SUBBAND AND WAVELET TRANSFORMS
Design and Applications

THE KLUWER INTERNATIONAL SERIES IN ENGINEERING AND COMPUTER SCIENCE

Distributors for North America:
Kluwer Academic Publishers
101 Philip Drive
Assinippi Park
Norwell, Massachusetts 02061 USA

Distributors for all other countries:
Kluwer Academic Publishers Group
Distribution Centre
Post Office Box 322
3300 AH Dordrecht, THE NETHERLANDS

Library of Congress Cataloging-in-Publication Data

A C.I.P. Catalogue record for this book is available
from the Library of Congress.

Printed on acid-free paper.

Printed in the United States of America

SUBBAND AND WAVELET TRANSFORMS
Design and Applications

edited by

Ali N. Akansu
New Jersey Institute of Technology

Mark J. T. Smith
Georgia Institute of Technology

Foreword by
James F. Kaiser

KLUWER ACADEMIC PUBLISHERS
Boston / Dordrecht / London

CONTENTS

CONTRIBUTORS

Mark J. T. Smith
Georgia Institute of Technology
Atlanta, Georgia, USA

Ali N. Akansu
New Jersey Institute of Technology
Newark, New Jersey, USA

Richard A. Haddad
Polytechnic University
Brooklyn, New York, USA

Kambiz Nayebi
Sharif University of Technology
Tehran, Iran

Thomas P. Barnwell
Georgia Institute of Technology
Atlanta, Georgia, USA

Rashid Ansari
University of Illinois
Chicago, Illinois, USA

Jean-Pierre Antoine
Université Catholique de Louvain
Louvain-la-Neuve, Belgium

Romain Murenzi
Clark Atlanta University
Atlanta, Georgia, USA

Albert Cohen
CEREMADE-Université Paris IX
Paris, France

Eric Séré
CEREMADE-Université Paris IX
Paris, France

Bernd Girod
Universität Erlangen-Nürnberg
Erlangen, Germany

Frank Hartung
Universität Erlangen-Nürnberg
Erlangen, Germany

Uwe Horn
Universität Erlangen-Nürnberg
Erlangen, Germany

Reginald L. Lagendijk
Delft University of Technology
Delft, The Netherlands

Frank Bosveld
Philips Sound and Vision
Eindhoven, The Netherlands

Jan Biemond
Delft University of Technology
Delft, The Netherlands

James D. Johnston
AT&T Bell Laboratories
Murray Hill, New Jersey, USA

Gary J. Saulnier
Rensselaer Polytechnic Institute
Troy, New York, USA

Michael J. Medley
Rome Laboratory
Griffiss Air Force Base, New York, USA

Pankaj K. Das
Rensselaer Polytechnic Institute
Troy, New York, USA

Phillip L. De León II
University of Colorado
Boulder, Colorado, USA

Delores M. Etter
University of Colorado
Boulder, Colorado, USA

Richard V. Cox
AT&T Bell Laboratories
Murray Hill, New Jersey, USA

Irving Kalet
RAKEFET Consulting Co.
Haifa, Israel

FOREWORD

The scientists and engineers of today are relentless in their continuing study and analysis of the world about us from the microcosm to the macrocosm. A central purpose of this study is to gain sufficient scientific information and insight to enable the development of both representative and useful models of the superabundance of physical processes that surround us. The engineers need these models and the associated insight in order to build the information processing systems and control systems that comprise these new and emerging technologies.

Much of the early modeling work that has been done on these systems has been based on the linear time-invariant system theory and its extensive use of Fourier transform theory for both continuous and discrete systems and signals. However many of the signals arising in nature and real systems are neither stationary nor linear but tend to be concentrated in both time and frequency. Hence a new methodology is needed to take these factors properly into account.

The wavelet approach, which has been around for more than ten years, is such a methodology which focuses locally both on time and scale. In the last five years it has attracted very active interest in the engineering and signal processing communities as evidenced by the plethora of papers and books that have appeared. One hallmark of this literature has been the showing of the relationship of the wavelet analysis approach to the subband representation of signals which has been fueled by practical applications in both the audio and video signal areas. It has been this interaction between theory and application that is leading to new and valuable insights into the wavelet methodology and its extensions. This edited book can therefore be viewed as a sampling of the fruits of this interaction showing both its extent and depth.

The past fifty years are unrivaled in the explosive nature of these new technologies which in large part have been fueled by the continuing development of integrated circuits and their progeny, the digital computing machines with all their ramifications. These computing machines have given us powerful new means for both devising and implementing these new wavelet analysis tools for our study of systems and their signals. Only time will reveal the extent and success of these endeavors. The seeds of discovery and application are all there.

James F. Kaiser

Department of Electrical Engineering,
Duke University, Durham, North Carolina.

PREFACE

The increasing volume of published literature on multirate signal processing and wavelets is a convincing indicator of the current interest surrounding these topics. Related research activities can be seen across the spectrum of technical journals, with most of the notably contributions coming from the engineering, science, and mathematics communities. Owing to the historical differences in notation, philosophy, and perspective, it is often difficult to follow all that is being done. Often even more difficult is appreciating the implications of these theoretical advances in various areas of applications. This book was conceived with the idea of trying to provide a useful reference volume, treating the fundamentals and discussing emerging applications in which filter banks and wavelets are thought to be important. It is intended for a broad audience of readers. We anticipate, however, that the book will be of particular interest to practicing engineers and scientists who wish to learn about utilization of these new developments, instructors teaching senior and graduate level engineering and applied mathematics courses in this field, and management personnel wishing to keep up with this rapidly developing technology.

The book is composed of chapter contributions from distinguished scholars in the field from prominent research laboratories and universities around the world. The first six chapters of the book are devoted to theory and design. Each has as a unique focus and perspective. The remaining seven chapters use the foundation built in the first part to address specific applications. Some of these application areas are somewhat mature, while others are just emerging. An extensive bibliography is provided at the end of the text, which is used as a reference list for all chapters. This should be very helpful to the interested reader for an expanded study on various aspects of this material.

Special words of thanks and appreciation are due Alen Docef for his many long hours and unflagging effort in the preparation and formatting of the manuscript's camera-ready version. We are particularly indebted to all of our chapter authors for their fine contributions to this volume. We have enjoyed long-standing associations and friendships with many of them and are grateful for their participation in this project. We are convinced that you, as readers of this edited volume, will benefit from their insights and perspectives as we have over the years.

Mark J. T. Smith
Ali N. Akansu

SUBBAND AND WAVELET TRANSFORMS
Design and Applications

1

INTRODUCTION AND OVERVIEW

Mark J. T. Smith and Ali N. Akansu*

Digital Signal Processing Laboratory, School of ECE, Georgia Institute of Technology, Atlanta, Georgia 30332-0250

**Department of Electrical and Computer Engineering, New Jersey Institute of Technology, Newark, New Jersey 07102*

1 INTRODUCTION

The technical community has witnessed recently a resurgence of activity involving the theory and design of linear systems for signal decomposition and reconstruction. Over the past few years, there have been considerable advances in multirate signal processing, with much of the attention being given to subband decompositions and wavelet transforms. These advances in algorithm theory combined with advances in microelectronics technology enable cost-effective digital computer processing of signals to be performed, and has attracted great interest in the signal processing and telecommunications areas. The influence of subband and wavelet representations can be seen in many important areas, such as teleconferencing, telemedicine, video transmission for unmanned autonomous vehicles, high definition television, picture phones, voice identification, speech recognition, secure voice transmission, signal storage, object detection and classification, and many others.

The chapters in this edited volume are devoted to the topic of subband representations and wavelet transforms, and their application to real-world problems. Subband signal representations are obtained from multirate filter banks. The input and output signals are all assumed to be digital and all processing is performed in the digital domain. Wavelet transforms, on the other hand, pertain to continuous or analog signals. The input signals are continuous, but the output of the transform may be discrete. Thus processing in the wavelet transform domain may also be discrete. The theory of multirate filter banks (which resides in the discrete-time domain) and wavelet theory (which resides in the continuous-time domain) share some commonalities, which will be discussed in detail later on in this book. These theories are rapidly making their way into a

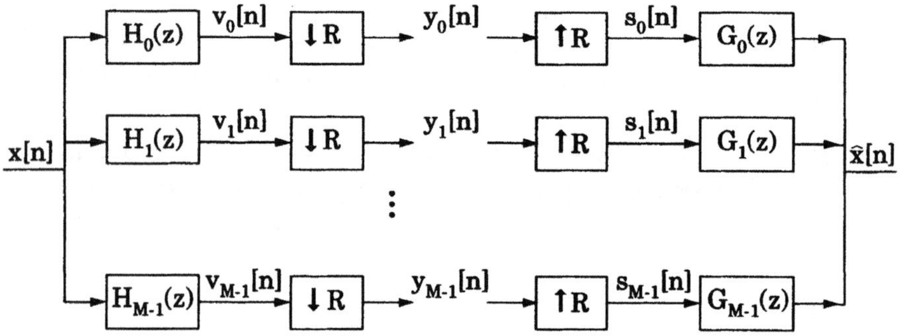

Figure 1 A 1-D uniform band analysis/synthesis filter bank.

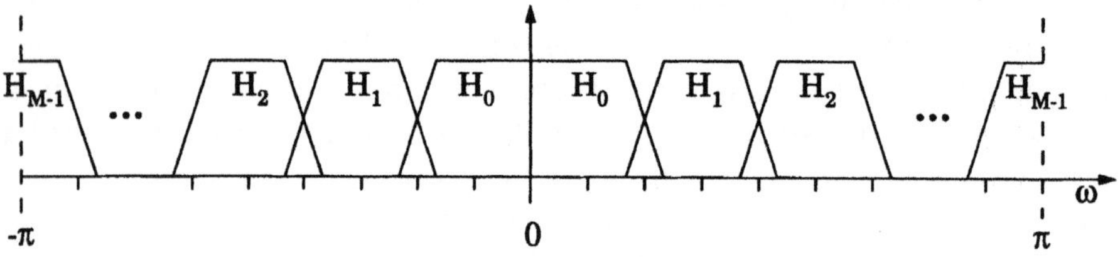

Figure 2 The spectral magnitude response corresponding to the analysis filters in Figure 1.

variety of applications. Most of the systems that are being developed for high visibility applications in which filter banks and wavelets are being considered are digital, i.e. they involve digital signals as inputs and perform processing in a digital signal environment. Thus a slightly greater emphasis is placed on the *digital* domain in this text, and we open with this as our starting point.

To begin the discussion for discrete signals, consider the *analysis/synthesis filter bank* shown in Figure 1. As can be seen, the input signal $x[n]$ is first filtered by a set of M filters, denoted $h_0[n], h_1[n], \ldots, h_{M-1}[n]$ or equivalently $H_0(z), H_1(z), \ldots, H_{M-1}(z)$ in the z-transform domain. This set is called the *analysis filter bank*. The individual filters are typically frequency selective bandpass filters of length L, with contiguous passbands that collectively cover the entire digital spectrum without gaps, as illustrated in Figure 2. The output of the filter bank may be described by the convolution

$$v_k[n] = h_k[n] * x[n] = \sum_{m=0}^{L-1} x[nR - m]h_k[m], \qquad k = 0, 1, \ldots, M-1,$$

where the asterisk $*$ denotes convolution. Since for every input sample, M output samples are produced, the filter bank has the inherent property of expanding the overall number of samples by a factor of M. To remove the redundancy, the individual outputs are downsampled by a factor of R $(1 \leq R \leq M)$. In most subband signal processing systems, R is set equal to M, which results in a ***maximally decimated*** or ***critically sampled*** filter bank. The R-to-one downsampling operation shown in the figure is described by the equation

$$y_k[n] = v_k[Rn], \tag{1.1}$$

and is denoted by the downarrow in the block diagram. Hence the outputs are given by

$$y_k[n] = \sum_{m=0}^{L-1} x[nR - m]h_k[m], \qquad k = 0, 1, \ldots, M-1, \tag{1.2}$$

where $y_0[n]$, $y_1[n]$, ..., $y_{M-1}[n]$ are called the ***subband*** signals (or channels)[1]. This part of the system is called the ***analysis section*** of the filter bank and performs a subband decomposition or transformation of the input. In typical applications, the subbands are processed in some way as dictated by the application. For example, in subband coding these outputs are quantized and then coded for transmission or storage. The motivation for performing the analysis decomposition is that improved performance can often be achieved by processing the subbands, $y_k[n]$, rather than the full-band input $x[n]$.

In the synthesis section, which follows, the signal is reconstructed from the subbands $y_k[n]$. This inverse operation consists of first upsampling[2] the subbands one-to-R as shown in Figure 1. Mathematically, ***upsampling*** by a factor of R may be described by the equation

$$s_k[n] = \begin{cases} y_k[n/R] & \text{for } n = 0, \pm R, \pm 2R, \pm 3R, \ldots \\ 0 & \text{otherwise.} \end{cases} \tag{1.3}$$

Descriptively, it may be interpreted as inserting $R-1$ zeros in between each of the samples $y_k[n]$. In the frequency domain, upsampling by R has the effect of

[1] The term ***subband*** (like many others) comes from the seminal work in subband speech coding by Crochiere, Weber, and Flanagan, [109].

[2] On occasion, the term ***interpolation*** has been used in reference to this operation. More commonly, interpolation refers to upsampling followed by filtering.

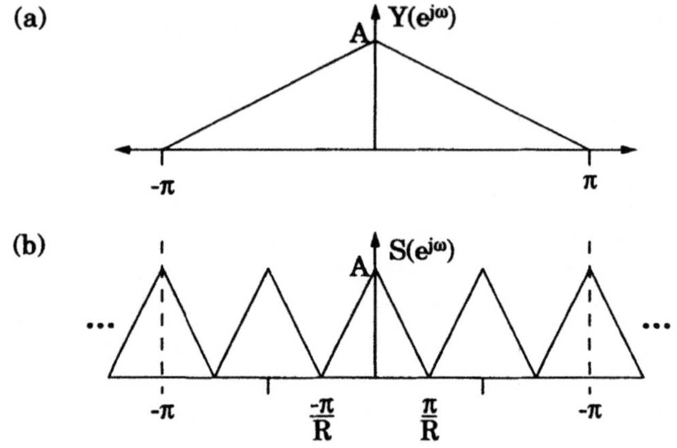

Figure 3 Frequency domain illustration of upsampling. The input signal $Y(e^{j\omega})$ is shown in (a). The output of one-to-R upsampling is shown in (b) for the case of $R = 4$. The effect is known as *imaging*.

shrinking the frequency axis R fold, bringing periodic replicas of the spectrum into the fundamental frequency range $-\pi \leq \omega \leq \pi$. This effect is called *imaging* and is illustrated in Figure 3. After the upsampling operation in the synthesis section, the signals are filtered by the synthesis filters, $g_0[n]$, $g_1[n]$, $\ldots$, $g_{M-1}[n]$, and summed together to produce $\hat{x}[n]$.

2 FILTER BANKS AND TRANSFORMS

Thus far, we have used the terms *filter banks* and *transforms* loosely, but said nothing of their distinctions and relationships. As it turns out filter banks and transforms are intimately related. A transform, in the general sense, implies any mapping from one signal into another, and so all filter banks are transforms. But in its common usage in signal processing, it often refers to a block transform given by

$$\mathbf{y} = \mathbf{H}\mathbf{x}, \tag{1.4}$$

where $\mathbf{y}$ and $\mathbf{x}$ are M-dimensional column vectors and $\mathbf{H}$ is an $M \times M$ matrix of transform basis vectors. Block transforms are pervasive in signal processing, notable examples of which are the DFT (discrete Fourier transform) and the DCT (discrete cosine transform), [2, 418, 7, 254, 108, 430]. These block transforms are a special case of the more general uniform band analysis filter bank in Figure 1. To see this, assume that the analysis filters in Equation (1.2) all have the same length with $L = M$. Then the filter bank may be interpreted as taking contiguous M-length blocks of the input, and applying to each block the transform in Equation (1.4). For example, consider the popular discrete cosine transform or DCT, in particular the DCT-II [333]. We may view this transform in the conventional way as a block transform, or alternatively as a filter bank. From the transform perspective, $\mathbf{H}$ is the M-point DCT matrix with elements $H_{k,n} = \cos[\frac{\pi}{M}(n + 1/2)k]$. In contrast, from the filter bank viewpoint, the analysis filters are the DCT-II basis vectors given by

$$h_k[n] = \cos[\frac{\pi}{M}(n + 1/2)k], \tag{1.5}$$

where $0 \leq n, k \leq M - 1$. This relationship between filter banks and block transforms is an important concept in multirate signal processing with synergistic benefits. Viewing a filter bank as a transform with extended basis vectors leads naturally to representations in terms of matrices of basis vectors, where the synthesis is the inverse matrix of the analysis. It is noteworthy that many of the recent advances in filter banks are based on matrix representations, which is a block transform perspective [254, 296]. The filter-bank-block-transform relationship has clearly had a major influence, motivating the use of FFT (fast Fourier transform) algorithms for filter banks in transmultiplexers [284, 359], the use of FFTs in the discrete short-time Fourier transform for time-scale modification and speech enhancement, [320, 339], and more recently, the introduction of efficient filter banks known as lapped transforms [254]. The filter bank perspective has also contributed much to the development of analysis/synthesis systems. Filter banks were historically viewed in the frequency domain via the spectral relationships associated with downsamplers and upsamplers [108]. But perhaps most important, the filter bank view is backed by an extensive collection of design tools and theory, both for FIR and IIR filters[3]. Recognizing these complementary perspectives, we can view this broad class of analysis/synthesis filter banks as being general linear ***subband*** transforms.

3 ANALYSIS/SYNTHESIS FILTER BANKS

The term ***analysis/synthesis***, used in the context of ***subband*** decomposition, comes from the early work in subband coding (SBC), and has the implication that the input signal is to be decomposed, processed and recombined. Thus, for convenience, analysis/synthesis filter banks can be divided into three basic

[3]FIR or ***finite impulse response*** refers to finite length filters. IIR refers to infinite length filters that are typically implemented recursively.

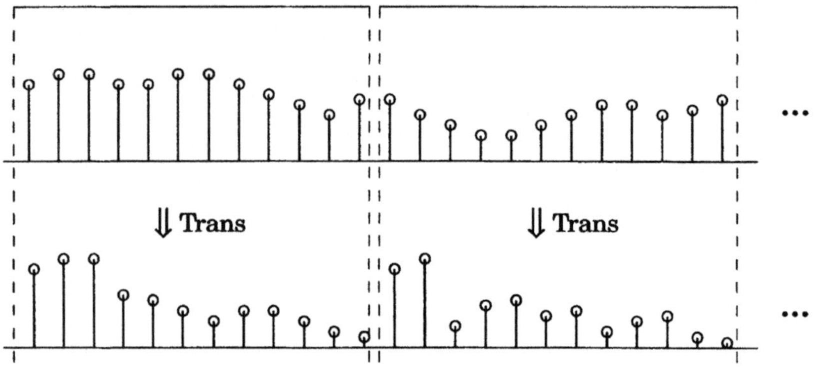

Figure 4 Illustration of the block transform interpretation of a filter bank with $R = M = L$.

parts for convenience of discussion: 1) the analysis filter bank; 2) the synthesis filter bank; and 3) the analysis/synthesis system. Each has its own set of associated design issues and applications as we discuss next.

Analysis Filter Banks. The analysis filter bank or analysis section as shown in Figure 1, performs a multirate transformation on the input. The input signal is decomposed into a set of subband signals $y_0[n]$, $y_1[n]$, ..., $y_{M-1}[n]$, which is used as a representation for extracting information. A very simple example of this is a discrete cosine transform (DCT) filter bank where the number of bands M and filter lengths L are the same. The analysis filter coefficients are as in Equation (1.5). Consider first the case where the downsampling rate R is equal to the filter length L, thus $L = M = R$. This filter bank is equivalent to taking an M-sample block ($0 \leq n \leq M-1$) of the input $x[n]$, and computing its DCT. The outputs $y_k[0]$ (where $0 \leq k \leq M-1$) are the DCT coefficients of the zeroth input block. Notice in particular that the filter bank output samples (in Figure 1) corresponding to the transform appear vertically in the variable k. Next we move to the following block of M samples ($M \leq n \leq 2M-1$), compute the DCT and obtain $y_k[1]$. Notice that again the DCT coefficients are in the variable k. This procedure is then repeated on a block-by-block basis, as depicted conceptually in Figure 4.

At the opposite extreme, consider the case where $R = 1$, which is equivalent to a filter bank with no decimation being performed on $v_0[n]$, $v_1[n]$, ..., $v_{M-1}[n]$. This case is equivalent to computing the DCT on the first block ($0 \leq n \leq M-1$) to obtain $y_k[0]$ as before. But for the next block, we only move one sample over, so the range of the block spans $1 \leq n \leq M$. Similarly the following block spans the range $2 \leq n \leq M+1$ and so on. This example of the DCT filter bank (with $R = 1$) is a short-time frequency representation that can be used to identify the frequency components in the input. It results in a two-dimensional time-frequency representation with frequency described in the variable k and time in the variable n. Subband transforms include a wide range of time-frequency representations governed by the filter length L, the number of bands M, and the decimation rate R. Such a time-frequency representation can be used to track embedded sinusoids or frequency components in a signal as they evolve in time and are commonly used in systems for speaker identification and verification, speech recognition, and in a variety of spectral analysis applications.

Synthesis Filter Banks. The synthesis filter bank or synthesis section as shown in Figure 1, transforms a set of multirate inputs into a single signal. To illustrate our discussion on synthesis filter banks, consider the case where $R = M$ and $L = 4M$. Observe that the filters are long with respect to the number of subbands. Consequently, the synthesis filters can be designed to have good spectral magnitude characteristics such as the ones shown in Figure 2. First we see that the input signals, $y_k[n]$ are upsampled by a factor of M and then passed to a bank of synthesis filters. Each output, $s_k[n]$, is a bandpass signal and is approximately orthogonal to the other channels $s_\ell[n]$, where $\ell \neq k$. Such a filter bank is a core component in translation multiplexers, where time division multiplexed (TDM) signals are converted to frequency division multiplexed (FDM) signals. The basic idea behind TDM-to-FDM trans-multiplexers can be appreciated by viewing the synthesis bank in terms of a telephone system, where the channels containing the signals $y_k[n]$ are telephone signals from M different speakers. In the TDM format, samples of the M signals are interleaved to form a single one. Conversion to an FDM format consist of first de-interleaving the signals, producing M baseband voice signals each bandlimited to less than π/M. These signals, which may be called $y_k[n]$ are input to the synthesis section. They are upsampled to the higher frequency, which produces spectral replicas of the signal in the frequency domain. Examine Figure 3 for an illustration of this effect. The synthesis filters then extract a unique non-overlapping spectral region for each channel. Summing these outputs together produces the FDM format. Thus we see that the synthesis section of the filter bank can be used to perform TDM-to-FDM conversion. Similarly, the analysis section can be used to perform FDM-to-TDM conversion.

Analysis/Synthesis Systems. Analysis/synthesis systems refer to the cascade of the analysis and synthesis filter banks as shown in Figure 1. The study and development of analysis/synthesis filter banks has been driven largely by subband coding—an application which will be discussed in detail in chapters

7, 8, and 9. For subband coding applications, the signal is split into subbands $y_k[n]$. As mentioned earlier, subband coding consists of quantizing and coding the subband signals $y_k[n]$ as opposed to the original input $x[n]$. For this important scenario, there are several crucial issues, some of which are not relevant for applications involving only analysis or synthesis individually. First, consider the choice of the number of subbands that are used in the system. This parameter implies a tradeoff between time and frequency resolution, and often between computational complexity and performance. In traditional subband speech coding applications, the number of bands typically ranges from 4 to 16, while for high quality audio the number of bands is often 100 or more.

A second issue is the bandwidths of the filters. To this point, we have only illustrated the simple uniform band filter bank where each of the bands has the same bandwidth. More common in subband coding, is the use of nonuniform band filter banks. This is particularly true in subband speech coding where the properties of auditory perception suggest that the filter bandwidths should match the so-called critical bands of the ear.

A third issue is the frequency domain characteristics of the individual analysis and synthesis filters. In applications like speech coding, the leakage between bands is undesirable. Therefore filters with narrow transition widths and high stopband attenuation are sought. To meet these frequency domain specifications the filter lengths are generally long relative to the number of subbands. For example, it is not uncommon for a four-band filter bank to have filters of length 32. This allows the filters to have good frequency domain characteristics—the longer the filters, the more closely they can approximate the ideals.

Fourth, consider the filter impulse response characteristics. In some applications, such as image coding for example, they can be traced to specific objectionable distortions in the output. By optimizing the filter coefficients with respect to constraints on impulse response behavior, these distortions can be mitigated to some degree [214, 237].

Fifth, the transfer function of the overall analysis/synthesis system should be equivalent to an identity system, or at least approximately so. The analysis and synthesis sections serve as a tool for more effective processing but are clearly counterproductive if their presence introduces unwanted distortion. Hence, it is generally important to design the analysis and synthesis systems to be effectively distortion free. Referring to Figure 1, this means that

$$\hat{x}[n] \approx x[n - n_0],$$

where n_0 is the system delay. By carefully examining the equations that describe the analysis and synthesis, we see that

$$\hat{x}[n] = \frac{1}{R} \sum_{m=-\infty}^{\infty} \sum_{\ell=-\infty}^{\infty} \sum_{k=0}^{M-1} x[m] h_k[\ell R - m] g_k[\ell - Rn], \tag{1.6}$$

which is a non-trivial relationship.

Analysis/synthesis systems have traditionally been analyzed in the frequency domain in terms of the discrete-time Fourier transform or DTFT. The corresponding relationship is

$$\hat{X}(e^{j\omega}) = \frac{1}{R} \sum_{k=0}^{M-1} \sum_{\ell=0}^{R-1} X(e^{j(\omega+2\pi\ell/R)}) H_k(e^{j(\omega+2\pi\ell/R)}) G_k(e^{j\omega}). \tag{1.7}$$

Notice that the summation term with the index k combines contributions from the M subband channels, while the sum with the index ℓ represents the aliasing components introduced by the downsampling. All terms involving k where $k \neq 0$ are aliasing components and should be canceled in the reconstruction process if possible. A great deal of attention has been given to addressing the design of analysis and synthesis filter banks that not only have good individual characteristics but also address the overall reconstruction property implied in the equation above.

The notion of overall system reconstruction leads naturally to the sixth issue for consideration, the system delay. The cascade of the analysis and synthesis sections may be thought of as a black-box system. As such it has a delay n_0 associated with it, i.e. $\hat{x}[n] \approx x[n - n_0]$. For some applications, as we will discuss in Chapter 3, large delays represent distortion and their negative impact should be considered in the analysis/synthesis system design.

The last issue we mention is computational complexity. Many applications, particularly those involving multidimensional signals, are very constrained by implementation complexity. Thus whatever computational processing can be saved in the analysis/synthesis, can be used to either lower the system cost or free-up processing power for another component of the system.

Before leaving the topic of subband transform issues, it is important to comment on the variety of inputs of interest. Inputs in this context are discrete-time signals that may be one-, two-, or multi-dimensional. The simplest input to consider is a one-dimensional (or 1-D) signal $x[n]$. In practice, such signals are typically, speech, music, or any audio signal that might be sent to a loudspeaker.

Typical two-dimensional (or 2-D) signals, which we denote by $x[n_1, n_2]$, are natural gray-scale images, and binary images such as written text. N-dimensional or multi-dimensional images, denoted $x[n_1, n_2, \ldots n_N]$ are also of general interest. The most common example is a video sequence, $x[n_1, n_2, n_3]$, where the n_1 and n_2 indices are the spatial coordinates associated the sequence frames, and n_3 is the time index that specifies the particular frame. Much of the theory can be understood using 1-D signals. In later chapters, the 1-D case is used as a foundation upon which to expand to higher dimensions.

4 TWO-BAND FILTER BANKS

The problem of designing analysis/synthesis filter banks is not a trivial task, due to the downsamplers and upsamplers. Often iterative algorithms are employed to optimize the filter coefficients subject to the constraints mentioned in the previous section. Interestingly, this formidable problem is greatly simplified for the special case where $M = 2$, leading to useful filter banks as we will soon see.

To open the discussion, consider the two-band filter bank illustrated in Figure 5. Again we emphasize that this is exactly the M-band filter bank with $M = R = 2$. The filters $H_0(e^{j\omega})$ and $G_0(e^{j\omega})$ are lowpass filters and $H_1(e^{j\omega})$ and $G_1(e^{j\omega})$ are highpass filters all with a nominal cutoff frequency of $\pi/2$. Using the frequency-domain equation given in (1.7), which describes the input-output behavior of M-band analysis/synthesis systems, observe that

$$\hat{X}(e^{j\omega}) = \frac{1}{2}X(e^{j\omega})\Big(H_0(e^{j\omega})G_0(e^{j\omega}) + H_1(e^{j\omega})G_1(e^{j\omega})\Big)$$
$$+\frac{1}{2}X(e^{j(\omega+\pi)})\Big(H_0(e^{j(\omega+\pi)})G_0(e^{j\omega}) + H_1(e^{j(\omega+\pi)})G_1(e^{j\omega})\Big). \quad (1.8)$$

Notice that the term on the second line of the equation above is the aliasing term. This is evident by the presence of the π frequency shift components in the DTFT signal arguments. The presence of aliasing is an undesirable by-product of the decimation. However, it can by removed by specifying the filters in the following way:

$$\begin{aligned} H_1(e^{j\omega}) &= H_0(e^{j(\omega+\pi)}) \\ G_0(e^{j\omega}) &= H_0(e^{j\omega}) \\ G_1(e^{j\omega}) &= -H_0(e^{j(\omega+\pi)}). \end{aligned} \quad (1.9)$$

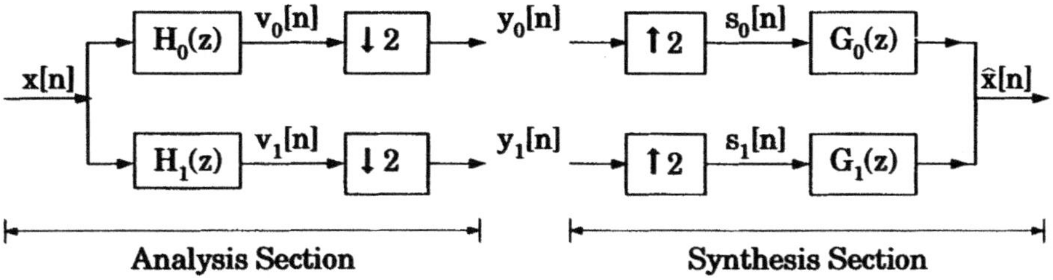

Figure 5 A two-band analysis/synthesis filter bank.

By direct substitution into Equation (1.8), we see that the aliasing term goes to zero, leaving

$$\hat{X}(e^{j\omega}) = \frac{1}{2}X(e^{j\omega})\left(H_0^2(e^{j\omega}) - H_1^2(e^{j\omega})\right). \tag{1.10}$$

Therefore the design task is to determine $H_0(e^{j\omega})$ such that the overall system frequency response,

$$\frac{1}{2}\left(H_0^2(e^{j\omega}) - H_1^2(e^{j\omega})\right),$$

approximates $e^{j\omega n_0}$, i.e. approximates an allpass function with constant group delay n_0. All four filters in the filter bank are specified by a length-L lowpass FIR filter, $h_0[n]$, where L is an even positive integer. This is the quadrature mirror filter (or QMF) solution that was introduced back in 1976 by Croisier, Esteban, and Galand [110, 140].

The beauty of this solution lies in its simplicity. The relationships among the filters as dictated by the QMF solution are very simple in the time domain:

$$\begin{aligned} h_1[n] &= (-1)^n h_0[n] \\ g_0[n] &= h_0[n] \\ g_1[n] &= -(-1)^n h_0[n]. \end{aligned} \tag{1.11}$$

Notice that the modulation by $(-1)^n$, (or equivalently $e^{j\pi n}$), converts the lowpass filter $h_0[n]$ into a highpass filter. In the spectral domain, this modulation corresponds to a frequency shift by π as seen in Equation (1.9). Johnston [216] has designed and published a set of QMFs with a variety of different filter lengths and stopband attenuations. All are optimized to have good stopband and transition band characteristics and very small reconstruction error. Johnston's QMFs are virtually a standard among users of two-band filter banks.

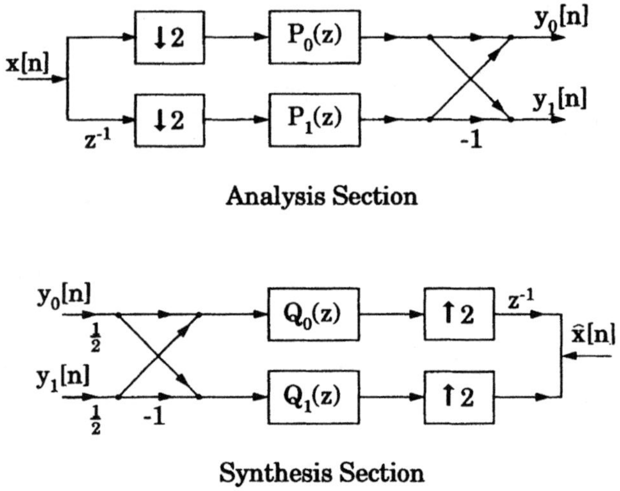

Figure 6 A two-band analysis/synthesis polyphase filter bank structure.

Two-band QMF filter banks can be implemented in direct form as suggested in Figure 5. However, they can be implemented more efficiently in polyphase form [43, 108]. The classical polyphase implementation structure for a two-band filter bank is shown in Figure 6[4]. The polyphase filters, $p_0[n]$ and $p_1[n]$, are obtained directly from the lowpass QMF $h_0[n]$. In particular, they are downsampled versions of the QMF impulse response given by

$$\begin{aligned} p_0[n] &= h_0[2n] \\ p_1[n] &= h_0[2n+1]. \end{aligned}$$

Similarly for the synthesis section, the synthesis polyphase filters are given by

$$\begin{aligned} q_0[n] &= h_0[2n+1] \\ q_1[n] &= h_0[2n]. \end{aligned}$$

The efficiency of this realization can be seen by recognizing two important properties of the structure. First, the filters $p_0[n]$ and $p_1[n]$, are of length $L/2$ as opposed to L, as is the case for QMFs. Second, the polyphase filters operate at half the sampling rate. Thus a savings of about four-to-one is realized

[4]The classical polyphase implementation can be generalized to form a matrix of filters, $E_{lk}(z)$. This is mentioned and utilized in Chapters 2–4.

over the direct implementation implied in Figure 5. It is evident, however, that complexity can be reduced by two in the direct form implementation by simply not computing the odd indexed $v_0[n]$ and $v_1[n]$ samples because they are discarded by the downsampling operation. Thus the polyphase implementation actually realizes a net two-to-one reduction in computational complexity over the direct form.

Due to the nature of the QMF solution, (see Equation (1.11)) it is *not* possible to make the reconstruction error exactly zero. The reconstruction will always contain some amount of distortion due to the two-band QMF filter bank. This distortion, however, can be removed by using the exact reconstruction two-band filter banks introduced in [373]. In the literature, these filters are sometimes called *Smith-Barnwell* filters or *conjugate quadrature filters (CQFs)*. They have the relationship

$$\begin{array}{lllllll} H_1(e^{j\omega}) & = & e^{-j\omega(L-1)}H_0(-e^{-j\omega}) & \text{or} & h_1[n] & = & (-1)^n h_0[L-1-n] \\ G_0(e^{j\omega}) & = & e^{-j\omega(L-1)}H_0(e^{-j\omega}) & \text{or} & g_0[n] & = & h_0[L-1-n] \\ G_1(e^{j\omega}) & = & -H_0(-e^{j\omega}) & \text{or} & g_1[n] & = & -(-1)^n h_0[n]. \end{array} \tag{1.12}$$

As in the QMF case, all filters are specified in terms of a single lowpass filter, $h_0[n]$, but $h_1[n]$, $g_0[n]$, and $g_1[n]$ are time reversed and/or $(-1)^n$-modulated versions of $h_0[n]$. Convenient and optimal design methods exist for constructing these exact reconstruction two-band filter banks. A detailed discussion of the theory can be found in [373, 371, 374, 274]. Interestingly these two-band perfect reconstruction filters are linked to the discrete wavelet transform, as discussed in Chapter 2. They have the advantage that they contribute no distortion and can have better frequency domain characteristics than QMFs [216], but cannot exploit the numerical utility of the classical polyphase form. As it turns out, the incompatibility with the polyphase structure is of small consequence, since there are many options now available for efficient implementation of exactly reconstructing two-band filter banks. A good example of an efficient two-band implementation structure is the lattice proposed in [421]. It supports perfect reconstruction with about the same number of multiplies and adds required for a QMF polyphase implementation. The lattice structure is structurally constrained to guarantee exact reconstruction and achieves good performance in a fixed point arithmetic environment. Filter banks can be designed within the lattice structure framework simply by optimizing the lattice coefficients. We will elaborate a little more on this point in Section 5, in the context of M-band filter banks, and also in Chapter 3.

Another useful option is provided by the binomial QMFs introduced by Akansu et al. [8]. These maximally flat magnitude square [184] paraunitary filters are *identical* to the wavelet filters devised by Daubechies [113]. The binomial structure can be implemented efficiently because all multiplication operations can be realized with a *small* number of binary shifts and adds [3]. Although the numerical efficiency *does* come at the expense of loss in stopband-passband-transition band quality, it can still be very attractive when implementation costs are weighed. This topic is revisited at a more basic level in Chapter 2.

The notion of multiplier-free filter coefficients, where convolution is reduced to simple shifts and adds, is often important for low-cost VLSI circuits. Filter bank design algorithms, most of which are iterative, can be made to optimize coefficients for easy implementation. An early example of a two-band filter bank with good numerical properties is the short-kernel filter set proposed by LeGall and Tabatabai [153]. These filters are short in length and thus have relatively poor spectral magnitude response characteristics. However, they are useful in the image coding application for which they are intended.

Finally it is appropriate to mention, at this point, the role of recursive *infinite impulse response* or IIR filters in two-band systems. IIR filters are known for their excellent spectral magnitude response characteristics relative to FIR filters of similar order. They have been studied in the context of two-band systems and shown to be very effective in some cases. In particular, for two-dimensional processing where causal filtering is not a necessity, IIR filter banks can provide remarkable efficiency and high quality [379, 22, 20]. In cases where causality is an issue, such as in subband speech coding, exact reconstruction is often in conflict with stability. Hence, strategies based on limiting the spectral magnitude and phase distortions to be within tolerable limits may be appropriate [41, 328, 268]. Indeed, the technical literature has addressed a variety of two-band filter banks with various properties. More in-depth discussions on IIR filter banks can be found in [418], [148], [7], and in Chapter 4 of this book.

5 TREE-STRUCTURED FILTER BANKS

Two-band filter banks are convenient, but subband applications generally require resolution greater than that given by two-band systems alone. To address this issue, two-band filter banks were typically embedded in tree structures. By cascading the two-band filter banks such that each band is split in two successively, an infinity of decompositions can be realized. For example, Figure 7 shows an eight-channel uniform-band analysis filter bank composed of two-band systems. Since the outputs from each two-band system can be reconstructed by

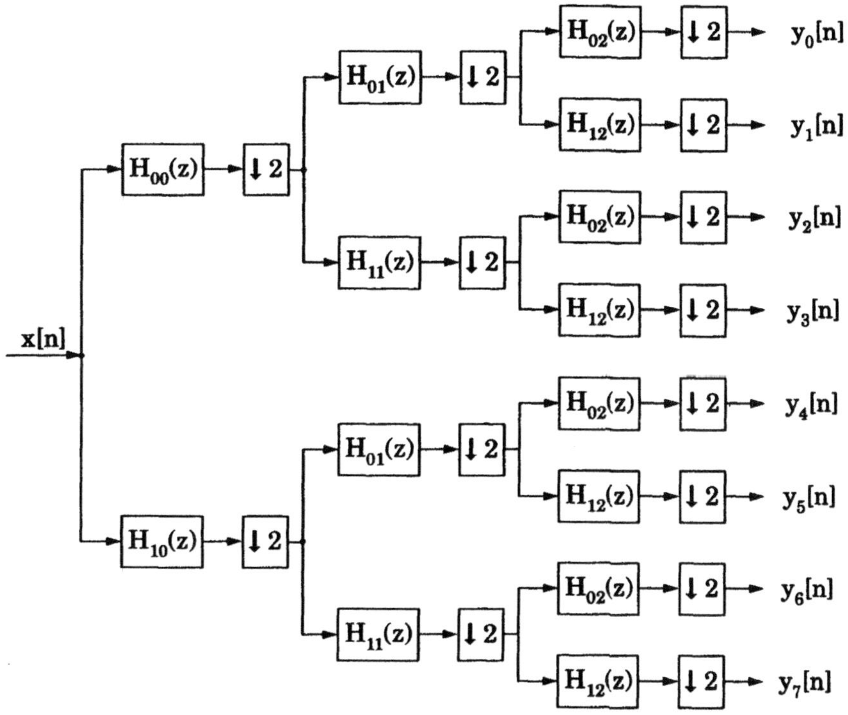

Figure 7 A uniform-band tree-structured filter bank.

its synthesis counterpart, constructing the synthesis tree-structured filter bank is trivial. Obviously a wide variety of tree structured decompositions exists. Note in particular, the octave-band tree structure shown in Figure 8.

Here the cutoff frequencies associated with the passbands occur at $\pi/2$, $\pi/4$, and $\pi/8$. Octave-band filter banks, which are prevalent in the 1990's, have been in vogue since the late 1970's [327] for speech coding. They were popular for speech coding because they provide a computationally efficient filter bank approximation to the critical bands of the ear. Octave-band filter banks also provide a structure for implementing the discrete wavelet transform, which we will discuss in Section 8. The infinite variety of tree structures arises from the infinite choices of filters at each stage in the tree, the lengths of the filters at the various stages, and the number of bands and tree configurations.

Tree structures can provide computationally efficient realizations of M-band filter banks. It is interesting to note that they can be implemented in an equivalent M-band form, often called a parallel form implementation. The

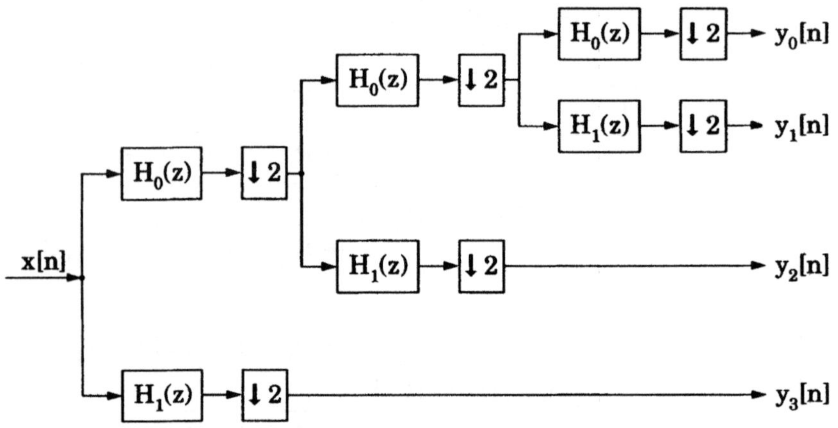

Figure 8 A four-channel octave-band tree-structured filter bank. The block diagram is shown in (a). The spectral magnitude response associated with the four channel filters is shown in (b).

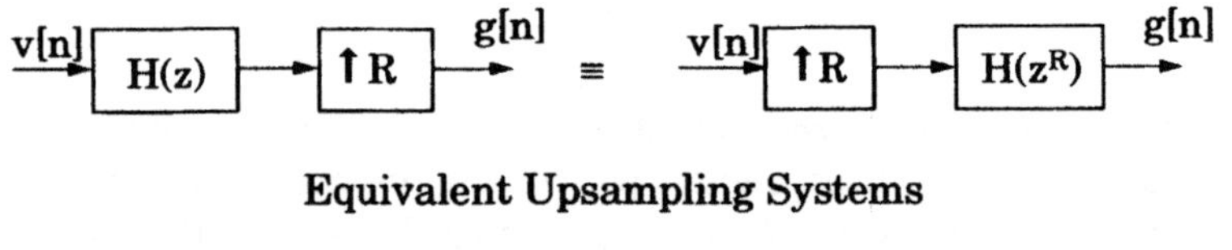

Figure 9 Identities for migrating filters across downsamplers and interpolators.

key to converting from a tree to parallel form is the notion of migrating filters forward and backward through the downsamplers and upsamplers as illustrated in Figure 9.

For example, the uniform band tree-structured filter bank in Figure 7 can be implemented in the form of the M-band filter bank (with $M = 8$) shown in

Figure 1, where

$$\begin{aligned} H_0(z) &= H_{0,0}(z)H_{0,1}(z^2)H_{0,2}(z^4) \\ H_1(z) &= H_{0,0}(z)H_{0,1}(z^2)H_{1,2}(z^4) \\ \vdots \quad &\vdots \quad \vdots \\ H_7(z) &= H_{1,0}(z)H_{1,1}(z^2)H_{1,2}(z^4). \end{aligned} \tag{1.13}$$

The disadvantage of implementing trees in parallel form is that the analysis and synthesis filters are very long and thus require many multiplies and adds. The concept of the parallel form equivalent representation can be useful in designing tree-structured filter banks. Improved system performance can often be achieved by optimizing the constituent two-band filter banks with respect to the characteristics of the parallel form equivalent, [41]. This point, which has recently been referenced as ***progressive optimality of subband trees***, is discussed further in Chapters 2 and 6. In addition, the configuration of the tree decomposition can be optimized to maximize overall performance. In other words, since there are many nonuniform band decompositions that are possible via a tree structure, we can choose the one that works best for our application. For speech, that decomposition is the one that matches the critical bands of the ear. For image coding, the decomposition may vary depending on the image or class of images. This idea was explored by Westerink [439] for image coding and later by other authors [10, 96] using different measures of optimality. Some authors have referred to these trees as ***wavelet packets*** [96]. In fact this idea is the major thrust of Chapter 6.

6 M-BAND FILTER BANKS

One of the clear disadvantages of tree-structured uniform filter banks is the long system delay associated with them. To better appreciate the system delay and complexity demands of these filter banks, consider the uniform tree in Figure 7. Let's assume that all the filters are linear phase QMFs, where in the first stage split, the filter lengths have length 32, in the second stage the lengths are 16, and in the third stage the lengths are 8. Assuming that the distortion introduced by the QMFs is negligible, the input will approximate the output delayed by the system delay n_0, i.e.

$$\hat{x}[n] \approx x[n - n_0].$$

For the specific tree-structured filter bank defined above, the effective length of the analysis and synthesis filters (or equivalently the length of the parallel form filters) is 90 and the system delay is 89 samples, all of which are very long. This requires more complex implementation hardware to handle the buffering associated with the long lengths. On the positive side, the number of multiplies per sample required for both analysis and synthesis (assuming a polyphase implementation) is just 56, which is relatively small.

In systems employing many channels, (e.g. for M greater than 32), M-band filter banks like the one shown in Figure 1 are often preferable, but with shorter filters. The delay requirements and consequently the number of registers needed can be much less for M-band systems. The viability of M-band filter banks is contingent on meeting the exact or near exact reconstruction conditions and on finding computationally efficient implementations.

Generalized polyphase and modulated filter banks are two classes of filter banks that are computationally efficient, have good reconstruction properties, and have reasonable system delays. The first of these, the lattice formulation, arises from a generalized polyphase interpretation, which in turn can be expressed as a cascade of simple lattice sections [417]. An important feature to spotlight is that the structure inherently guarantees perfect reconstruction. More on this point is given in Chapter 3.

The modulated filter banks, mentioned above, are perhaps the most attractive of the FIR uniform band filter banks. They are based on the idea of modulating a single baseband filter to form a bank of bandpass filters. To better illustrate the concept, consider a lowpass baseband filter $h[n]$ with frequency response shown in Figure 10a. By modulating this baseband filter by a cosine function, the filter bank

$$h_k[n] = h[n] \cos[\frac{\pi}{M}(n+\alpha)(k+\beta)] \tag{1.14}$$

is obtained, and is illustrated in Figure 10b. The constants α and β allow the modulation function to realize a variety of different forms. These filter banks can be implemented efficiently using structures such as the ones introduced by Malvar [255, 254], an example of which is shown in Figure 11.

7 MULTIDIMENSIONAL FILTER BANKS

Filter banks have been studied extensively for one-dimensional signals. More recently, multidimensional filter banks have been studied, sparked by the in-

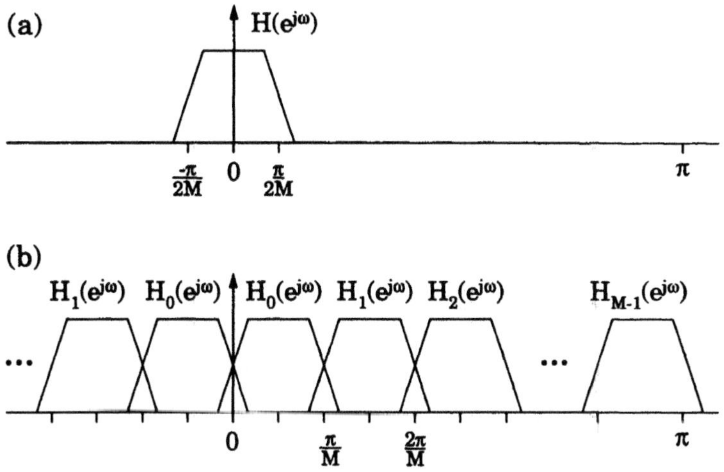

Figure 10 Modulated filter bank illustration. A baseband lowpass filter $H(e^{j\omega})$ is shown in (a). The cosine modulated filter banks $H_0(e^{j\omega})$, $H_1(e^{j\omega})$, $\ldots$, $H_{M-1}(e^{j\omega})$ are shown in (b).

terest in processing images and video signals. For multiple dimensions—most notably two dimensions—it is common to employ separable implementations where filter banks are applied to the dimension indices individually. For example, for a square image $x[n_1, n_2]$, a separable subband decomposition might consist of first filtering the rows of the image (i.e. in the n_1 dimension) with a two-band filter bank resulting in two rectangular images as shown in Figure 12a and b. The columns of the result may then be filtered with the same two-band filter bank resulting in four square subband images as shown in Figure 12c. As can be seen, each of the four resulting images is derived from a different filtering process. The subband image labeled A, is obtained by lowpass filtering and downsampling the rows and columns. Hence it may be interpreted as the low-low subband. The subband labeled B, is obtained by lowpass filtering the rows and highpass filtering the columns, resulting in a low-high subband. Similarly, subband images C and D may be interpreted as high-low and high-high subband images respectively. As more and more subband splits of the image are taken, the number of subbands clearly increases and their sizes become smaller in general.

A unique size expansion phenomenon generally occurs when 2-D and multi-D signals are the subject. This is because most multi-D signals of interest are limited in size. Consider the 2-D example just discussed and illustrated in Figure 12. If we assume that the image size is $N \times N$ samples (or *pixels* as they are routinely called) and that the analysis filters are of length L, then each subband image on average is of size $(N + L - 1)/2 \times (N + L - 1)/2$. This is

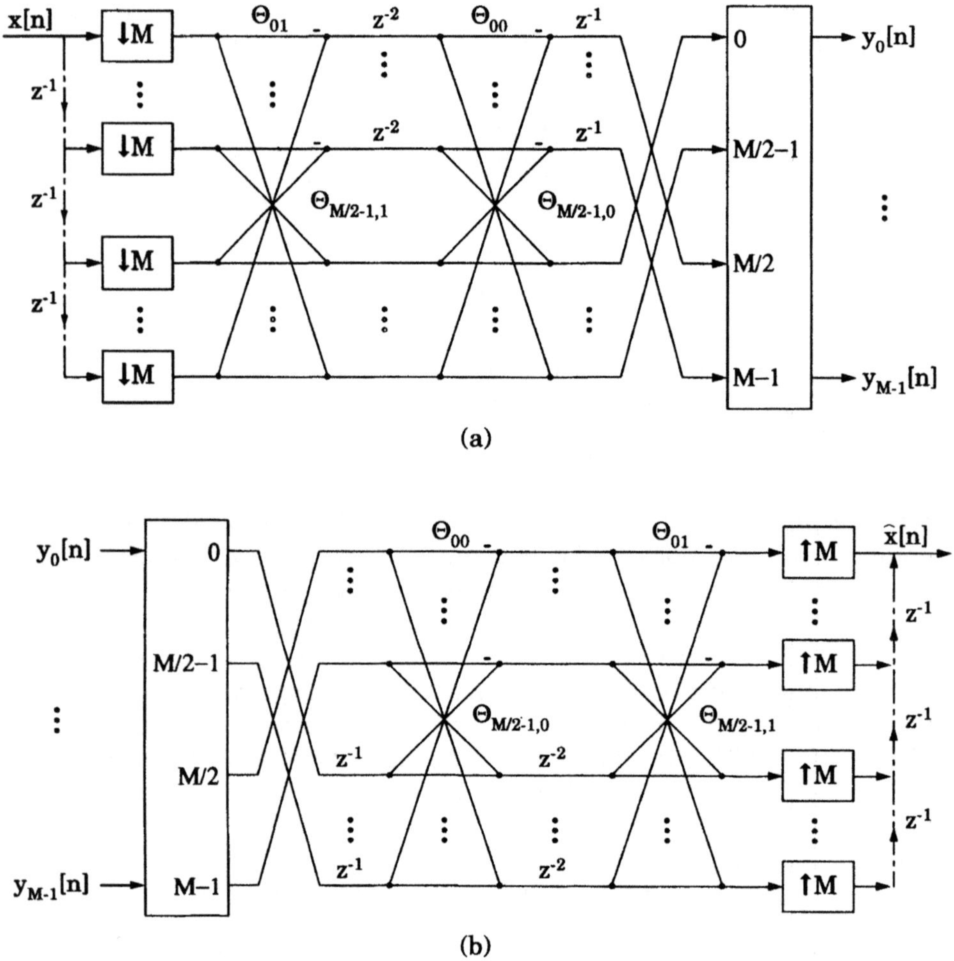

Figure 11 Block diagram of implementation structure introduced by Malvar for cosine modulated DCT-IV filter banks.

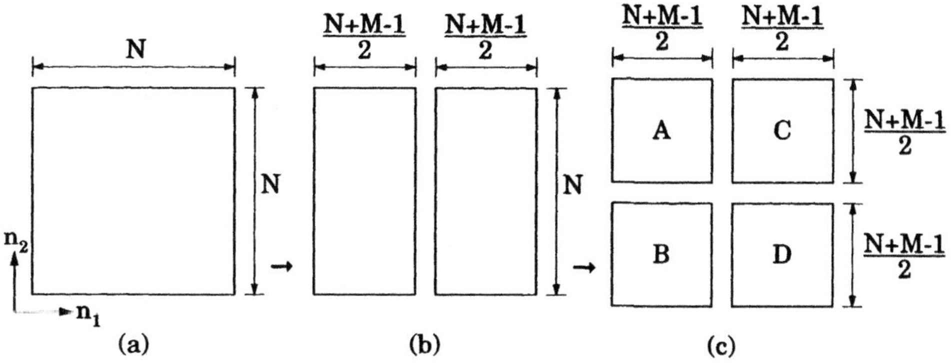

Figure 12 Illustration of a separable filter bank applied to a 2-D image. The original image size is shown in (a). The result after filtering and downsampling the rows is shown in (b). The result after filtering and downsampling the columns of the subband images in (b) is shown in (c).

because linear convolution of two finite sequences of length N and L results in an output length of $N+L-1$. The downsampling by a factor of two effectively reduces this length by two, resulting in the size mentioned above. The total number of output samples in all subbands is $(N+L-1)\times(N+L-1)$, which is larger than the input. This inherent expansion in size can pose a problem in applications like coding where the goal is to reduce, not expand, the data. Solutions to this problem have been proposed and may be grouped into two general classes at this point. The first family of solutions is based on periodically extending the input in various ways, filtering the image with conventional filter banks, and then exploiting the symmetry in the output to avoid expanding the overall number of samples [378, 222]. Numerous papers have appeared in the literature discussing this approach. The second approach is based on performing the subband analysis decomposition as before but discarding the pixels at the borders associated with the image expansion. Exact reconstruction is achieved by solving a set of boundary equations to determine the missing initial conditions for the synthesis filter bank. Discussions of this approach may be found in [121, 301, 302].

Interpretation of these bands can be very important in a variety of applications. The individual subband images represent spatial frequency components of the original image. Within the image, it is valid to interpret the samples as being a spatial domain (time domain) representation. However across the set of subband images, a frequency domain interpretation is valid. As the number of subband images is increased, the size of the subband images decreases and the overall representation approaches that of a block transform, (i.e. a purely spatial-frequency domain representation).

This manner of extending 1-D filter banks to multiple dimensions is called a separable implementation. It has the advantage of being conceptually very simple, easily extendible to many dimensions, and computationally efficient. Nonseparable filter banks can also be considered. They may be conceptualized as a k-dimensional extension of the 1-D case from first principles. If, for example, we generalize n to the k-D vector $\mathbf{n}$ and $\downarrow M$ to $\downarrow\mathbf{M}$, where $\mathbf{M}$ is now a matrix, the equations derived previously in the context of 1-D signals resemble the equations for the k-dimensional extension. We hasten to caution against concluding that multi-dimensional filter banks are trivial extensions of the 1-D case. This is far from the truth. Rather we simply point out that the form of the equations have many similarities.

Before leaving this topic, it is appropriate to say a few words about multi-dimensional filtering, downsampling and upsampling, as these are the marrow of the filter banks. A k-dimensional FIR filter $h[\mathbf{n}]$, which can also be expressed equivalently as $h[n_1, n_2, \ldots, n_k]$, is a k-dimensional hypercube of filter coefficients, which is convolved with the k-dimensional input. Computational complexity is typically the major problem associated with filters of this type. In some select cases, efficiency can be achieved by using non-rectangularly separable implementations [426, 22, 37], but in general high complexity and multi-dimensional filters go hand-in-hand.

For the case of the downsamplers and upsamplers, in multiple dimensions, we have matrices that define the process. As an example, consider the 2-D downsampling matrix given by

$$\mathbf{M} = \begin{pmatrix} 1 & -1 \\ 1 & 1 \end{pmatrix}.$$

The 2-D downsampling operation

$$y[\mathbf{n}] = x[\mathbf{Mn}] \tag{1.15}$$

defined by this matrix reduces the number of samples by a factor of two as illustrated in Figure 13. The downsampling matrix discards the unshaded points in the figure, leaving only the pixels shown as black dots. The resulting pattern of pixels is known as a quincunx lattice. It is a simple matter to verify point-by-point that the 2-D downsampler in Equation (1.15) produces this result. The determinant of the downsampling matrix $\mathbf{M}$ indicates the factor by which the

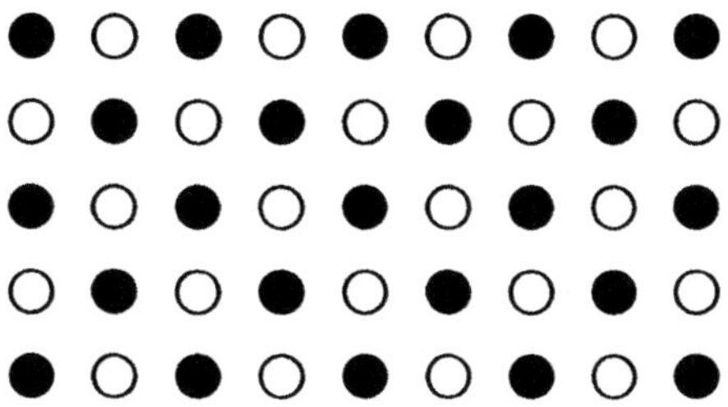

Figure 13 Illustration of the quincunx sampling lattice that results from 2-D downsampling with a quincunx downsampling matrix.

number of inputs samples is reduced, in this case 2. The theory of multidimensional digital signal processing is extensive; good treatments of the topic can be found in [130] and [246]. Good treatments of the theory of multidimensional filter banks can be found in [156, 223, 425, 15]. A brief discussion is also given in Chapter 4 of this book.

8 WAVELET TRANSFORMS

The wavelet transform is a relatively new linear transform for continuous-time signals. It is often compared to and contrasted with the classical short-time Fourier transform (STFT). Owing to the discrete-time focus of the previous sections, we introduce the STFT and wavelet transform as *continuous-time* analogs of the digital filter banks just discussed. In particular, one may view the STFT as the continuous-time analog of an M-band filter bank, and the wavelet transform as being the continuous-time analog of an octave band filter bank. To see this more clearly, consider the STFT, defined by

$$X(t,\Omega) = \int_{-\infty}^{\infty} h(t-\tau)x(\tau)e^{-j\Omega\tau}\,d\tau, \tag{1.16}$$

where $h(t)$ is a window function (which is typically a finite length bell-shaped curve). The window function may be interpreted as a lowpass filter, since its frequency response has a lowpass characteristic. Writing Equation (1.16) as

$$\begin{aligned} X(t,\Omega) &= e^{j\Omega t}\int_{-\infty}^{\infty} h(\tau)x(t-\tau)e^{-j\Omega\tau}\,d\tau \\ &= e^{j\Omega t}[x(t) * h(t)e^{-j\Omega t}], \end{aligned} \tag{1.17}$$

it is evident that $X(t,\Omega)$ is the convolution of the input $x(t)$ with a modulated lowpass filter $h(t)e^{-j\Omega t}$. When $\Omega = 0$, $h(t)e^{-j\Omega t}$ is the lowpass filter $h(t)$. For values of Ω greater than zero, we have an infinite set of bandpass filters, all with the same bandwidth and a continuum of center frequencies. The exponential term that precedes the convolution in (1.17) merely modulates the bandpass outputs down to baseband (i.e. to $\Omega = 0$). Hence the STFT in many ways resembles an M-band filter bank in continuous time. The bandwidth of the baseband filter, $h(t)$, defines the frequency resolution of the STFT, while the width of $h(t)$ in time, defines the time resolution. Clearly what we have is a time-frequency representation with uniform resolution in time and in frequency.

In contrast, the wavelet transform, $W_x(a,b)$ is defined as

$$W_x(a,b) = \frac{1}{\sqrt{a}} \int_{-\infty}^{\infty} \psi(\frac{t-b}{a}) x(t) dt, \tag{1.18}$$

where $a \in R^+$, $b \in R$, and $\psi(\frac{t-b}{a})$ is the wavelet basis. The basis functions are based on scaling by a and translating by b a mother wavelet $\psi(t)$. Like with $h(t)$, we may also picture $\psi(t)$ as a time-localized function (but in this case interpret it as a ***bandpass*** filter). The scaled and shifted versions of this mother wavelet correspond to bandpass filters with different bandwidths and different time durations. This is evident from the scaling property, which yields the Fourier transform pair

$$\psi(t) \leftrightarrow \Psi(\Omega). \tag{1.19}$$

Scaling the argument results in

$$\frac{1}{\sqrt{a}} \psi(\frac{t}{a}) \leftrightarrow \sqrt{a} \Psi(a\Omega), \tag{1.20}$$

where $a > 0$ is a continuous parameter. It is seen from Equation (1.20) that a contraction in one domain implies a dilation in the other. This implies that as a increases, so do the bandwidth and center frequency of $\psi(\frac{t}{a})$. The parameter b controls the time shifting of the function and behaves like the convolution variable. This leads to a nonuniform partitioning of the time-frequency space. Frequency resolution is good in the low frequency range but poor at high frequencies.

Looking at the behavior of $\psi(\frac{t-b}{a})$ in the time domain, we observe that it has good time resolution (i.e. it has a short time duration) at high frequencies, but poor time resolution at low frequencies. Hence it is analogous to the octave band filter bank in discrete time.

Since the wavelet transform, as described, is a function of two *continuous* parameters, this transform is more formally called the *continuous wavelet transform* or CWT. For computer implementations of the CWT, the parameters a and b are often oversampled to preserve accuracy, which yields a redundant representation. However, if sampled appropriately, the redundancy in the transform can be removed. If, for instance, the transform parameters a and b are sampled as

$$a = a_0^m \qquad b = nb_0a_0^m$$

so that

$$\frac{1}{\sqrt{a}}\psi(\frac{t-b}{a}) \leftrightarrow a_0^{-m/2}\psi(a_0^{-m}t - nb_0)$$

where $m, n \in Z$, the representation is efficient. This is called the *discrete wavelet transform* (DWT) because the parameters a and b are now discrete, but the signal it represents is continuous. As it turns out, the digital octave band filter bank can be used to generate the coefficients in the DWT. The relationships here are very interesting and are presented in the next chapter.

The scientific literature can be confusing with respect to wavelet related terminology. Often the liberal use of the term "wavelets" results in contradictions with basic definitions. Usage is a delicate issue at this point in time, due to the many conflicting opinions on this subject. As a case in point, the reader perusing the literature may find examples where authors have used the term *wavelet* as a descriptor for uniform band decompositions. The same reader may find other examples where wavelets are being contrasted with uniform decompositions as an illustration of its advantages (this latter position being consistent with the basic definitions). These contradictions can make this subject difficult to teach and the literature difficult to read. In this book we have tried to present wavelets and filter banks in a way that is consistent with their principal definitions. We speculate that wavelet terminology will be treated with more technical precision and care in the future as fine grain distinctions between representations become more important for clarity.

9 HISTORICAL OVERVIEW

Many ideas are now being exchanged on the theory of wavelets and analysis/synthesis filter banks, with attention being given to applications. Due to the large volume of contributions from many authors over many years, it would take an entire book to properly treat these contributions in a comprehensive manner. To this point, we have only mentioned the rudimentary elements of multirate filter banks and wavelet theory as a foundation for the discussions to come. In the remainder of this section, we try to fill in some of the other mainstream contributions to give a sense for how the area has evolved, and conclude with an overview of the topics discussed in subsequent chapters. As we begin this journey, some of the early methods for signal decomposition and the applications that shaped the course of research development are highlighted first.

The true genesis of filter banks might best be considered as a matter of opinion. For our purposes we will start by mentioning the early work in spectrum analysis for speech processing. The sound spectrograph is an analog device that was popular in the 1930's for generating time-frequency representations of speech. The device is based on using the input speech signal to modulate a variable frequency oscillator, the output of which is filtered by a bandpass filter. The average short-term energy is plotted over time and frequency (i.e. oscillator frequency) to give the time-frequency representation we know as the spectrogram. The spectrograph may be viewed, in a liberal way, as a form of filter bank.

Also during this time period was the introduction of the channel vocoder by Dudley [131]. It is the oldest type of digital speech coder and contains a cosine-modulated filter bank of the form of Equation (1.14). The filter bank outputs $y_k[n]$ are full-wave rectified, which is equivalent to taking the absolute value of the output samples. Next, each channel is lowpass filtered to achieve an approximation of the short-time spectrum. These output are used with voicing and pitch information to represent the speech signal at a low bit rate.

The discrete short-time Fourier transform[5], which is a uniform DFT filter bank, is another noteworthy example of a commonly used representation. It was popular in speech processing during the 1970s and early 80s and employs complex modulators as shown in Figure 14. The input is modulated by a complex exponential that effectively shifts high frequency regions of the spectrum to baseband. This is followed by a lowpass filter that attenuates all but the frequency region centered at the origin. Because spectral regions are translated to the origin, they can be lowpass filtered by $h[n]$ (which is called the analysis window or filter) and downsampled to produce a critically sampled representation. Typically the same analysis filter is used for all the channels, resulting in a uniform band DFT filter bank. Since the modulators are complex exponentials, the filter bank outputs are complex valued as are the outputs of the

[5]The discrete short-time Fourier transform to which we refer is a time-frequency decomposition for digital signals.

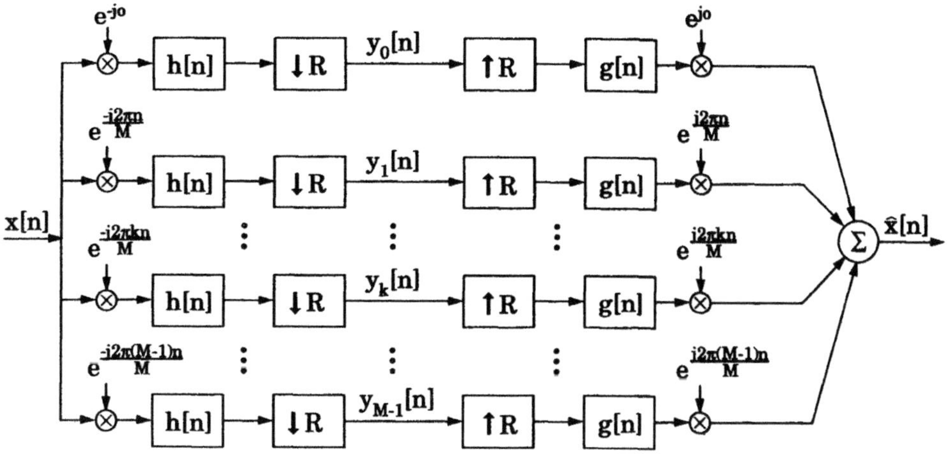

Figure 14 Block diagram of the discrete short-time Fourier transform. The analysis section is shown on the left. The synthesis section is on the right.

DFT. Pioneering work in this area was done by Portnoff [321], who among other things explored the relationship between the filter bank and transform interpretations of the discrete STFT, and derived the analysis/synthesis conditions for reconstruction.

By closely examining the figure, the analysis output equation for the M-channel discrete STFT can be seen to be

$$y_k[n] = \sum_{m=-\infty}^{\infty} h[nR - m]x[m]e^{-j2\pi kn/M} \quad k = 0, 1, \ldots, M-1.$$

The synthesis section associated with the discrete STFT is also shown in Figure 14. The reconstruction equation is given by

$$\hat{x}[n] = \sum_{m=-\infty}^{\infty} g[n - mR]\frac{1}{M}\sum_{k=0}^{M-1} y_k[m]e^{j2\pi kn/M}.$$

Baseband analysis and synthesis filters ($h[n]$ and $g[n]$) could be designed using the conditions derived in [321] to allow for approximate reconstruction. In cases where redundant time-frequency representations do not pose a problem, the full redundant representation (i.e. where $R = 1$) can be used. However, for the more common case where complexity and memory must be economized, the discrete STFT is typically decimated to its maximum (i.e. R=M), to achieve a minimally sampled representation. In this form, it may be implemented efficiently using a polyphase structure and the FFT algorithm.

As the theory of analysis/synthesis filter banks matured, more attention was given to systems employing aliasing cancellation, like two-band QMFs, octave-band QMF tree-structures, and the efficient filter banks we use today. The decade of the 1980s was a very active time for development. It was during this period that perfect reconstruction (PR) filter banks, and modern FIR modulated filter banks emerged. Interestingly, the development paths for PR and modulated filter banks have merged and crossed in several places. As pointed out in Section 3, the work in PR filter banks began with the solution for the two-band system [373]. A few years later, solutions for the M-band case were introduced [417, 296]. Also during the 1983-1984 time frame, the modulated filter banks of the form shown in Equation (1.14) were introduced. The first treatment of this appearing in the open literature was by Rothweiler who presented his work at ICASSP83 [345][6]. These modulated filter banks are similar, in a way, to the ones developed for transmultiplexers [284, 359], but differ in that they explicitly address analysis/synthesis reconstruction. This class of modulated filter banks cannot attain exact reconstruction, but is acceptable for many speech applications [85, 105].

The overlap between modulated filter banks and PR systems began with a solution introduced by Princen and Bradley [324], which they called time domain aliasing cancellation (TDAC). These filter banks are restricted to having the filter length L be less than or equal to twice the number of bands M, i.e. $L \leq 2M$. Malvar, Staelin, and Cassereau approached the problem from a transform perspective and arrived at essentially the same system, which they appropriately call lapped transforms, [255, 73]. Unique to this body of work is an elegant FIR implementation structure with very high computational efficiency, [254]. These filter banks are limited, however, in that the filter lengths are restricted to $L \leq 2M$, which is a fundamental limitation on the passband-stopband-transition band properties of the filter bank. As time marched on, it was discovered how to design PR modulated filter banks without this length constraint [253, 329, 292, 233]. With these new developments in theory and design tools, PR filter banks could now be viewed in a more unified light, with cosine modulation being a simple efficiency constraint that could be imposed on the system.

[6] Related work was done by Nussbaumer, as reported in the IBM company technical bulletin [303].

Concurrent with the modulated filter bank activity was work in IIR filter banks. The recursive nature of these filter banks leads naturally to computational efficiency. However, the reconstruction and stability constraints can be more difficult to handle [41, 328, 145], particularly for non-causal systems.

A major turning point in filter bank research came with the introduction of subband image coding by Woods and O'Neil [447] in 1986. New issues and questioned surfaced that had to be addressed as mentioned in Section 6. Toward the end of the 1980's many useful filter banks were available and effective design tools were in place. It was at this time that wavelets starting receiving some attention in the signal processing community. The theory of wavelet transforms, the ***continuous wavelet transform*** (CWT) and the ***discrete wavelet transform*** (DWT), are continuous signal representations, much like the Fourier transform and the Fourier series. However, by projecting these ideas into the discrete-time signal case, wavelets have been referenced in the digital processing context. As was mentioned in Section 8, there are some fundamental problems with adopting this perspective. What is clear, however, is that there is much uncharted territory to explore for both the continuous and discrete signal cases.

On the continuous signal side of the coin, there is mounting interest in applying wavelet transforms to a variety of analysis and communications problems, some of which are mentioned in Chapters 5, 6, and 10. On the discrete signal side of the coin, some of the new frontiers are time-varying filter banks, [294, 382, 120], designs that incorporate signal quantization, [177, 235, 386, 70, 224], systems based on non-conventional decimation, [295], and low delay filter banks [297, 403], just to mention a few. Some of the important challenges ahead lie in video and image transmission, data storage, detection and recognition, and enhancement. The chapters that follow discuss some of these as well as others.

In the early chapters of this book, we discuss theory and design, beginning with the fundamentals of linear time-frequency representations in Chapter 2. The constraint imposed by the uncertainty principle is presented. Issues such as energy compaction, entropy minimization, aliasing, phase and step response, flatness, MSE quantization, and computational efficiency are discussed in fair detail. Finally, the relationships and distinctions among subband trees, fast wavelet transforms, and wavelet packets are addressed.

In the following chapter, the emphasis shifts to the design of filter banks and wavelets. In particular, a time domain formulation is presented, which provides both a flexible analysis framework and a design methodology for a wide variety of different filter banks. In addition to the traditional M-band, tree, and modulated filter banks, several relatively new analysis/synthesis systems are also discussed, like low delay, nonuniform band, and time-varying filter banks.

Chapter 4 also discusses theory and design but focuses on IIR filter banks. Designs based on classical Butterworth and elliptic filters are discussed, as well as efficient two-dimensional decompositions. It is also shown how wavelets may be designed with the classical analog Butterworth design technique.

The next two chapter take on a different flavor, presenting wavelets from the mathematics community perspective. In particular, Chapter 5 presents the fundamentals of the CWT, and discusses properties and interpretations. This in addition to the reconstruction formula, discretization of the CWT, and extension of the CWT to two and three dimensions form the body of the chapter. The use of classical 2-D analyzing wavelets and the notion of directionally oriented analysis is discussed as well. A few applications are mentioned at the end and some general conclusions and perspectives are given. In contrast, Chapter 6 presents a more advanced treatment of wavelets. It focuses on a generalization of wavelet packets in which different filters are used at each level in the tree.

The remaining chapters of the book are bent toward applications and the use of subband filter banks and wavelets in practical situations. This half of the text begins with subband image coding, starting with the information theoretical motivation for the approach and perceptual aspects of the problem. The presentation not only treats the analysis/synthesis system issues, but also the hard-core compression issues that have a deep impact on the system performance, such as bit allocation, quantization, and entropy coding.

Chapter 8 is related to the previous one but differs in that it treats digital video and the issues unique to this 3-D signal. Traditional approaches to coding video are based on DCT transform coding of motion compensated prediction error frames. Chapter 8 discusses the use of subband coding in lieu of transform coding. Important issues such as how to effectively utilize motion compensation, how to extend 2-D coding to 3-D subband coding, and how to achieve resolution scalability are addressed.

Chapter 9 focuses on the application of the subband coding concept to high quality audio. Properties of the audition and the characteristics of the audio signal are examined, and used to formulate the analysis/synthesis filter bank and compression strategy. Issues such as pre-echo distortion, noise masking, and robustness to channel errors are all topics of discussion. It is quite remarkable how effective subband coding is for high quality audio.

Chapter 10 examines the application of filter banks to spread spectrum communications. This chapter begins with an overview of spread spectrum systems and the motivation for using such systems. A major part of the text focuses on interference suppression and the use of filter banks to accomplish this goal. In the last part of the chapter, waveform coding in subbands is discussed.

Acoustic echo cancellation is the topic of Chapter 11. The approach presented involves using analysis/synthesis filter banks to perform adaptive filtering. The principles of linear FIR adaptive filtering are presented first. Then the acoustic echo cancellation is discussed. Full-band adaptive filter architectures are contrasted with those of subband adaptive systems. Performance comparisons are presented in terms of mean square error, speed, and computational complexity.

The following chapter addresses the problem of speech scrambling for voice transmission. The chapter opens with an introduction to simple frequency and time domain techniques for analog speech scramblers. The notion of combined time-frequency scramblers is presented. It is shown how subband filter banks can be employed to achieve high levels of cryptanalytical strength.

The last chapter treats multitone modulation and the role DFT filter banks can play in this application. The chapter begins with an introduction to the concept of multitone transmission. This is followed by discussions on QAM multitone modulation, twisted pair channel issues, and the implementation of discrete multitone transceivers.

It is evident at this point that subband filter banks and wavelet transforms represent an active area of study. Although the applications discussed in this book do not represent a comprehensive description of the ways in which these transforms are being employed, they do represent some of the more visible uses of this technology. For further discussion of theory and applications, the reader is advised to consult the bibliography given at the end of the book.

2

FUNDAMENTALS AND OPTIMAL DESIGN OF SUBBAND AND WAVELET TRANSFORMS

Ali N. Akansu and Richard A. Haddad*

Department of Electrical and Computer Engineering, New Jersey Institute of Technology, Newark, New Jersey

**School of Electrical Engineering and Computer Science Polytechnic University, Brooklyn, New York*

1 INTRODUCTION

As noted in Chapter 1, a major thrust of research in subband filter banks has focussed on methods for achieving perfect reconstruction (PR), defined as the property wherein the reconstructed output signal is simply a delayed version of the input signal. There are several established categories of PR filter banks, the most popular of which are the paraunitary and the biorthogonal varieties, as well as a spate of others [7, 418, 430, 371]. While all of these insure PR, there is literally an infinite number of solutions, even for the simplest of filter structures. The space of feasible solutions can be narrowed by design procedures that optimize certain objective measures while maintaining perfect reconstruction as a constraint.

In this Chapter, we review the theory of PR filter banks as a prelude to a detailed discussion of various measures of optimality for parallel realizations and hierarchical trees.

2 FUNDAMENTALS

In this section, we present a review of the more salient aspects of signal decomposition by subband filter banks. The reader is invited to consult [7, 418, 430] for a more detailed treatment.

The basic objective of subband transforms is to separate the signal into uncorrelated frequency subbands by a filter bank prior to application specific processing. Each subband signal occupies only a fraction of the original frequency bandwidth and consequently can be downsampled to the Nyquist rate without loss of information. The most popular application is signal coding where subband signals are quantized and coded preferentially to achieve high quality performance.

2.1 Decimation and Interpolation in Multirate Filter Banks

In a multirate system, the signal samples are manipulated at different clock rates at various points in the configuration. First, the analog signal is bandlimited by an analog anti-aliasing filter to $f_s/2$ and then sampled at the Nyquist rate f_s to generate the full-band signal with spectral content on $[0, \frac{f_s}{2}]$. This signal is then filtered by a digital band-limiting filter so that it occupies only a fraction of the original full band spectrum. The filtered signal can then be downsampled by a factor, say M, so that it occupies a full band, but at the slower sampling rate $\frac{f_s}{M}$. The decimation process therefore consists of a band-limiting filter (e.g. lowpass, bandpass, or highpass) followed by a downsampler, as shown in Figure 1a.

The downsampler retains every M^{th} sample of $x(n)$ and re-labels the time index axis, so that

$$y(n) = v(Mn), \qquad \text{downsampler,} \tag{2.1}$$

$$y(n) = \sum_k h(Mn-k)x(k) \qquad \text{decimator.} \tag{2.2}$$

It is shown in [7, 418, 430] that the transform behaves as

$$Y(z) = \frac{1}{M}\sum_{k=0}^{M-1} V(z^{1/M}W^k), \qquad W = e^{-j2\pi/M} \tag{2.3}$$

or

$$Y(e^{j\omega}) = \frac{1}{M}\sum_{k=0}^{M-1} V(e^{j(\frac{\omega-2\pi k}{M})}). \tag{2.4}$$

Figure 1 illustrates the time and frequency domain consequences of downsampling. In this illustration, the downsampler operates at exactly the Nyquist rate

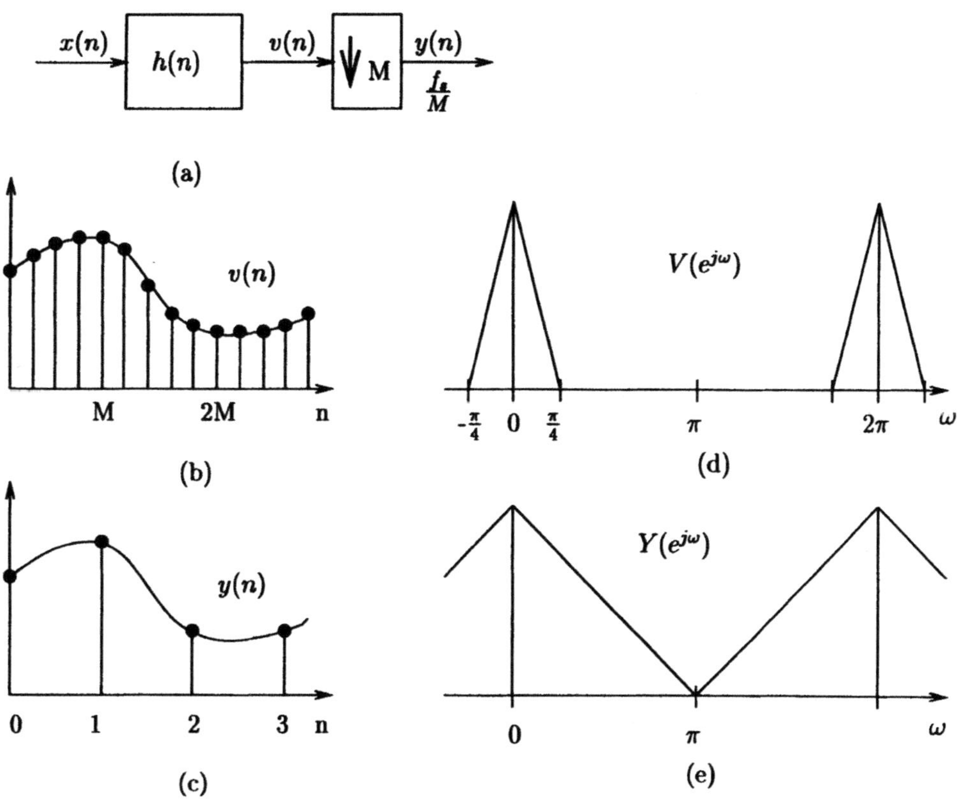

Figure 1 The decimation operation: (a) composite filter and downsampler; (b) filtered signal at fast clock rate; (c) downsampled signal at slow clock rate; (d) signal spectrum occupying $\frac{1}{4}$ of full band at fast clock rate, f_s; (e) spectrum of signal downsampled by 4, at slow clock rate, $f_s/4$.

for the signal shown. If $v(n)$ were downsampled at a lower rate, say $M = 5$ in this example, the triangular segments would overlap causing the distortion known as *aliasing*.

Interpolation is the process of increasing the sampling rate of a signal by an integer M. This is achieved by the upsampler, $\uparrow M$, and interpolating filter, $g(n)$. The upsampler inserts $(M-1)$ zeros between samples and re-indexes the time scale, thereby effectively increasing the clock rate by M. The operator is defined by

$$y(n) = \begin{cases} x(n/M), & n = 0, \pm M, \pm 2M, \ldots \\ 0, & \text{otherwise,} \end{cases} \tag{2.5}$$

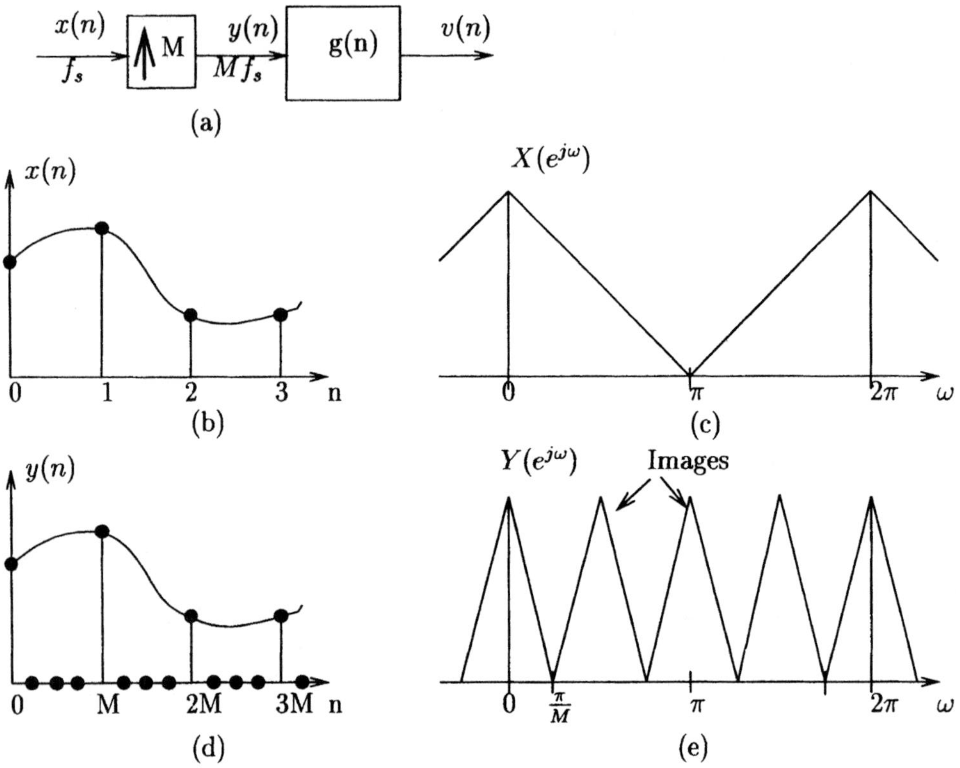

Figure 2 The interpolation operation: (a) upsampling and filtering, (b) input $x(n)$ in the time domain, (c) input $X(e^{j\omega})$ in the frequency domain, (d) upsampled signal $y(n)$ in time domain for $M = 4$, (e) upsampled signal $Y(e^{j\omega})$ in the frequency domain for $M = 4$.

and the interpolated output is

$$v(n) = \sum_k g(n - Mk)x(k). \tag{2.6}$$

It is easy to show that

$$Y(z) = X(z^M), \qquad Y(e^{j\omega}) = X(e^{j\omega M}). \tag{2.7}$$

As indicated in Figure 2, upsampling has two effects: first, stretching the time axis induces a compression in frequency; second, forcing the upsampled signal to pass through zero between samples of $x(n)$ generates high frequency signals

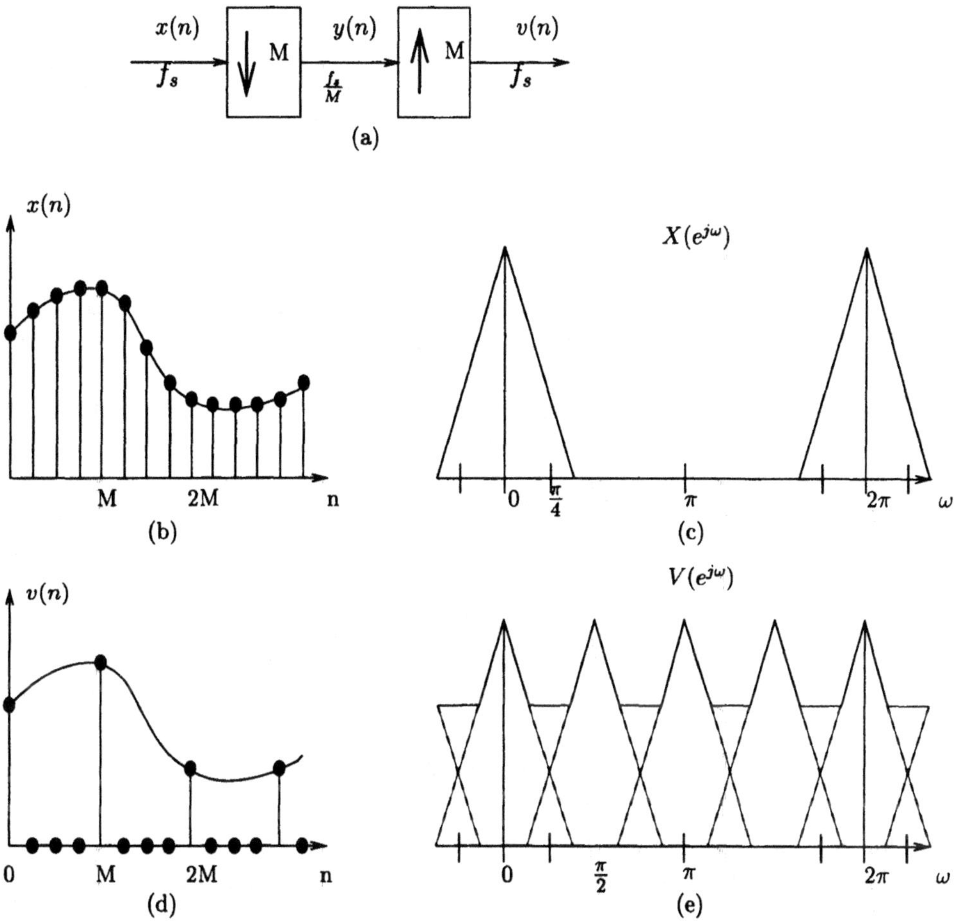

Figure 3 Downsampler/upsampler cascade: (a) the cascade combination; (b) input $x(n)$ in the time domain; (c) input $X(e^{j\omega})$ in the frequency domain; (d) output $v(n)$ in the time domain; (e) output $V(e^{j\omega})$ in the frequency domain.

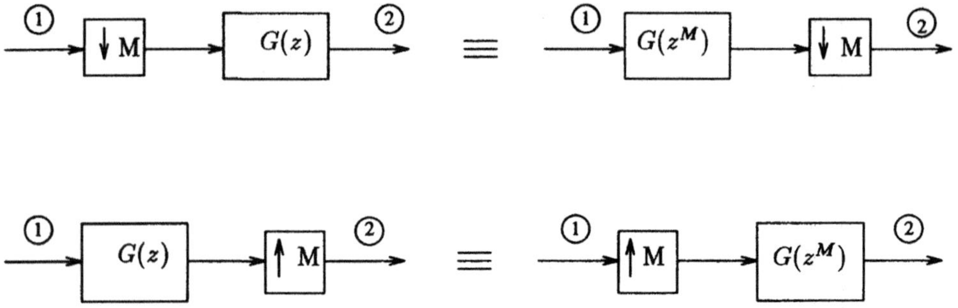

Figure 4 Equivalent structures.

or *images*, as shown in Figure 2e. The purpose of the interpolating filter, $g(n)$, is to eliminate these images by smoothing the upsampled signals.

Back-to-back downsampling-upsampling is shown in Figure 3 along with time and frequency domain plots of the signals. It is evident that

$$v(n) = \begin{cases} x(n) & n = 0, \pm M, \pm 2M, \ldots \\ 0 & \text{otherwise,} \end{cases} \tag{2.8}$$

and that

$$V(z) = \frac{1}{M} \sum_{k=0}^{M-1} X(zW^k) \tag{2.9}$$

$$V(e^{j\omega}) = \frac{1}{M} \sum_{k=0}^{M-1} X(e^{j(\omega - \frac{2\pi k}{M})}). \tag{2.10}$$

This latter form shows that $V(e^{j\omega})$ is simply the sum of M replicas of $X(e^{j\omega})$ spaced $\frac{2\pi}{M}$ apart. Figure 3e illustrates both aliasing and imaging.

Transfer functions can be moved across samplers as indicated by the equivalent structures shown in Figure 4. Manipulations based on these identities are used in the subsequent sections.

3 POLYPHASE DECOMPOSITION

A polyphase decomposition is a regrouping of terms in the z-transform of a signal which is useful in multirate systems. It also can lead to more computa-

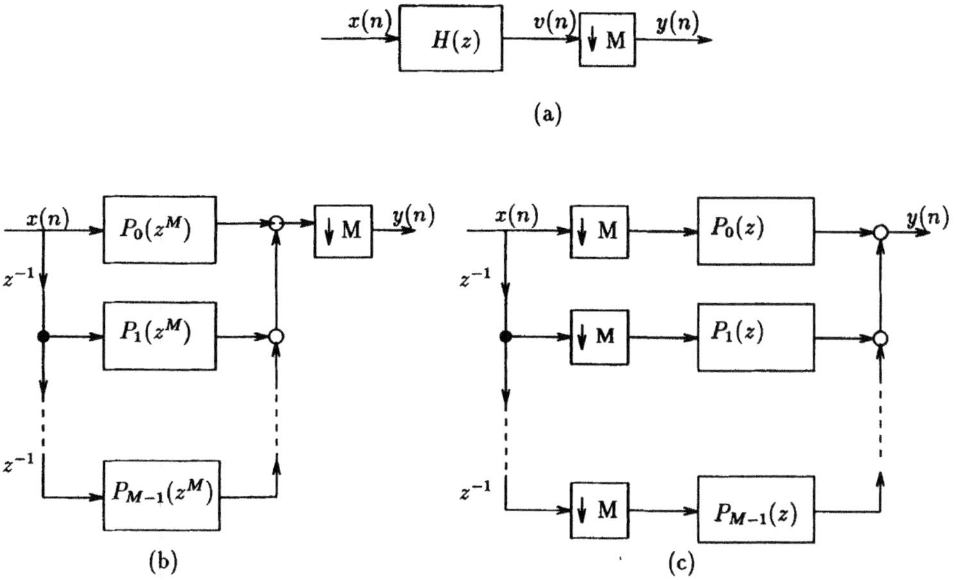

Figure 5 Polyphase decomposition of decimator: (a) decimator; (b) polyphase representation; (c) equivalent structure after transposing the down-samplers.

tionally efficient filter realizations [108]. For any filter we can write

$$\begin{aligned} H(z) &= \sum_{k=-\infty}^{\infty} h(k)z^{-k} \\ &= (\ldots + h(-M)z^{M} + h(0) + h(M)z^{-M} + h(2M)z^{-2M} + \ldots) + \\ &\quad z^{-1}(\ldots + h(-M+1)z^{M} + h(1) + h(M+1)z^{M} + h(2M+1)z^{-2M} + \ldots) + \\ &\quad \ldots + z^{-(M-1)}(\ldots + h(-1)z^{M} + h(M-1) + h(2M-1)z^{-2M} + \ldots) \\ &= \sum_{r=0}^{M-1} z^{-r} P_r(z^M) \end{aligned} \tag{2.11}$$

where

$$P_r(z) = \sum_{\ell=-\infty}^{\infty} h(r+\ell M)z^{-\ell}, \qquad r = 0, 1, \ldots, M-1. \tag{2.12}$$

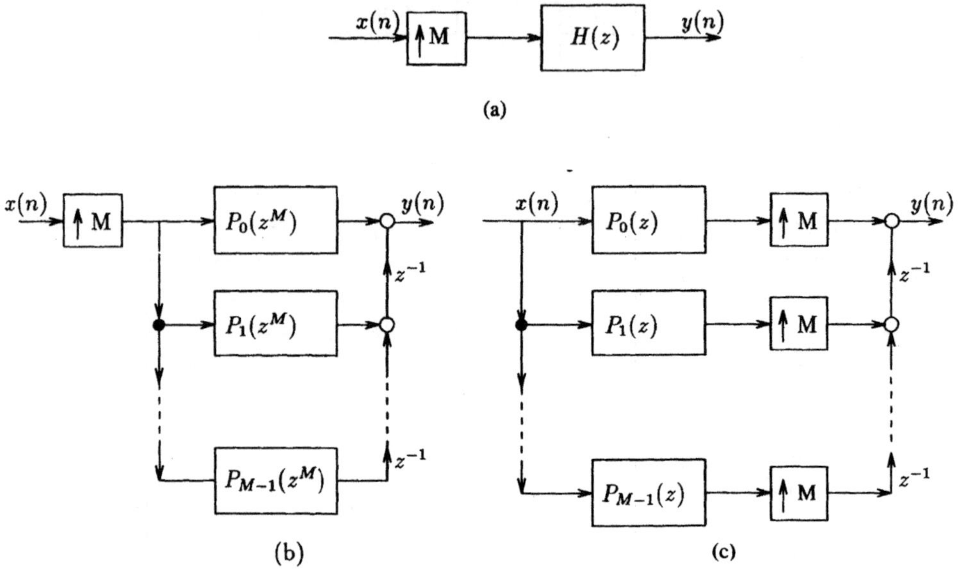

Figure 6 Polyphase decomposition of interpolator: (a) interpolator; (b) polyphase representation of filters; (c) equivalent structures after transposing the upsampler.

The polyphase filters $P_r(z)$ are recognized as shifted and downsampled versions of the filter $h(n)$. Therefore, from Equation (2.3), it follows that

$$P_r(z) \quad = \quad \frac{z^{r/M}}{M} \sum_{m=0}^{M-1} W^{mr} H(z^{1/M} W^m). \tag{2.13}$$

The three equations (2.11), (2.12), (2.13) define the polyphase analysis and synthesis equations. The polyphase decompositions of a decimator and interpolator are illustrated in Figures 5 and 6. Each filter is replaced by its polyphase equivalent. Then the downsamplers and upsamplers are transposed with these polyphase components in accord with the equivalences suggested by Figure 4.

4 UNIFORM M-BAND FILTER BANKS

The equal bandwidth M-band analysis/synthesis filter bank considered in this chapter is shown in Figure 7. The bank of filters $\{H_k(z)\}$ constitutes the analysis filters. Each filtered signal is downsampled by the same decimation

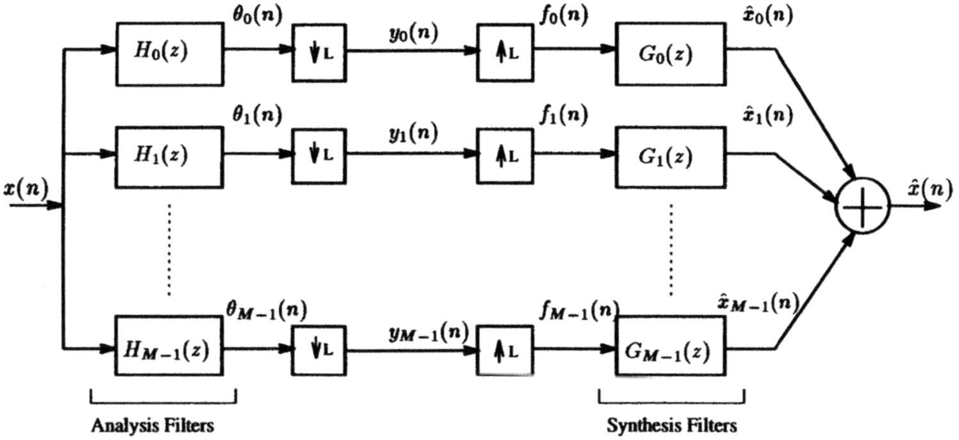

Figure 7 Uniform M-band analysis/synthesis filter bank.

ratio L. The bank of upsamplers/synthesis filters reconstructs the signal. The decimation factor satisfies $L \leq M$. For critical sampling $L = M$, and the total data rate in samples per second is unaltered from input $x(n)$ to the set of downsampled signals $\{y_k(n)\}$. The analysis and synthesis filters are FIR, of length N, and N' respectively.

Nonuniform filter banks can have variable decimation factors, which depend on the subband. Perfect reconstruction and design procedures for such nonuniform structures are described in references [299] and [418].

Our objective here is to analyze the structure of Figure 7 using the "aliasing component" (or AC) matrix, and the polyphase matrix approaches [371, 416]. From these, we can derive perfect reconstruction requirements, and paraunitary and biorthogonal solutions.

The AC matrix formulation

From Figure 7, and Equation (2.9) we have

$$\theta_k(z) = H_k(z)X(z), \qquad k = 0, 1, \ldots, M-1 \qquad \text{analysis,} \qquad (2.14)$$

$$\hat{X}(z) = \sum_{k=0}^{M-1} G_k(z)F_k(z), \qquad \text{synthesis,} \qquad (2.15)$$

$$F_k(z) = \frac{1}{M}\sum_{r=0}^{M-1} \theta_k(zW^r) \qquad \text{down/upsampler.} \tag{2.16}$$

Combining these gives

$$\hat{X}(z) = \frac{1}{M}\sum_{r=0}^{M-1}\sum_{k=0}^{M-1} X(zW^r)H_k(zW^r)G_k(z). \tag{2.17}$$

We can write this last equation compactly as

$$\begin{aligned}\hat{X}(z) &= \frac{1}{M}[X(z),\ldots,X(zW^{M-1})] \\ &\cdot \begin{bmatrix} H_0(z) & \ldots & H_{M-1}(z) \\ H_0(zW) & \ldots & H_{M-1}(zW) \\ \vdots & \vdots & \vdots \\ H_0(zW^{M-1}) & \ldots & H_{M-1}(zW^{M-1}) \end{bmatrix} \begin{bmatrix} G_0(z) \\ G_1(z) \\ \vdots \\ G_{M-1}(z) \end{bmatrix} \\ &= \frac{1}{M}\underline{X}^T H_{AC}(z)\underline{g}(z), \end{aligned} \tag{2.18}$$

where H_{AC} is the "aliasing component" or AC matrix, and $\underline{h}(z)$ and $\underline{g}(z)$ are vectors representing the analysis and synthesis filters

$$\begin{aligned}\underline{h}^T(z) &= [H_0(z),\ldots,H_{M-1}(z)], \\ \underline{g}^T(z) &= [G_0(z),\ldots,G_{M-1}(z)], \\ &\text{and} \\ \underline{X}^T(z) &= [X(z),\ldots,X(zW^{M-1})]. \end{aligned} \tag{2.19}$$

Two kinds of distortion terms can be deduced from Equation (2.18).

(i) Aliasing distortion. The sampling operations in the filter bank make the system time-varying and produce aliasing and imaging. These can be eliminated if we impose

$$H_{AC}\cdot\underline{\mathrm{g}}(z) = \begin{bmatrix} MT(z) \\ 0 \\ \vdots \\ 0 \end{bmatrix}. \tag{2.20}$$

In this case, the input-output relation reduces to that of a time-invariant system

$$\begin{aligned}\hat{X}(z) &= \left[\sum_k G_k(z)H_k(z)\right]X(z) \\ &= T(z)X(z). \end{aligned} \tag{2.21}$$

(ii) Magnitude and Phase Distortion. Having constrained $H_k(z)$ and $G_k(z)$ to eliminate aliasing via Equation (2.20), we are left with classical magnitude and phase distortions implicit in $T(e^{j\omega})$,

$$T(e^{j\omega}) = |T(e^{j\omega})|e^{j\phi(\omega)}. \tag{2.22}$$

Perfect reconstruction requires $T(z) = z^{-n_0}$, a pure delay, or

$$T(e^{j\omega}) = e^{-jn_0\omega}. \tag{2.23}$$

Deviation of $|T(e^{j\omega})|$ from unity and $\phi(\omega)$ from linearity constitutes magnitude and phase distortion, respectively. Thus perfect reconstruction requires that all three errors be zero.

The Polyphase Representation. The reconstruction formula using the AC matrix, Equation (2.18) can be formulated in an alternative way using polyphase representations. Following Equation (2.11), each analysis filter can be decomposed as

$$H_r(z) = \sum_{k=0}^{M-1} z^{-k} E_{r,k}(z^M), \qquad r = 0, 1, \ldots, M-1. \tag{2.24}$$

When this is done for each analysis filter, we can stack the results to obtain

$$\begin{bmatrix} \theta_0(z) \\ \theta_1(z) \\ \vdots \\ \theta_{M-1}(z) \end{bmatrix} = \begin{bmatrix} E_{00}(z^M) & \cdots & E_{0,M-1}(z^M) \\ E_{10}(z^M) & \cdots & E_{1,M-1}(z^M) \\ \vdots & & \\ E_{M-1,0}(z^M) & \cdots & E_{M-1,M-1}(z^M) \end{bmatrix} \begin{bmatrix} 1 \\ z^{-1} \\ \vdots \\ z^{-(M-1)} \end{bmatrix} X(z)$$

or

$$\underline{\theta}(z) = [\mathcal{E}_p(z^M)\underline{Z}_M]X(z) = \underline{h}(z)X(z) \tag{2.25}$$

where $\mathcal{E}_p(z)$ is the polyphase matrix, and $\underline{Z}_M$ is a vector of delays.

Similarly, we can represent the synthesis filters by

$$G_s(z) = \sum_{k=0}^{M-1} z^{-k} R_{s,k}(z^M), \tag{2.26}$$

$$R_{s,k}(z) = \sum_{n=-\infty}^{\infty} g_s(k + Mn)z^{-n}. \tag{2.27}$$

With the synthesis polyphase matrix $\mathcal{R}_p$ defined in the same way as $\mathcal{E}_p(z)$ above, we can replace Figure 7 by its equivalent in Figure 8a, where

$$J\mathcal{R}_p^T(z) \triangleq \mathcal{R}_p'(z) \tag{2.28}$$

and J is the *counter-identity* matrix

$$J = \begin{bmatrix} 0 & 0 & \dots & 0 & 1 \\ 0 & 0 & \dots & 1 & 0 \\ \vdots & & & & \\ 1 & 0 & \dots & 0 & 0 \end{bmatrix}. \tag{2.29}$$

Defining $\underline{L}(z)$ as the vector $\underline{L}^T(z) = [L_0(z), L_1(z), \dots, L_{M-1}(z)]$, it is easy to verify that

$$\begin{aligned} \hat{X}(z) &= \underline{L}^T(z)J\underline{Z}_M \quad \text{and} \\ \underline{L}(z) &= \mathcal{R}_p'(z^M)\underline{F}(z). \end{aligned} \tag{2.30}$$

The synthesis equations become

$$\hat{X}(z) = \underline{F}^T(z)\mathcal{R}_p(z^M)\underline{Z}_M = \underline{F}^T(z)\underline{g}(z). \tag{2.31}$$

Note that we have defined the analysis and synthesis polyphase matrices in the same way so as to result in

$$\begin{aligned} \underline{h}(z) &= \mathcal{E}_p(z^M)\underline{Z}_M, \\ \underline{g}(z) &= \mathcal{R}_p(z^M)\underline{Z}_M. \end{aligned} \tag{2.32}$$

Shifting the downsamplers and upsamplers to the left and the right of the respective polyphase matrices, and replacing z^M by z in these arguments gives the equivalent structure of Figure 8b. Note that the structure from $\underline{\xi}(n)$ to $\underline{\eta}(n)$ operating at the slow clock rate is linear and time-invariant. The vector matrix equivalent of this time-invariant structure is shown in Figure 8c,

$$\underline{\eta}(z) = [\mathcal{R}_p'(z)\mathcal{E}_p(z)]\underline{\xi}(z) = \mathcal{Q}(z)\underline{\xi}(z), \tag{2.33}$$

from which we may discern the roles of the polyphase matrices.

$\mathcal{Q}(z)$ is an $M \times M$ matrix, each component of which consists of sums of products of FIR polyphase components, and hence is itself FIR. A sufficient condition for perfect reconstruction is $\mathcal{Q}(z) = z^{-\mu}I$, μ is an integer, and I the identity matrix. This implies [7], $\hat{x}(n) = x(n - n_0)$, $n_0 = \mu M + M - 1$.

Equation (2.33) also implies a time-domain, block filtering interpretation. It is convenient for us to now denote the lengths of the analysis and synthesis filters by NM and $N'M$, respectively. Then $\mathcal{R}'_p(z)$, and $\mathcal{E}_p(z)$ can be expanded as

$$\mathcal{E}_p(z) = \sum_{k=0}^{N-1} E_{p,k} z^{-k}, \qquad \mathcal{R}'_p(z) = \sum_{k=0}^{N'-1} R'_{p,k} z^{-k} \tag{2.34}$$

where $(E_{p,k}, R'_{p,k})$ are $M \times M$ coefficient matrices in the time-domain. The block filtering operation is

$$\underline{\eta}(n) = \sum_k Q_{n-k} \underline{\xi}(k), \qquad Q_n = \sum_{k=0}^{N'-1} R'_{p,k} E_{p,n-k}. \tag{2.35}$$

The PR requirements in the time domain is simply, $Q_n = I\delta(n-\mu)$.

The connection between polyphase and AC matrices can be derived by substituting the polyphase expansion for $H_k(z)$ of Equation (2.24) into the definition of $H_{AC}(z)$ of Equation (2.18). This leads to

$$H_{AC}(z) \quad = \quad \mathcal{W} A(z) \mathcal{E}_p^T(z^M) \tag{2.36}$$

where $\mathcal{W}$ is the DFT matrix $[W^{nk}]$, $W = e^{j2\pi/N}$, and

$$A(z) \quad = \quad diag\{1, z^{-1}, \ldots, z^{-(M-1)}\}. \tag{2.37}$$

The literature on PR solutions is extensive [7, 274, 373, 416, 418, 430, 428]. For the needs of this chapter, we confine the discussion to two classes of solutions: the paraunitary or orthonormal, and the biorthogonal. Recall that the sufficient condition for PR in terms of the AC matrix is

$$\frac{1}{M} H_{AC} \underline{g}(z) \quad = \quad \begin{bmatrix} z^{-n_0} \\ 0 \\ \vdots \\ 0 \end{bmatrix}, \tag{2.38}$$

and for the polyphase formulation

$$\mathcal{R}'_p(z)\mathcal{E}_p(z) \quad = \quad z^{-\mu} I, \tag{2.39}$$

where μ is selected to make the FIR filter causal. Without loss of generality the delay μ could be made zero.

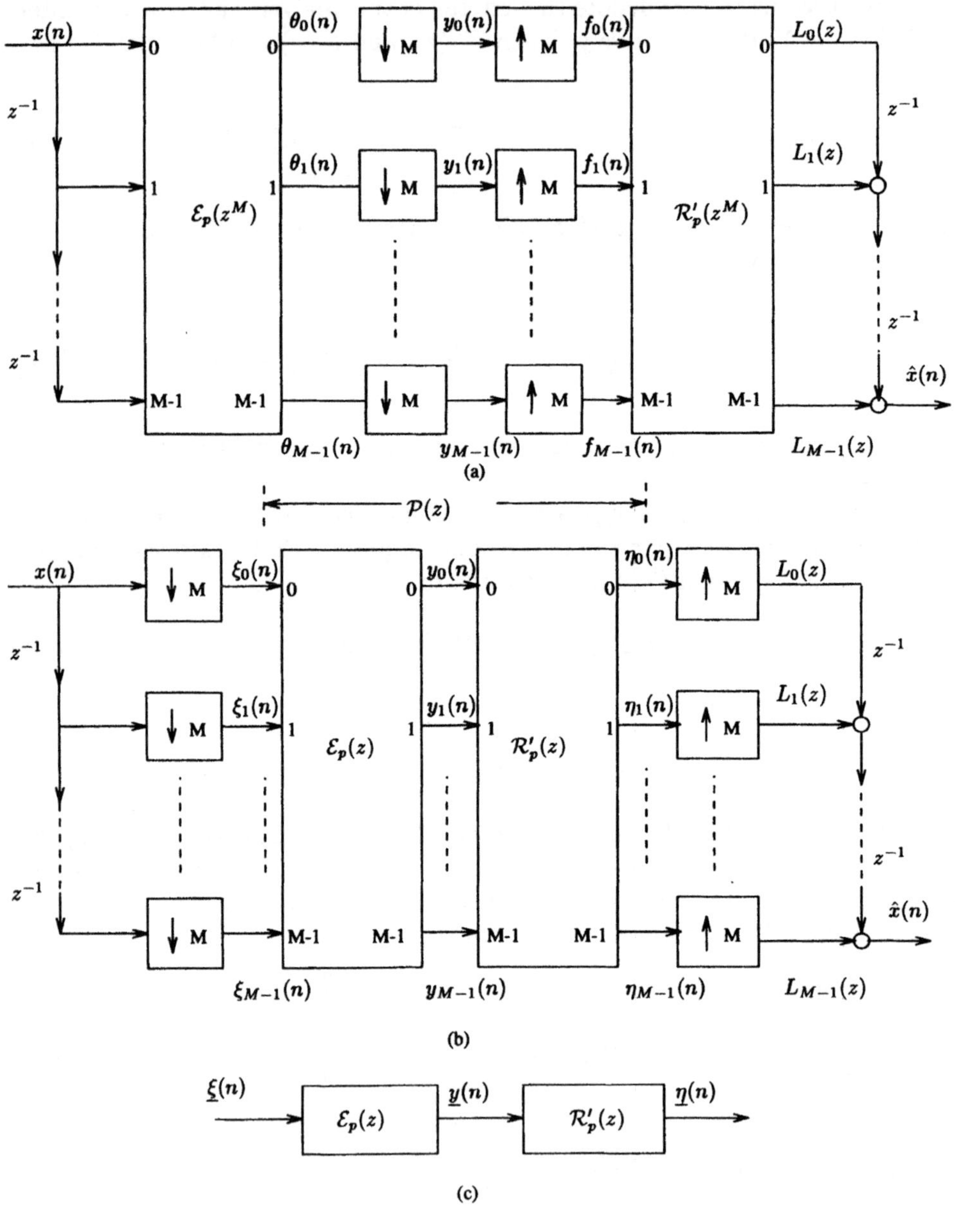

Figure 8 (a) Polyphase representation of filter bank; (b) equivalent polyphase structure; (c) slow clock, time-invariant, vector matrix representation.

The Orthonormal Solution. For this case, we choose $\mathcal{E}_p(z)$ to be paraunitary,

$$\begin{aligned} \tilde{\mathcal{E}}_p(z)\mathcal{E}_p(z) &= I, \\ \tilde{\mathcal{E}}_p(z) &\triangleq \mathcal{E}_p^T(z^{-1}). \end{aligned} \tag{2.40}$$

And, if this condition is met, we can simply choose

$$\mathcal{R}_p'(z) = z^{-\mu}\mathcal{E}_p^T(z^{-1}) = z^{-\mu}\tilde{\mathcal{E}}_p(z). \tag{2.41}$$

From Equation (2.32), this in turn implies that each synthesis filter is just a time-reversed version of the analysis filter

$$G_k(z) = z^{-\mu}H_k(z^{-1}) = z^{-\mu}\tilde{H}_k(z). \tag{2.42}$$

Using Equation (2.36), we can show that a paraunitary $\mathcal{E}_p(z)$ implies a paraunitary $H_{AC}(z)$, and conversely, so that

$$\frac{1}{M}\tilde{H}_{AC}(z)H_{AC}(z) = I, \tag{2.43}$$

where $\tilde{H}_{AC}(z) \triangleq \tilde{H}_{AC*}^T(z^{-1})$, and the subscripted asterisk implies conjugation of the coefficients in the matrix. In [7], it is shown that the necessary and sufficient conditions for a paraunitary filter bank are as follows[1]: let

$$\Phi_{rs}(z) \triangleq \tilde{H}_r(z)H_s(z) \leftrightarrow \rho_{rs}(n) = h_r(-n) * h_s(n). \tag{2.44}$$

Then

$$\boxed{\rho_{rs}(Mn) = \textstyle\sum_k h_r(k)h_s(k+Mn) = \delta(r-s)\delta(n)} \tag{2.45}$$

For $r = s$, we see that $\rho_{rr}(Mn) = \delta(n)$ which implies that $\Phi_{rr}(z) = H_r(z^{-1})H_r(z)$ is the transfer function of an Mth band filter and $H_r(z)$ must be a spectral factor of $\Phi_{rr}(z)$. In the time domain, Equation (2.45) implies

$$(1) \quad \textstyle\sum_k |h_r(k)|^2 = 1, \qquad \text{normalization} \tag{2.46}$$

$$(2) \quad \textstyle\sum_k h_r(k)h_r(k+Mn) = 0, \qquad n \neq 0 \tag{2.47}$$

$$(3) \quad \textstyle\sum_k h_r(k)h_s(k+Mn) = 0, \qquad r \neq s. \tag{2.48}$$

Equation (2.47) asserts that $h_r(k)$ is orthogonal to its translates shifted by M, while Equation (2.48) for $r \neq s$, implies $h_r(k)$ is orthogonal to all M translates

[1] We assume real-valued FIR filter coefficients.

of $h_s(k)$. This is the time domain equivalent of aliasing cancellation. Another consequence of the paraunitary AC matrix is that the filter bank is power complementary, i.e.

$$\frac{1}{M}\sum_{r=0}^{M-1} |H_r(e^{j\omega})|^2 = 1. \tag{2.49}$$

By Equation (2.42), the same paraunitary requirements are imposed on the synthesis filters as well.

For the two-band case, it is easy to demonstrate that paraunitary filters can not have linear phase except for trivial cases, such as $H_0(z) = (1+z^{-1})/\sqrt{2}$, $H_1(z) = (1-z^{-1})/\sqrt{2}$. PR linear phase analysis and synthesis filters in the two-band case can be realized by relaxing the paraunitary constraint, leading to the so-called biorthogonal solutions. In this solution, orthogonality is distributed across the analysis and synthesis sections [7, 430]. For the M-band case, the result is a generalization of Equation (2.45)

$$\rho_{rs}(Mn) = \sum_k h_r(k) g_s(Mn-k) = \delta(r-s)\delta(n). \tag{2.50}$$

Equation (2.50) permits wider latitude in the selection of the filter coefficients, and, in particular, it allows linear phase filters in the two-band case.

It is emphasized, however, that the paraunitary and the biorthogonal requirements for perfect reconstruction really constitute a set of constraints. The solutions to Equations (2.45) and (2.50) are not unique. There are more filter coefficients than constraint equations, and the excess of coefficients can be used to satisfy appropriate optimality measures as described in Section 8.

5 THE TWO-BAND CASE AND TREE EXTENSIONS

The two-band case, shown in of Figure 9 is important in its own right, and serves as a springboard for canonic tree structures. These filter banks serve as one of the prototype decomposition cells in adaptively structured subband trees [10]. Additionally, two-band paraunitary filter banks play a vital role in the wavelet theory which will be discussed later in detail. We will review the paraunitary and biorthogonal realizations. The paraunitary PR requirements are summarized below:

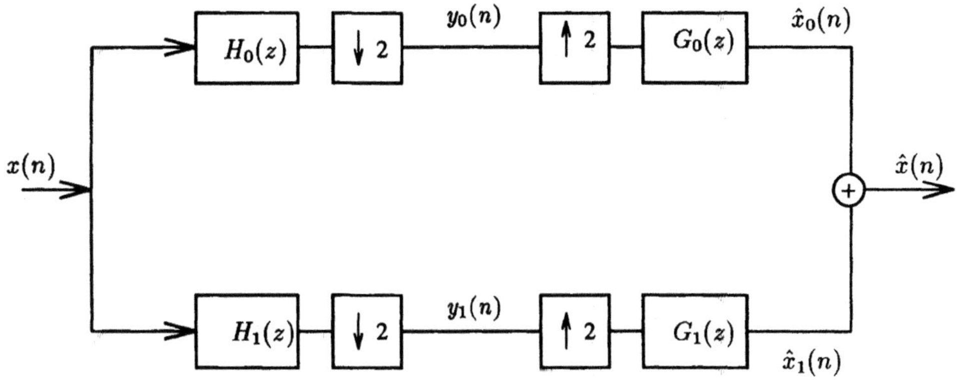

Figure 9 Two-band filter bank.

All filters are N tap and N is even:

$$\begin{aligned} \sum_k h_0(k)h_1(k+2n) &= 0 \\ \sum_k h_i(k)h_i(k+2n) &= \delta(n), \qquad i=0,1 \end{aligned} \tag{2.51}$$

$$\begin{aligned} H_i(z^{-1})H_i(z) + H_i(-z^{-1})H_i(-z) &= 2, \qquad i=0,1 \\ H_0(z^{-1})H_1(z) + H_0(-z^{-1})H_1(-z) &= 0 \end{aligned} \tag{2.52}$$

$$\begin{aligned} H_1(z) &= z^{-(N-1)}H_0(-z^{-1}) \\ G_0(z) &= z^{-(N-1)}H_0(z^{-1}) \\ G_1(z) &= H_0(-z) \end{aligned} \tag{2.53}$$

$$|H_0(e^{j\omega})|^2 + |H_1(e^{j\omega})|^2 = 1, \quad \text{power complementary.} \tag{2.54}$$

If we assume $x(n) \in L^2(Z)$, i.e. $\sum_n |x(n)|^2 < \infty$, we can view the two-band analysis/synthesis filter bank as an orthonormal decomposition and reconstruction of $x(n)$. The sequences $h_0(n)$, and $h_1(n)$ are orthonormal sequences spanning complementary orthogonal subspaces of $L^2(Z)$. That is, $h_0(n)$, $h_1(n)$ and their even translates form an orthonormal basis for $L^2(Z)$.

In the top branch of Figure 9, the signal $x(n)$ is filtered by the lowpass filter $h_0(n)$ and downsampled by 2 to produce the coarse half-resolution approximation $y_0(n)$. Upsampling by 2 and interpolating via lowpass $g_0(n) = \tilde{h}_0(n)$

gives $\hat{x}_0(n)$, a coarse approximation (or projection of $x(n)$ onto the subspace spanned by $h_0(n)$. Similarly, passing $x(n)$ through the highpass $h_1(n)$ and downsampling by 2 gives $y_1(n)$, the highpass or *detail* signal at half resolution. Upsampling and interpolating via $g_1(n) = \tilde{h}_1(n)$ gives $\hat{x}_1(n)$, the projection of $x(n)$ onto the subspace spanned by $h_1(n)$. The sum of $\hat{x}_0(n)$ and $\hat{x}_1(n)$ gives $\hat{x}(n) = x(n - n_0)$.

The decomposition of a signal into two bands or signals could be continued by successive decomposition of each of the half-resolution subband signals. Several tree structures capable of differing decompositions have been studied [10, 11].

The *binary* tree structure of Figure 10a decomposes the signal into four subband signals $y_0(n)$, ..., $y_3(n)$, each of which has energy concentrated in four equal subbands. The binary tree can be developed as a succession of nested two-band PR realizations. In Figure 10b, the requirements are that the nests $\{H_0, G_0, H_1, G_1\}$, $\{H_0', G_0', H_1', G_1'\}$, $\{H_0'', G_0'', H_1'', G_1''\}$ are each two-band PR. Using the equivalent structures of Figure 4, we can construct the equivalent four-band structure depicted in Figure 10b. Each filter in this bank is realized as the *product filter* shown. Single-level M-band paraunitary filter banks can be constructed this way, using two-band designs. Since $H_i(z)$ and $H_i'(z)$, $i = 0, 1$ need not be the same, there is wide latitude in designing such structures. However, linear phase cannot be realized, and this filter bank is more constrained than a freely chosen paraunitary M-band configuration. Figure 11 shows the equivalent four-band split where identical eight-tap Binomial QMF filters are used in each of the two-band steps.

An L-level binary tree splits the frequency band into 2^L equal-width bands and provides a maximum frequency resolution of $\pi/2^L$. Reference [7] further elaborates on these points.

The *dyadic* or *octave*-band tree splits the lower half of the spectrum into two equal bands at any level of the tree. The higher half-band component of the signal at any level of the tree is decomposed no further.

The three-level dyadic tree of Figure 12a provides the octave frequency band split shown in Figure 12b, and a multiresolution decomposition at each node. Observe that this structure is critically sampled so that the sum of the data rates at the four end nodes equals f_s. The resolution at each node is also indicated by the data rate at that node. The resolution of the input $x(n)$ is 1 corresponding to f_s, while the resolution at the output nodes is 1/8, 1/8, 1/4, 1/2, corresponding to frequency bands LLL, LLH, LH, H. Every downsampling by 2 reduces the resolution by that factor.

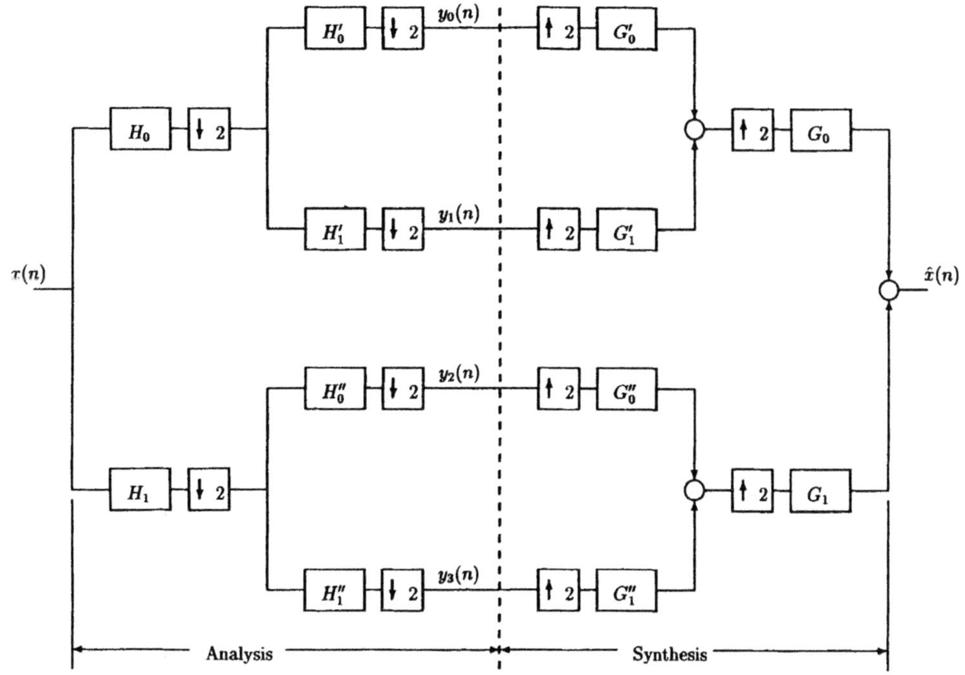

(a)

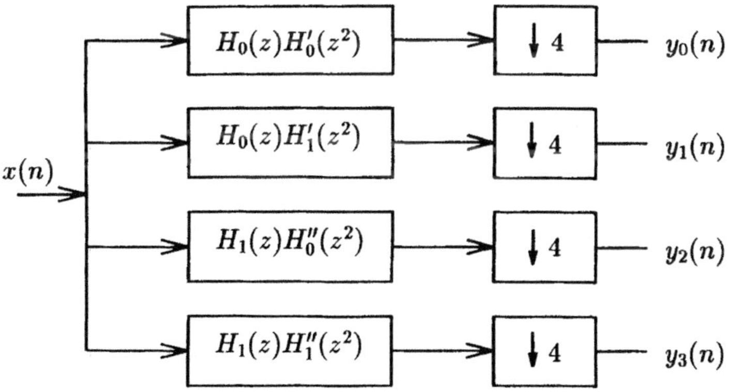

(b)

Figure 10 (a) Four-band two-level binary analysis/synthesis tree structure; (b) four-band single level equivalent filter bank.

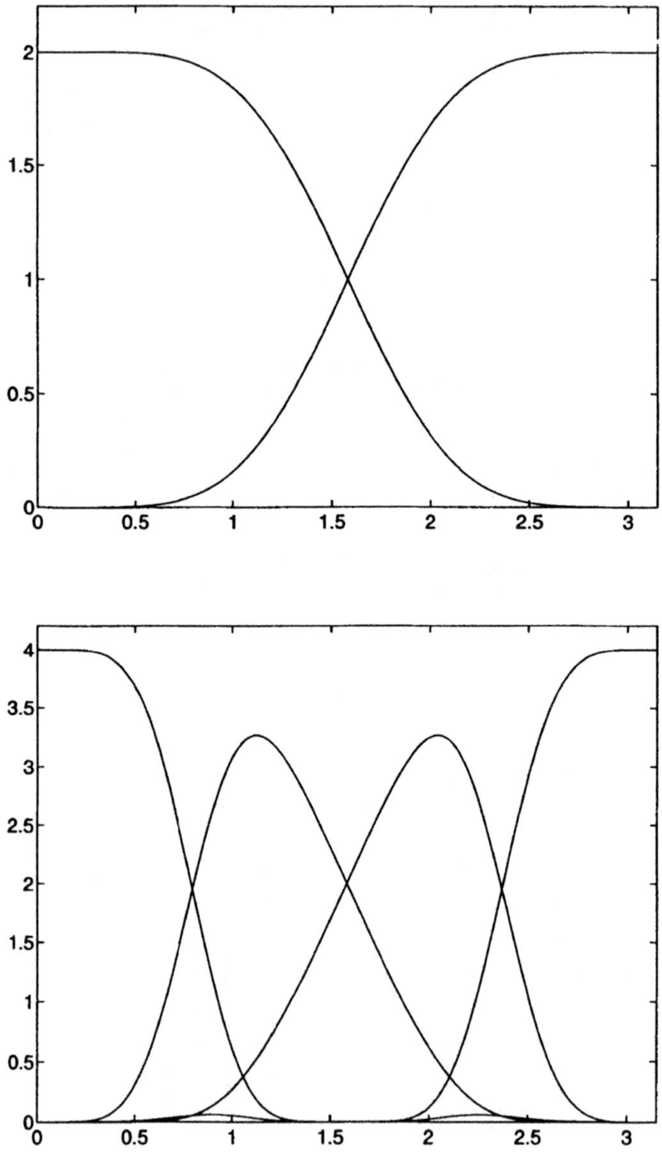

Figure 11 Frequency responses of 8-tap binomial-QMFs for a four-band binary tree using the same filters throughout.

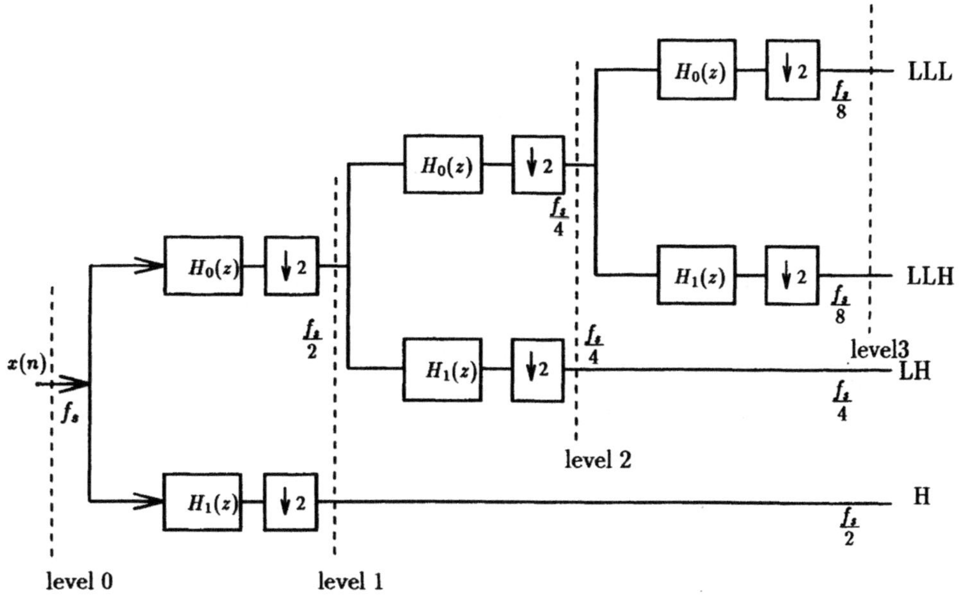

(a)

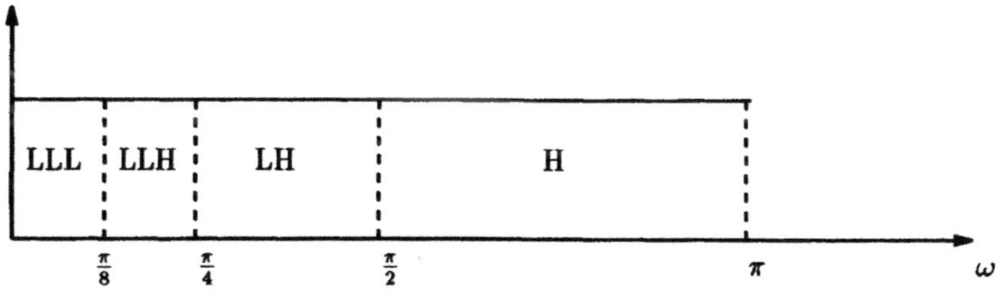

(b)

Figure 12 (a) Three level dyadic tree; (b) idealized frequency band decomposition.

At any level in the tree the signal is represented by a bandpass component, which may be viewed as a particular resolution or scale of the signal, (Figure 12a). At any level, the low frequency component is the "coarse" approximation and the high frequency term may be viewed as the "detail" at that resolution. We can view the dyadic tree as generating an octave-band signal decomposition in the frequency domain, or as a multiresolution decomposition in time. The latter aspect follows from the signal decomposition onto orthonormal basis sequences $\{h_0(n)\}$, $\{h_1(n)\}$ as discussed earlier. The decomposed signal can be reassembled using the nested paraunitary synthesis filters shown schematically in Figure 13.

As mentioned in Chapter 1, the idea of using different filters in a subband tree has a long history in the signal processing field. A simple *tree-structuring algorithm* was introduced in the signal processing field in [10]. A similar method was also independently proposed as the fashionable *wavelet packets* in [95]. The tree-structuring algorithm tailors the filter bank to the given input spectrum. Therefore, it is signal-dependent and adaptive. Its merits have been proven in several applications areas such as spread spectrum communications and discrete multitone transceivers [397, 50].

The two-channel biorthogonal PR constraints are obtained from Equation (2.50) with $M = 2$ [7, 430],

$$\begin{aligned} \sum_k h_v(k) g_v(2n-k) &= \delta(n), \qquad \nu = 0,1 \\ \sum_k h_0(k) g_1(2n-k) &= 0, \\ \sum_k h_1(k) g_0(2n-k) &= 0. \end{aligned} \tag{2.55}$$

In the transform domain, aliasing is eliminated with

$$\begin{aligned} G_0(z) &= -H_1(-z), \\ G_1(z) &= H_0(-z) \end{aligned} \tag{2.56}$$

and perfect reconstruction is achieved via

$$[-H_0(z)H_1(z) + H_0(-z)H_1(z)] = 2z^{-n_0},$$

$$[G_0(z)G_1(-z) - G_0(-z)G_1(z)] = 2z^{-n_0}. \tag{2.57}$$

This biorthogonal structure is a generalization of the paraunitary filter bank.

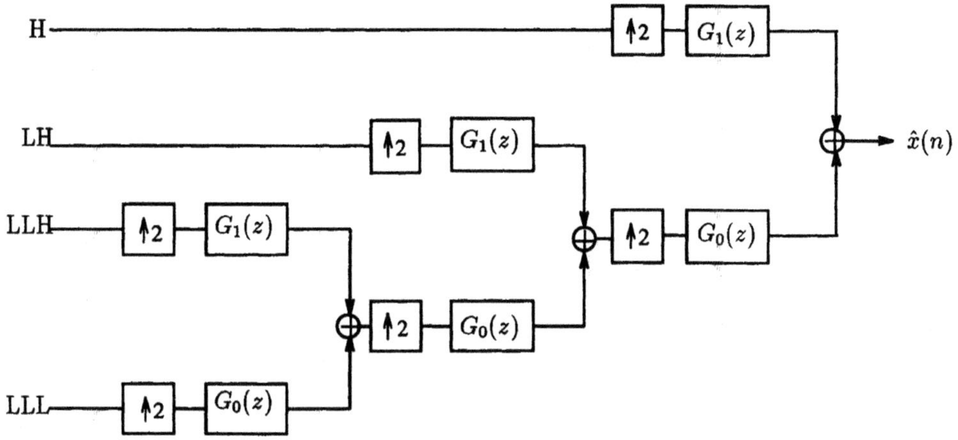

Figure 13 Dyadic synthesis tree.

This decomposition and reassembly can also be used in the binary and dyadic tree expansions of Figures 10, 12, and 13. Design procedures for linear phase biorthogonal filters can be found in [430].

6 THE WAVELET TRANSFORM AND DYADIC FILTER BANKS

Wavelet transforms have received much visibility recently in the technical literature. They are linear multiresolution decompositions for continuous-time signals whose basis functions are wavelets. Interestingly, there is a connection between orthonormal wavelet bases for continuous-time signals and dyadic discrete-time filter banks. Here we elaborate on this important connection. The discussion that follows serves as both an introduction to wavelets and as a prelude to the time-frequency localization properties described in the next section.

6.1 The Continuous Wavelet Transform

The wavelet transform is a time-frequency decomposition that partitions the time-frequency plane into non-uniform regions. It is the *analog* counterpart of the ***digital*** octave band (or dyadic) filter bank. The scaling property of the Fourier transform plays a significant role in wavelet theory. Let us assume a Fourier transform pair

$$\psi(t) \leftrightarrow \Psi(\Omega). \tag{2.58}$$

Scaling the argument results in

$$\frac{1}{\sqrt{a}}\psi(\frac{t}{a}) \leftrightarrow \sqrt{a}\Psi(a\Omega) \tag{2.59}$$

where $a > 0$ is a continuous parameter. It is seen from Equation (2.59) that a contraction in one domain implies a dilation in the other. As we shall see, this property leads to an octave-band tiling of the time-frequency plane.

The wavelet transform basis $\psi_{ab}(t)$ is defined by the scale and shift parameters a and b where

$$\psi_{ab}(t) = \frac{1}{\sqrt{a}}\psi(\frac{t-b}{a}) \tag{2.60}$$

and its Fourier transform is given by

$$\Psi_{ab}(\Omega) = \sqrt{a}\Psi(a\Omega)e^{-jb\Omega}.$$

The wavelet coefficients associated with the input $x(t)$ are obtained by the inner product

$$W_x(a,b) = \int_{-\infty}^{\infty} \psi_{ab}(t)x(t)dt \quad = \quad <\psi_{ab}, x> \tag{2.61}$$

where $a \in R^+$, $b \in R$. Thus we see that the wavelet transform maps a function $x(t)$ into time-scale space (a, b). This transform can be inverted provided $\psi(t)$ satisfies the ***admissibility*** condition [174], (see Chapter 5, Section 2.1). Given this, the inverse can be expressed as

$$x(t) = \frac{1}{C}\int_{-\infty}^{\infty}\int_{0}^{\infty} \frac{W(a,b)\psi_{ab}(t)}{a^2} dadb \tag{2.62}$$

where

$$C = \int_{0}^{\infty} \frac{|\Psi(\Omega)|^2}{\Omega} d\Omega. \tag{2.63}$$

This implies that $\Psi(0)$ must be zero; consequently $\psi(t)$ resembles a bandpass filter.

6.2 The Discrete Wavelet Transform

The wavelet parameters, a and b, in $W_x(a,b)$ are continuous. Hence the transform is called the *continuous wavelet transform* (CWT). Since the coefficients are continuous, the CWT is not well suited to computer implementations. Therefore, a and b are often sampled as

$$a = a_0^m \qquad b = nb_0a_0^m$$

so that $\psi_{ab}(t)$ goes to

$$\psi_{mn}(t) = a_0^{-m/2}\psi(a_0^{-m}t - nb_0) \tag{2.64}$$

where $m, n \in Z$. This is called the *discrete wavelet transform* (DWT). Note that the time variable of the *discrete* wavelet transform is *continuous.*

Any $x(t) \in L^2(R)$ can be represented as

$$x(t) = \sum_m \sum_n d(m,n)\psi_{mn}(t) \tag{2.65}$$

where the wavelet coefficients are the inner product

$$\begin{aligned} d(m,n) &= <\psi_{mn}(t), x(t)> \\ &= \frac{1}{a_0^{m/2}} \int x(t)\psi(a_0^{-m}t - nb_0)dt \end{aligned} \tag{2.66}$$

if the wavelet family $\{\psi_{mn}(t)\}$ is complete. Since the wavelet $\psi(t)$ is a zero-mean, bandpass function, the superposition in Equation (2.65) requires an infinite number of terms, which is not practical. Therefore, a lowpass, complementary scaling basis $\{\phi_{mn}(t)\}$ plays a major role in wavelet theory.

We can express a function $x(t)$ using the scaling basis as

$$x(t) = \sum c(0,n)\phi(t-n). \tag{2.67}$$

A finite *L-resolution* wavelet decomposition of the function $x(t)$ can be expressed by jointly using the complementary scaling basis of resolution-L as

$$x(t) = \sum_{n=-\infty}^{\infty} c(L,n)2^{-L/2}\phi(\frac{t}{2^L} - n) + \sum_{m=1}^{L}\sum_{n=-\infty}^{\infty} d(m,n)2^{-m/2}\psi(\frac{t}{2^m} - n). \tag{2.68}$$

In a finite resolution wavelet representation, $x(t)$ is lowpass approximated at scale L by using the corresponding scaling basis and the detail components of the L different resolutions are represented by the wavelets.

The multiresolution signal decomposition of Equation (2.65) is founded on a sequence of closed subspaces $\{V_m | m \in Z\}$ (where Z is the set of integers) of $L^2(R)$ with the following properties [251, 7]:

(i) Containment.

$$\ldots V_2 \subset V_1 \subset V_0 \subset V_{-1} \subset V_{-2} \ldots \tag{2.69}$$

(ii) Completeness.

$$\bigcap_{m \in Z} V_m = \{0\} \qquad \bigcup_{m \in Z} V_m = L^2(R).$$

(iii) Scaling Property.

$$x(t) \in V_m \Longleftrightarrow x(2t) \in V_{m-1} \text{ for any function } f \in L^2(R).$$

(iv) Basis/Frame Property. There exists a scaling function $\phi(t) \in V_0$, such that $\{\phi(t-n)\}$ is an orthonormal basis for V_0. From *(iii)*, it follows that $\phi_{mn}(t) = 2^{-m/2}\phi(2^{-m}t - n)$ is an orthonormal basis for V_m.

(v) Orthogonal Complement. Let W_m be the orthogonal complement of V_m in V_{m-1}, i.e. $V_{m-1} = V_m \oplus W_m$, implying that if $x(t) \in V_m$, and $y(t) \in W_m$, then $< x, y >= 0$.

In applications it is often desirable to have the wavelets be orthogonal and of compact support. Daubechies has shown the linkages of compactly supported orthonormal wavelet transform bases and paraunitary two-band FIR PR-QMFs in her seminal paper [113]. She has suggested a wavelet design procedure which starts with the design of two-band FIR PR-QMF having a zero-mean highpass filter. We will briefly present next the connections between wavelet and subband transforms as well as their distinctions. Since $V_m \subset V_{m-1}$, and $W_m \subset V_{m-1}$, then basis functions $\{\phi_{m,n}(t)\}$, and $\{\psi_{m,n}(t)\}$ of V_m and W_m respectively must be linear combinations of $\{\phi_{m-1,n}(t)\}$, the basis functions of V_{m-1}.

Hence, for $m = 0$, we can write

$$\phi(t) = 2\sum_n h_0(n)\phi(2t - n), \tag{2.70}$$

$$\psi(t) = 2\sum_n h_1(n)\phi(2t-n), \tag{2.71}$$

where $\phi(t)$, $\psi(t)$ are the scaling and wavelet functions, respectively. The coefficient sets $\{h_0(n)\}$, and $\{h_1(n)\}$ are ***interscale basis coefficients***. The orthogonality of V_m and W_m demand

$$<\phi(t), \psi(t+n)> = 0 \Rightarrow \sum_n h_0(n)h_1(n+2k) = 0. \tag{2.72}$$

Since the bases for V_m, and W_m are orthonormal, we must have $<\phi(t), \phi(t-n)> = \delta(n)$ and $<\psi(t), \psi(t-n)> = \delta(n)$. These in turn lead to the sequence orthogonality

$$\begin{aligned} &<h_0(n), h_0(n+2k)> = \delta(k), \\ &<h_1(n), h_1(n+2k)> = \delta(k). \end{aligned} \tag{2.73}$$

Equations (2.72) and (2.73) are just a replay of the two-band paraunitary filter bank requirements, Equation (2.51).

There remains the construction of $\phi(t)$, $\psi(t)$ from $\{h_0(n), h_1(n)\}$. Taking the Fourier transform of Equation (2.70), and using $\phi(2t - nT_0)$ to remove the time normalization, gives

$$\begin{aligned} \Phi(\Omega) &= \sum_n h_0(n)e^{-j\frac{\Omega}{2}nT_0}\Phi(\frac{\Omega}{2}) \\ &= H_0(e^{j\frac{\omega}{2}})\Phi(\frac{\Omega}{2}), \qquad \omega = \Omega T_0 \\ &= \Phi(0)\prod_{k=1}^{\infty} H_0(e^{j\frac{\omega}{2^k}}), \end{aligned} \tag{2.74}$$

where T_0 is the sampling period associated with the discrete-time filter representation. Similarly, from Equation (2.71), we get

$$\Psi(\Omega) = H_1(e^{j\frac{\omega}{2}})\prod_{k=2}^{\infty} H_0(e^{j\frac{\omega}{2^k}}). \tag{2.75}$$

The requirement for convergence of these infinite products to a smooth function is a ***regularity*** condition, namely $H_0(e^{j\omega}) = 0$, at $\omega = \pi$, i.e. $H_0(z)$ has at least one zero at $z = -1$. It also can be shown that $\phi(t)$, $\psi(t)$ are of compact support if $h_0(n)$, $h_1(n)$ are FIR. Thus Equations (2.74), (2.75) provide formulas for the construction of compactly supported wavelet families from two-band paraunitary filter pairs, with only a modest regularity requirement.

6.3 Wavelet Expansion and the Dyadic Subband Tree

The main result here is that we can use a paraunitary dyadic subband tree to compute the coefficients $d(m,n)$ in the wavelet expansion of a continuous-time function. The basic construction is as follows: suppose $x(t)$ can be represented by a sum of orthonormal scaling functions at full resolution, i.e.,

$$x(t) = \sum_n c(0,n)\phi(t-n) = \sum c(0,n)\phi_{0n}(t) \tag{2.76}$$

where the scaling coefficients are

$$c(0,n) = < x(t), \phi_{0n}(t) > . \tag{2.77}$$

This is the starting point for the wavelet decomposition. Now suppose that $c(0,n)$ is the input to the dyadic subband tree shown in Figure 12a, and $h_0(n)$, $h_1(n)$ are the paraunitary filters that generated $\Phi(\Omega)$ and $\Psi(\Omega)$ in Equations (2.74) and (2.75).

By the finite-resolution wavelet representation, $x(t)$ can be decomposed into a (coarse) lower resolution scaling expansion, and a lower resolution wavelet expansion which carries the detail:

$$x(t) = \sum_n d(1,n)\psi_{1n}(t) + \sum_n c(1,n)\phi_{1n}(t). \tag{2.78}$$

The coarse approximation at the half-resolution can in turn be decomposed into yet a coarser approximation at the quarter-resolution and a detail component

$$x(t) = \sum_n d(1,n)\psi_{1n}(t) + \sum_n d(2,n)\psi_{2n}(t) + \sum_n c(2,n)\phi_{2n}(t). \tag{2.79}$$

Continuing in this manner, we see that $x(t)$ can be represented as a lowpass approximation at the $1/L$-resolution plus a sum of L detail wavelet components at successively finer resolutions. This decomposition can be continued indefinitely. The coefficients $c(l,n)$, and $d(l,n)$ in this pyramid decomposition are the coarse and detail outputs at level l of the paraunitary dyadic tree of Figure 12a when $x(n) = c(0,n)$ is the input. Hence the terminology "Fast Wavelet Transform" was coined to suggest the computation of the coefficients in Equation (2.78) by this dyadic tree.

The fly in the ointment here is the initialization or feeding the subband tree by $c(0,n)$. If this starting point, Equation (2.77), is only an approximation, then the expansion which follows is itself an approximation [113, 7, 128]. As a case in point, suppose $x(t)$ is a band-limited signal. Then the set

$$\phi_{0n}(t) = \frac{\sin \pi(t-n)}{\pi(t-n)}$$

constitutes an orthonormal basis, and

$$x(t) = \sum x(n) \frac{\sin \pi(t-n)}{\pi(t-n)}.$$

In this case $h_0(n)$ and $h_1(n)$ must be ideal lowpass and highpass filters. If $x(n)$ itself is the input to the dyadic subband tree with filters that only approximate the ideal filters, then the resulting coefficients or subband signals $d(m,n)$ and $c(m,n)$ are themselves only approximations to the exact values. This problem becomes even worse in the case of nonstationary wavelet packets discussed in Chapter 6.

The theoretical precision of wavelet transforms has long been overlooked by many signal processing engineers and researchers. The terms ***wavelet transform***, ***fast wavelet transform*** and ***wavelet packets*** have been loosely used without considering the approximation error introduced by the implicit assumption that

$$c(0,n) = x[n] \tag{2.80}$$

The validity of the ***wavelet transform*** and its variations requires that

$$x[n] = \int x(t)\phi(t-n)dt,$$

which in general is only satisfied when $\phi(t)$ is an ideal *sinc* function. The implication of this condition is that $h_0[n]$ and $h_1[n]$ must be ideal lowpass and highpass filters, which unfortunately are infinite in duration and unstable. Thus there is an implicit error inherent in the wavelet transforms discussed in the literature, which is largely ignored. A thorough analysis and evalutation of the effects of this approximation error caused by (2.80) is a topic for future study.

On the other hand, analysis-synthesis filter banks (or subband transforms) perfectly represent the sampled signals without any such implicit errors. Moreover, they do not impose any zero-mean requirements on the filters, which may or may not be appropriate for applications.

It is expected that this delicate issue will be treated with more technical precision and care in the future and that loosely used terminologies in the field will be refined to avoid confusion. The real potential of wavelet transform theory in signal processing is not yet clear but will become so in the future.

7 TIME-FREQUENCY LOCALIZATION

In many signal processing applications the time-frequency localization property of the decomposition technique is an important consideration. While bandwidth compression of the signals require kernels with good localization in frequency, signal domain features demand a high degree of localization in the time or spatial variables. These compete with each other and one is secured at the expense of the other. The classical *uncertainty principle* provides the back drop for this tradeoff. The ideal operator can extract local features of a signal in both time and frequency domains.

Let the transformation kernel and its transform be $\phi(t) \leftrightarrow \Phi(\Omega)$. Then for any $\phi(t)$, with $\sqrt{t}\phi(t) \to 0$, as $t \to \pm\infty$, the uncertainty principle [310] asserts

$$\sigma_T \sigma_\Omega \geq 1/2 \tag{2.81}$$

where σ_T, σ_Ω are the RMS spread of $\phi(t)$ and $\Phi(\Omega)$ around the center values. That is,

$$\begin{aligned} \sigma_T^2 &= \frac{\int (t-\bar{t})^2 |\phi(t)|^2 dt}{E}, \\ \sigma_\Omega^2 &= \frac{\frac{1}{2\pi}\int_{-\infty}^{\infty} (\Omega-\bar{\Omega})^2 |\Phi(\Omega)|^2 d\Omega}{E} \end{aligned} \tag{2.82}$$

where E is the energy in the signal

$$E = \int_{-\infty}^{\infty} |\phi(t)|^2 dt = \frac{1}{2\pi}\int_{-\infty}^{\infty} |\Phi(\Omega)|^2 d\Omega \tag{2.83}$$

and $\bar{t}, \bar{\Omega}$ are the centers of mass of these kernels:

$$\begin{aligned} \bar{t} &= \frac{\int_{-\infty}^{\infty} t |\phi(t)|^2 dt}{E}, \\ \bar{\Omega} &= \frac{\frac{1}{2\pi}\int_{-\infty}^{\infty} \Omega |\Phi(\Omega)|^2 d\Omega}{E}. \end{aligned} \tag{2.84}$$

The time-bandwidth product $\sigma_T \sigma_\Omega$ is called the resolution cell. The equal sign holds in Equation (2.81) if and only if $\phi(t)$ is Gaussian of the form $exp(-at^2)$, [310]. The short-time Fourier transform (STFT) positions a window $g(t)$ at some point τ on the time axis and calculates the Fourier transform of the signal within that window:

$$X(\beta, \tau) = \int_{-\infty}^{\infty} x(t) g(t-\tau) e^{-j\beta t} dt. \tag{2.85}$$

When $g(t)$ is Gaussian, the STFT is termed a Gabor transform. The basis functions are generated by ***modulation*** and ***translation*** of the window function by parameters β and τ respectively. Observe that as τ increases, the kernel simply translates in time while keeping the spread of the window fixed.

Let $g(t) \leftrightarrow G(\Omega)$ be a Fourier transform pair, and assume $\bar{t} = 0, \bar{\Omega} = 0$. Then the translated, modulated kernel pair is

$$g_{\tau,\beta}(t) = g(t-\tau)e^{-j\beta t} \leftrightarrow G_{\tau,\beta}(\Omega) = e^{-j(\Omega-\beta)\tau}G(\Omega-\beta). \tag{2.86}$$

This two-parameter family is centered at (τ, β) in the time-frequency plane, i.e.,

$$\bar{t}_{\tau,\beta} = \tau, \qquad \bar{\Omega}_{\tau,\beta} = \beta.$$

Therefore it follows that the spread of the shifted modulated kernel is constant in both domains, i.e.,

$$\sigma^2_{\Omega_{(\tau,\beta)}} = \sigma^2_{\Omega} \quad \text{and} \quad \sigma^2_{T_{(\tau,\beta)}} = \sigma^2_{T} \tag{2.87}$$

where σ_Ω and σ_T are the RMS spreads of the unmodulated, untranslated kernel. This implies that the resolution cell has a constant shape as well as area in the time-frequency plane shown in Figure 14a.

The time-frequency localization of the wavelet kernel, Equation (2.60), is distinctly different from that of the STFT. Define $\bar{t}$ and $\bar{\Omega}$ as the centers of mass of $\psi(t)$ and $\Psi(\Omega)$ on $(-\infty, \infty)$ and $[0, \infty)$, respectively. The spreads are then

$$\begin{aligned} \sigma^2_\Omega &= \frac{\frac{1}{\pi}\int_0^\infty (\Omega-\bar{\Omega})^2 |\Psi(\Omega)|^2 d\Omega}{E}, \\ \sigma^2_T &= \frac{\int_{-\infty}^\infty (t-\bar{t})^2 |\psi(t)|^2 dt}{E}. \end{aligned} \tag{2.88}$$

The prototype kernel is centered at $\bar{t}$ and $\bar{\Omega}$ in the time-frequency plane with spreads σ_T and σ_Ω. It follows that $\psi_{ab}(t) \leftrightarrow \Psi_{ab}(\Omega)$ is centered at $(\bar{t}_{ab}, \bar{\Omega}_{ab}) = (a\bar{t} + b, \bar{\Omega}/a)$ with spreads

$$\sigma^2_{\Omega_{(a,b)}} = \frac{1}{a^2}\sigma^2_\Omega, \qquad \sigma^2_{T_{(a,b)}} = a^2\sigma^2_T \tag{2.89}$$

and time-frequency resolution cell $\sigma_{T_{(a,b)}}\sigma_{\Omega_{(a,b)}} = \sigma_\Omega \sigma_T \geq 1/2$. The resolution, i.e., localization in time and frequency depends on the scale parameter a. The area of the resolution cell is constant but the shape depends on a, as shown in Figure 14b.

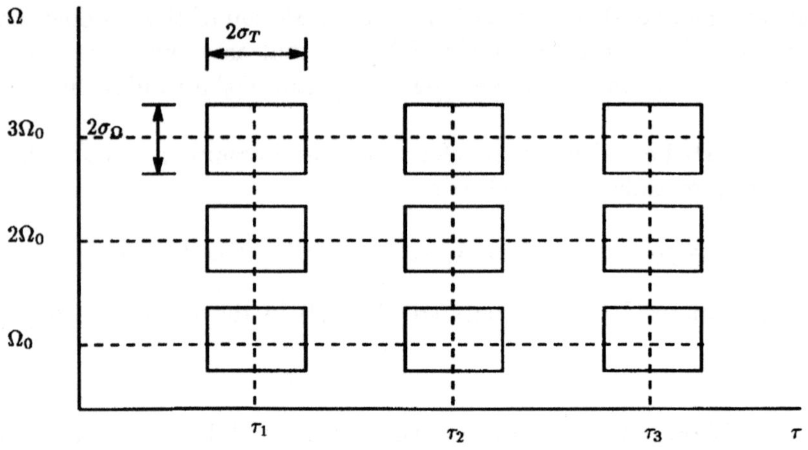

(a)

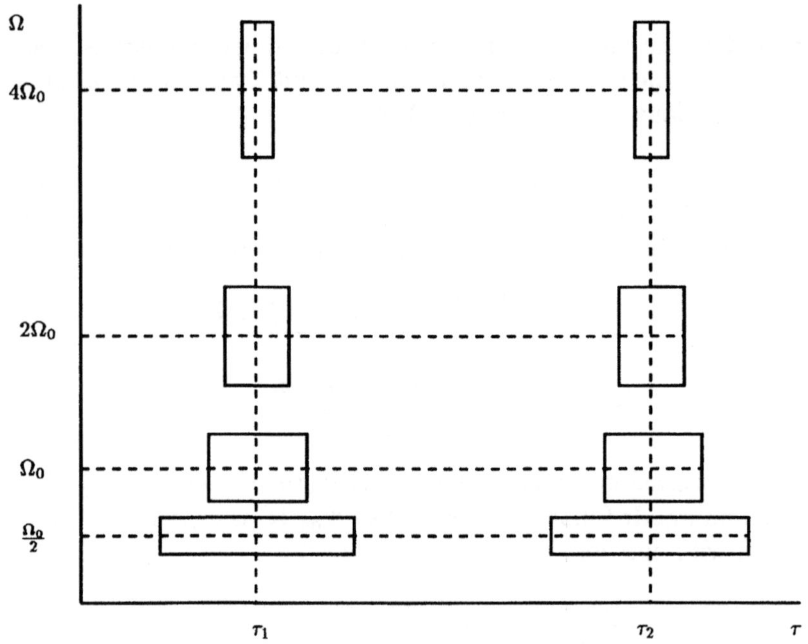

(b)

Figure 14 Time-frequency plane showing resolution cells: (a) STFT cells; (b) resolution cells for the wavelet transform.

An often quoted example of a wavelet is the second derivative of a Gaussian, the so-called "Mexican hat," given by

$$\psi(t) = (1-t^2)e^{-t^2/2} \leftrightarrow \Psi(\Omega) = \sqrt{2\pi}\Omega^2 e^{-\Omega^2/2}.$$

This kernel has excellent localization in time and frequency and satisfies the admissibility condition. But it is not of compact support and is not complete. For this kernel, we find that ($\bar{t} = 0, \sigma_T^2 = 7/6$), and ($\bar{\Omega} = 1.505, \sigma_\Omega^2 = 0.23646$), resulting in $\sigma_T\sigma_\Omega = 0.525$ [176].

The discrete-time version of the uncertainty principle is as follows [176, 71]. Let $x(n) \leftrightarrow X(e^{j\omega})$ be a discrete-time Fourier transform pair

$$X(e^{j\omega}) = \sum_{n=-\infty}^{\infty} x(n)e^{-jn\omega} \leftrightarrow x(n) = \frac{1}{2\pi}\int_{-\pi}^{\pi} X(e^{j\omega})e^{jn\omega}d\omega. \tag{2.90}$$

By Parseval's theorem, the energy is

$$E = \sum_{n=-\infty}^{\infty} |x(n)|^2 = \frac{1}{2\pi}\int_{-\pi}^{\pi} |X(e^{j\omega})|^2 d\omega.$$

We define the means

$$\begin{aligned}\bar{\omega} &= \frac{\frac{1}{2\pi}\int_{-\pi}^{\pi}\omega|X(e^{j\omega})|^2 d\omega}{E},\\ \bar{n} &= \frac{\sum_{-\infty}^{\infty} n|x(n)|^2}{E}.\end{aligned} \tag{2.91}$$

The spread in time and frequency is then

$$\begin{aligned}\sigma_n^2 &= \frac{\sum_{-\infty}^{\infty}(n-\bar{n})^2|x(n)|^2}{E},\\ \sigma_\omega^2 &= \frac{\frac{1}{2\pi}\int_{-\pi}^{\pi}(\omega-\bar{\omega})^2|X(e^{j\omega})|^2 d\omega}{E}.\end{aligned} \tag{2.92}$$

For any real signal $\bar{\omega} = 0$, and without loss of generality we can make $\bar{n} = 0$. It can be shown [176], that the time-frequency product or resolution cell is given by

$$\begin{aligned}\sigma_n\sigma_\omega &\geq \frac{|1-\mu|}{2},\\ \mu &\triangleq \frac{|X(e^{j\omega})_{\omega=\pi}|^2}{E} = \frac{|X(-1)|^2}{E}.\end{aligned} \tag{2.93}$$

For bandpass functions with frequency peaks centered at $\pm\hat{\omega}$, the frequency measure in Equation (2.91) is not suitable. In this case, we are concerned about the spread around $\hat{\omega}$. Therefore, we define frequency spread on $[0,\pi]$ rather than $[-\pi,\pi]$. In this case, we take

$$\hat{\omega} = \frac{\frac{1}{\pi}\int_0^{\pi}\omega|X(e^{j\omega})|^2 d\omega}{\frac{1}{\pi}\int_0^{\pi}|X(e^{j\omega})|^2 d\omega},$$
$$\hat{\sigma}_{\omega}^2 = \frac{\frac{1}{\pi}\int_0^{\pi}(\omega-\hat{\omega})^2|X(e^{j\omega})|^2 d\omega}{\frac{1}{\pi}\int_0^{\pi}|X(e^{j\omega})|^2 d\omega} \tag{2.94}$$

and σ_n^2 remains unchanged. It can be shown [176] that

$$\hat{\sigma}_{\omega}\sigma_n \geq \frac{1}{2}|1-\mu'|,$$
$$\mu' = \frac{\hat{\omega}}{\pi}\frac{|X(1)|^2}{E} + (1-\frac{\hat{\omega}}{\pi})\frac{|X(-1)|^2}{E}. \tag{2.95}$$

For bandpass signals with zero DC gain, $X(1) = 0$, and zero high frequency gain, $X(-1) = 0$, Equation (2.95) reduces to

$$\hat{\sigma}_{\omega}\sigma_n \geq 1/2.$$

For lowpass filters, $X(e^{j\omega})|_{max}$ occurs at $\omega = 0$ and Equations (2.92) are used. In this case, there are two classes of filters

$$\text{Class I:} \qquad X(-1) = 0 \rightarrow \sigma_n\sigma_{\omega} \geq \frac{1}{2}, \tag{2.96}$$

$$\text{Class II:} \qquad X(-1) \neq 0 \rightarrow \sigma_n\sigma_{\omega} \geq \frac{|1-\mu|}{2}. \tag{2.97}$$

It can be shown that the optimum is reached, (i.e., the equal sign holds in Equation (2.93)), when $\frac{dF}{d\omega} = -K\omega F^*$. The solution to this differential equation is a Gaussian $e^{-K\omega^2/2}$. But this Gaussian cannot satisfy the Class I boundary condition, $X(-1) = 0$. In this case, the lower bound cannot be attained, and the strict inequality holds. For the Class II set of functions, the Gaussian can satisfy both the differential equation and the boundary condition, resulting in $\sigma_n\sigma_{\omega} = \frac{1}{2}|1-\mu|$. In reference [176], it is shown that the binomial filter is an excellent FIR approximation to the Gaussian that matches the zero boundary condition at $\omega = \pi$.

The time-frequency resolution for several filter banks and wavelet families are compared in [176]. Table 1 compares the resolutions of scaling and wavelet

Table 1 Time-frequency localizations of 6-tap wavelet filters, and corresponding scaling and wavelet functions [176].

		Daubechies [113][9]	Most regular [112]	Coiflet [113]
Scaling Function	σ_T^2	0.134	0.143	0.086
	σ_Ω^2	5.22	5.77	11.86
	$\sigma_T^2\sigma_\Omega^2$	0.699	0.825	1.02
Wavelet Function	σ_T^2	0.178	0.188	0.108
	σ_Ω^2	8.97	11.70	39.36
	$\sigma_T^2\sigma_\Omega^2$	1.596	2.199	4.25
Lowpass PR-QMF	σ_n^2	0.453	0.470	0.305
	σ_ω^2	0.987	0.996	1.059
Highpass PR-QMF	σ_n^2	0.453	0.470	0.305
	σ_ω^2	0.987	0.996	1.059

functions for three wavelet families generated by 6-tap paraunitary filters—the Daubechies [113], the *most regular* [112], and Coiflet [113]—along with the localization properties of the progenitor discrete-time filters. (The progenitor filters for the Daubechies wavelets are the binomial QMF filters described in [9].) This table demonstrates some of the tradeoffs between time and frequency resolution. It also demonstrates that among the three, the Daubechies wavelets and scaling functions have the smallest resolution cells.

8 OPTIMAL DESIGN AND PERFORMANCE

It was shown in Section 7 that a function cannot localize perfectly in both the time and frequency domains. Figure 15 shows that there are infinite possible filter bank solutions. Each point in that figure represents a lowpass PR-QMF solution and its associated time and frequency localizations. The resolution tradeoffs are explicitly evident, and clearly non-trivial. This section emphasizes certain subsets of the solution space that meet the requirements of the design criteria selected [70].

The desired features of the design depend on the application under consideration. The optimality concepts and their measures are constrained by the limits of time and frequency localization and completeness requirements.

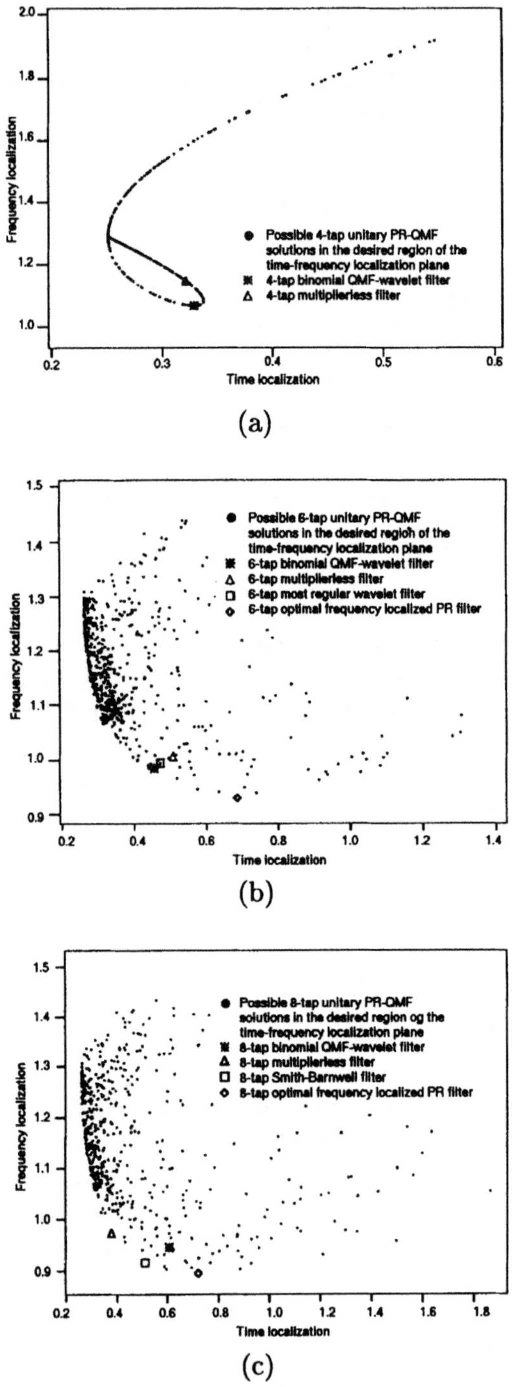

Figure 15 The time-frequency localizations of possible regular unitary two-band QMFs for (a) 4-tap, (b) 6-tap, and (c) 8-tap cases along with some other of the known PR-QMFs in the literature.

The optimal PR-QMF design techniques discussed in this section consider several measures of practical significance. The optimal criteria might be based on a single design measure or a set of measures. Some of the measures imply a primarily frequency domain localization of filter functions, while the others focus on a time domain localization. A few of the optimality measures will be given in the next section with their practical merits in signal processing applications. We also describe a criterion and design procedure based on a mean-square error(MSE) metric that minimizes the mean-square signal reconstruction error in the presence of subband signal quantization. In our discussions, the two-band PR-QMF bank serves as the platform for evaluating system performance.

8.1 Measures of Optimality

Energy Compaction: This is a desired performance measure for most signal decomposition techniques. The energy compaction measure is derived from rate-distortion theory and merges with the entropy minimization measure for Gaussian sources. The complete derivation of this measure can be found in [213].

Basically, the measure compares the mean square reconstruction error σ_ϵ^2 for different signal coding schemes with that of pulse code modulation (PCM), when each is used in conjunction with *pdf-optimized* quantizers, and optimal bit allocation [213]. This ratio emerges as an energy compaction measure which for unitary transforms (or transform coders, TC), is

$$G_{TC} = \frac{\sigma_\epsilon^2(PCM)}{\sigma_\epsilon^2(TC)} = \frac{\frac{1}{N}\sum_{j=0}^{N-1}\sigma_j^2}{[\prod_{j=1}^{N-1}\sigma_j^2]^{1/N}}.$$

This is recognized as the ratio of the arithmetic mean of the energy, $\{\sigma_j^2\}$, in each subband to the geometric mean. This measure is widely quoted in evaluating the performance of block and subband transforms.

The output energy of the prototype lowpass filter $\underline{h}^T = [h_0, h_1, \ldots, h_{N-1}]$ can be expressed in matrix form as

$$\sigma_L^2 = \underline{h}^T R_{xx} \underline{h} \tag{2.98}$$

where R_{xx} is the covariance matrix of a zero-mean input. It is sufficient to maximize σ_L^2 in order to maximize the energy compaction of a two-band PR-

QMF bank, which reduces to

$$G_{TC} = \frac{\sigma_x^2}{[\sigma_L^2 \sigma_H^2]^{1/2}}, \tag{2.99}$$

where σ_x^2 and σ_H^2 are the input and highpass band variances, respectively. The unitary property in two-band PR-QMF banks is expressed as

$$\sigma_x^2 = \frac{1}{2}(\sigma_L^2 + \sigma_H^2). \tag{2.100}$$

This measure shapes the frequency responses of the filter functions for the given input spectrum. Hence, the solution is a matched filter bank.

Aliasing Energy: Any realizable lossless decomposition technique performs an aliasing cancellation in order to achieve perfect reconstruction. Because of the quantization or discarding of some of the subbands, non-cancelled aliasing signal components might exist in the reconstructed signal in practice. The significance of the energy of aliased components has been observed in speech applications and more recently in low bit rate image coding and signal excision in spread spectrum communications [397, 41].

The aliasing energy component at the reconstructed lowpass filter output, Figure 9, is expressed as

$$\sigma_{LA}^2 = \frac{1}{2}\int_{-\pi}^{\pi} |H(e^{-j\omega})|^2 S_{xx}(e^{j(\omega+\pi)})|H(e^{j(\omega+\pi)})|^2 d\omega \tag{2.101}$$

where $S_{xx}(e^{j\omega})$ is the input spectrum. The time domain version of Equation (2.101) is given by

$$\sigma_{LA}^2 = \sum_k [\rho(n) * (-1)^n \rho(n)] R_{xx}(k) \tag{2.102}$$

where $R_{xx}(k)$ is the autocorrelation sequence of the input and $\rho(n)$ is the autocorrelation sequence of the filter $h(n)$ defined as

$$\rho(n) = h(n) * h(-n). \tag{2.103}$$

The optimal solution based on this measure minimizes the aliasing energy component of the lowpass branch. This measure somehow complements the energy compaction measure. The latter does not pay any attention to the spillover of subband energies. In contrast, the aliasing component of the output energy is carefully monitored by this measure. Therefore, the non-aliasing energy ratio (NER) was proposed [6] as a new measure to complement the energy compaction measure. The aliasing energy measure focuses on the frequency localization of the filter functions for the given input spectrum.

Most of the two-band biorthogonal filter banks proposed in the literature overlooked this concept. Therefore, their performance did not meet expectations although they have linear phase responses. For less aliasing in biorthogonal filter banks, one should minimize the differences in filter bandwidths, an aspect which is inherent in biorthogonal structures.

Unit Step Response: The time-local features of signals can be efficiently represented with bases which consists of functions well localized in the time domain. The unit-step response of the filter is widely used as a measure of time localization and defined as

$$a(n) = h(n) * u(n) \tag{2.104}$$

where $u(n)$ is the unit step function. The energy difference between the unit step response $a(n)$ and the unit step function $u(n)$ is expressed as

$$E_s = \sum_{k=0}^{N-1} \left[\sum_{n=0}^{k} h(n) - 1 \right]^2 . \tag{2.105}$$

It is seen from Equation (2.105) that E_s becomes zero when $h(n) = \delta(n)$. This is the best time-local solution with the worst frequency selectivity. This measure is a mathematical tool to monitor the time domain properties of the designed filter.

Constrained Non-Linearity in Phase Response: The orthonormality conditions presented in Section 5 indicate that there cannot be any linear phase two-band PR-QMF solution. But, it is also well-known that a severe phase non-linearity degrades the performance in many applications. Therefore, the level of phase non-linearity is monitored by a phase measure defined as

$$E_p = \sum_{n} [h(n) - h(N-1-n)]^2 . \tag{2.106}$$

This measure simply monitors the non-symmetry of the filter's unit-sample response.

Zero-Mean Bandpass and Highpass Filters: Many practical signal sources have a significant spectral component at zero frequency. Therefore, an efficient decomposition technique should be able to represent the DC component with only one basis function. Following this argument one should constrain the highpass filter of two-band PR-QMF banks to have zero-mean or DC gain via

$$\sum_n (-1)^n h(n) = 0. \tag{2.107}$$

This requirement implies that there must be at least one zero of the lowpass prototype filter $H(e^{j\omega})$ at $\omega = \pi$. That implies some degree of regularity in the context of wavelet transform theory [113].

Uncorrelated Subband Coefficients: The Karhunen-Loève Transform (KLT) is the unique example of block transforms with perfectly uncorrelated transform coefficients for the given input statistics. The uncorrelatedness and maximum energy compaction requirements merge in the KLT solution. But, this is not the case with filter banks.

The cross-correlation of the two subband signals $y_0(n)$, $y_1(n)$ in Figure 9, at the same time index is

$$R_{LH}(0) = \sum_n \left[\sum_l h(l)(-1)^l h(n-l) \right] R_{xx}(n), \quad \forall m. \tag{2.108}$$

In general, there is more than one filter solution that satisfies the perfect decorrelation condition, $R_{LH}(0) = 0$. The one that maximizes the objective function for optimization is the desired solution.

More recently, the cross-correlations $R_{LH}(k)$ of the subband signals in different time slots, $k = 0, 1, \ldots, p$ were constrained to be zero in reference [262]. This is an extension of the decorrelation measure discussed here. In the ultimate case, the ideal filter bank decorrelates perfectly all subband samples that are not positioned in the same sampling point, i.e. $R_{LH}(k) = 0$, for all k.

Maximally Flat Frequency Response: The fundamental question of the degree of flatness in the frequency response vs. the sharpness of the transition band has been investigated since the inception of filter design in the signal processing field. The concepts of time-frequency localization, aliasing energy and the smoothness of the magnitude response of the filter are implied in this question. Therefore, there is not a unique answer.

This basic concept also has been extended to filter banks and transform bases. The most common mathematical tool for flatness is the number of z-plane zeros in the filter at $\omega = \pi$. This condition forces the filter bank solution to have zero-mean for the bandpass and highpass functions. This degree of smoothness might be necessary for some signal processing applications.

The linkages between subband and wavelet transforms were shown in Section 6. Wavelet regularity was proposed as the measure of wavelet basis design. The number of zeros at $z = -1$ in the lowpass prototype filter in a two-band PR-QMF bank was used as the design tool for wavelet and scaling bases. Daubechies proposed a wavelet basis obtained by placing the maximum possible number of zeros of the lowpass filter at $\omega = \pi$ [113]. Interestingly enough, it was shown in reference [69] that this filter bank solution has the unique maximally flat magnitude square function that was studied by Herrmann in 1971 [184].

Time-Frequency Localizations: All of the measures of design discussed in this section merely shape the time and frequency features of the filters or basis functions. Therefore, the time and frequency centers and spreads defined in Equation (2.92) can be directly used as the design criteria. The time and frequency localizations of all possible lowpass PR-QMF filters are displayed in Figure 15 for 4-, 6-, and 8-tap two-band cases.

Computational Efficiency: Computationally efficient filter bank solutions are always of practical interest. An example of such an efficient two-band filter bank is the ***multiplierless PR-QMF*** proposed in [3]. The values of these filter coefficients are limited to powers of two or can be obtained by a shift-add operation. These filters are unique orthonormal solutions in the space of infinitely many solutions.

Optimum (MSE) Design of Quantized M-Band Filter Banks: The quantizer is the key compression element in subband coding. Yet the effects of quantization in the design of filter banks have received only scant attention in the literature. In this section we review an approach to optimal filter design based on a MSE performance measure that accounts explicitly for quantization errors [177, 415, 178, 312].

It is shown in [213] that the Lloyd-Max ***pdf-optimized*** quantizer of Figure 16a has the properties that the quantization error $\tilde{x}$ is unbiased and orthogonal to the quantized output $\hat{x}$, but is, in fact, ***correlated*** with the input x. Jayant and Noll [213] have shown that the network of Figure 16b consisting of nonlinear gain α and additive fictitious noise $r(n)$ ***uncorrelated*** with $x(n)$ is equivalent to the optimum quantizer, provided one chooses $\alpha = 1 - \sigma_{\tilde{x}}^2/\sigma_x^2$, $\sigma_r^2 = \alpha(1-\alpha)\sigma_x^2$. Furthermore, from rate-distortion theory, we know that $\sigma_{\tilde{x}}^2 = \beta(R)2^{-2R}\sigma_x^2$ where $\beta(R)$ depends on the ***pdf*** of the input signal and R, the number of bits

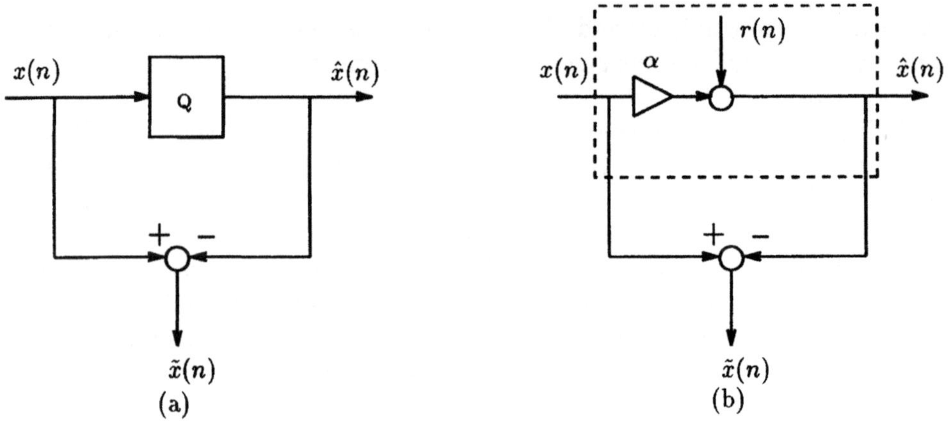

Figure 16 (a) Pdf-optimized quantizer; (b) nonlinear gain plus additive noise model.

in the quantizer. $\beta(R)$ can be measured experimentally for any normalized *pdf* [213]. The result is a nonlinear gain, $\alpha = 1 - \beta(R)2^{-R}$.

Placing quantization noise models (Figure 16) in all the subbands of an M-band filter bank (Figure 7), and replacing the analysis and synthesis filters by equivalent polyphase matrices, we can obtain the vector-matrix block diagram of Figure 17. This structure running at the slow clock rate is nonlinear, but time-invariant. A and S are diagonal gain and (proposed) compensation matrices, respectively and $\underline{\xi}(n)$ and $\underline{r}(n)$ are uncorrelated by construction. Hence, the total quantization error in the reconstructed signal is the sum of two uncorrelated components: a distortion term $\underline{e}_D$ due to A, and a noise term $\underline{e}_N$ due to $\underline{r}(n)$,

$$\begin{aligned} \underline{e}_D(z) &= \mathcal{R}_p^{'}[SA - I]\mathcal{E}_p(z)\underline{\xi}(z) \\ \underline{e}_N(z) &= \mathcal{R}_p^{'}(z)S\underline{R}(z). \end{aligned} \tag{2.109}$$

The mean-square quantization error at the filter bank output (averaged over M fast clock pulses) can then be represented as

$$\sigma_Q^2 = \sigma_D^2 + \sigma_N^2. \tag{2.110}$$

This decomposition allows us to analyze and treat each component separately. In most circumstances, σ_N^2 dominates σ_D^2. Explicit time-domain expansions of Equation (2.110) reported in [415, 178] give the output MSE in terms of

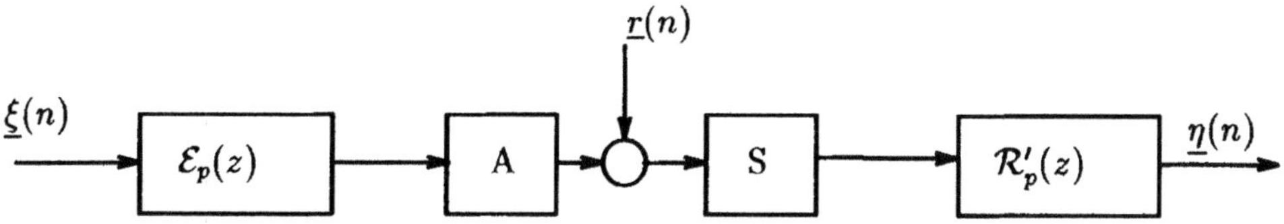

Figure 17 Slow clock equivalent of filter bank with quantizers and compensators.

FIR analysis/synthesis filters, the input signal correlation, the non-linear gains $\{\alpha_i\}$, and compensators $\{s_i\}$.

The optimization algorithm determines the filter coefficients and compensators for a given input correlation and bitrate subject to the PR requirement. The compensation scenarios are as follows:

(i) Let $s_i = 1$; there is no compensation.
(ii) Minimize σ_Q^2 by jointly minimizing the sum in Equation (2.110) for a given input signal. This is the ***optimal*** solution.
(iii) Let $s_i = 1/\alpha_i$. This is the ***null*** solution.

This ***null*** compensation completely eliminates the distortion term σ_D^2, but it increases σ_N^2, and consequently σ_Q^2 as well. In defense of this measure, it is found [177, 312] that as the input signal correlation changes from nominal design values, the null compensated system is superior to the ***optimal*** one.

For the paraunitary PR constraint, the optimal compensation is $s_i = 1$; yet again, the null compensated $s_i = 1/\alpha_i$ is more robust over a wide range of operating conditions. The paraunitary filter coefficients are also insensitive to average bit rate R and input correlation ρ.

Optimal designs for equal-length analysis/synthesis biorthogonal structures are reported in [177]. These filter coefficients are very sensitive to changes in R and ρ. The general conclusion is that the null-compensated paraunitary system is the robust choice over a wide operating range.

8.2 Design Examples

The set of measures of Section 8.1 can be combined in an objective function that also incorporates constraints of interest. Since there are infinitely many solutions the purpose here is to select the criteria suitable to the subset of desired features. We present a few two-band optimal PR-QMF design examples in the next section in order to illustrate the issues discussed earlier.

8.3 Optimal Design Based on Energy Compaction

We now formulate the objective function J, which is to be maximized, as

$$max\{J\} = \underline{h}^T R_{xx} \underline{h} \tag{2.111}$$

with the perfect reconstruction and zero-mean highpass filter constraints as [70]

$$\sum_n h(n)h(n+2k) = \delta(k),$$
$$\sum_n (-1)^n h(n) = 0, \tag{2.112}$$

where $h(n)$ is the prototype lowpass filter and $\underline{h}^T$ is its vector form representation. In this formulation, the objective function implies the optimal shaping of the filter frequency response in order to maximize its output for the given input spectrum. The M-band extensions of this design were reported by Soman and Vaidyanathan in reference [386].

8.4 Optimal Design Based on an Extended Set of Variables

The measures and constraints of energy compaction, aliasing energy, unit step response, constrained non-linear phase response, zero-mean highpass filter, and uncorrelated subband signals are all combined in this optimal design. The optimization problem is now formulated as [70]

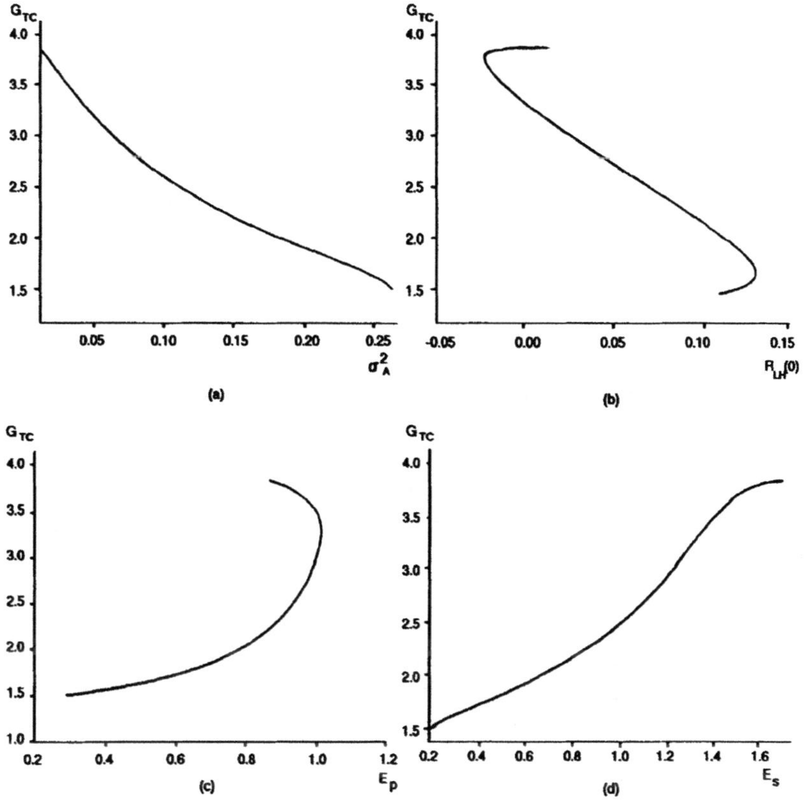

Figure 18 The relations of (a) G_{TC} versus σ_A^2; (b) G_{TC} versus $R_{LH}(0)$, (c) G_{TC} versus E_p, and (d) G_{TC} versus E_s of 8-tap, two-band PR-QMFs for AR(1), $\rho = 0.95$ source.

$$max\{J\} = \underline{h}^T R_{xx}\underline{h} - \alpha \sum_k [\rho(n) * (-1)^n \rho(n)] R_{xx}(k)$$

$$-\beta \sum_{k=0}^{2N-1} [\sum_{n=0}^{k} h(n) - 1]^2 - \gamma \sum_n [h(n) - h(2N-1-n)]^2, \qquad (2.113)$$

with the constraints of perfect reconstruction, zero-mean highpass filter, and uncorrelated subband signals given by

$$\begin{aligned} \sum_n h(n)h(n+2k) &= \delta(k), \\ \sum_n (-1)^n h(n) &= 0, \\ R_{LH}(0) &= 0. \end{aligned} \qquad (2.114)$$

This is a general optimization problem. The objective function consists of conflicting requirements. Figure 18 displays the interrelations of several measures discussed in this section for two-band PR-QMF banks. These relationships are not straightforward but can be explained by the fundamentals of signal processing. Therefore, the values of optimization parameters, namely α, β, and γ are very important. They serve to fine tune the time and frequency domain features of the filter under consideration. The optimal basis design should be performed hand-in-hand with experimental studies. The significance of the optimization variables should be quantified for the application of interest. The following section evaluates the objective performance of some known filter bank families.

8.5 Objective Performance Evaluation

We present the objective performance of a few known two-band filter bank families in this section. Table 2 displays the values of the performance measures defined earlier in this chapter for 4-, 6-, 8-tap binomial QMF-wavelet filters, the 8-tap Smith-Barnwell filter, and the 6-tap maximally regular wavelet filter designed by Daubechies. The time and frequency domain features of these filter bank families are easily interpreted from this table. It also provides the means for the interpretation and assessment of wavelet transform concepts like regularity from a signal processing perspective.

Filter	G_{TC}	σ_A^2	$R_{LH}(0)$	Mean	E_p	E_s
Binomial-QMF (8-tap)	3.8109	0.0147	0.0003	0.0000	0.9085	1.8647
Binomial-QMF (6-tap)	3.7588	0.0177	-0.0233	0.0000	1.2368	1.3550
Binomial-QMF (4-tap)	3.6426	0.0240	-0.0422	0.0000	0.7500	0.8365
Smith-Barnwell (8-tap)	3.8391	0.0127	0.0302	0.0142	0.6008	4.2500
Maxregular (6-tap)	3.7447	0.0185	-0.0103	0.0000	1.2703	1.4446

Table 2 Objective performance of two-band binomial QMF-wavelet filters [8], Smith-Barnwell CQF [373], and Maxregular wavelet filters [112] for AR(1) source with $\rho = 0.95$.

Figure 15 displays the possible two-band PR-QMF solutions from the time-frequency localization point of view. Some of the well-known filter bank families are also highlighted in these plots for comparison purposes.

The solution space of orthonormal filter banks is very rich. Lately, the efforts in the filter bank field have been focussed mostly on design and performance issues. More studies on the subjective performance of subband transforms are expected in the future [348].

9 PROGRESSIVE OPTIMALITY IN SUBBAND TREES AND WAVELETS

Subband trees employ the prototype two-band (or M-band in general) non-ideal PR-QMF bank repetitively in the tree in their most popular version. The tree can grow in a symmetrical or irregular fashion. Hence, different tree types are obtained. The concept of using different PR-QMFs at different nodes of a tree for performance improvements is discussed in this section. In the most general case, it is not necessary to limit the tree structure with two-band splits only. An arbitrary spectral split is possible.

Figure 10a displays a two-stage, four-band, binary subband tree. The equivalent subband product filters of this tree are shown in Figure 10b, and are given by

$$\begin{aligned} H_1(z) &= H_0(z)H_0'(z^2) \\ H_2(z) &= H_0(z)H_1'(z^2) \\ H_3(z) &= H_1(z)H_0''(z^2) \end{aligned}$$

$$H_4(z) = H_1(z)H_1''(z^2). \tag{2.115}$$

The product filters here are the special four-band PR-QMF bank solutions with an additional factorization feature. They provide the hierarchical as well as the direct split of the input spectrum.

The repetitive use of an optimal two-band PR-QMF bank in a tree does not yield an optimum subband structure. Hence, progressive optimality plays a role for fine tuning of the product filters with respect to the features of interest. The design procedure of a progressively optimized subband tree is summarized for the present four-band case as the following. First, an optimal two-band PR-QMF solution (H_0, H_1) for the first stage of the tree is searched based on some design criterion. Then, the second stage two-band PR-QMF solutions (H_0', H_1') and (H_0'', H_1'') for each node that optimizes the four product filters are searched.

The optimization problem of hierarchical filter banks can also be formulated in the direct form in order to search for globally optimal solutions. The direct form with the factorization properties, Equation (2.115), has a much more difficult search procedure than the progressively optimized case. The latter breaks the optimization problem into smaller subproblems at the expense of a sure globally optimal solution. Our experience with the available optimization software suggests that the progressive approach is the more viable and a better way of solving the problem.

It was reported in [395] that performance improvements are possible by using different PR-QMF banks at different nodes in a subband tree.

9.1 A Note on Optimal Wavelets

The wavelet and scaling functions are expressed as an infinite product of the same prototype subband functions as given in Equations (2.75) and (2.74), respectively. However, as discussed earlier the repetitive use of an optimal two-band PR-QMF solution does not yield an optimal product filter from a subband transform perspective. Therefore, optimal product filters require the use of different two-band PR-QMFs at the different scales. This is in contradiction with the optimal wavelet bases designed in the literature based on a repetitive use of the same optimal two-band PR-QMF. Cohen and Sere also point at the same issue in Chapter 6 from the perspective of wavelet transform theory.

10 DISCUSSIONS AND FUTURE RESEARCH

The theory of subband transforms is now a mature subject currently utilized in many signal processing problems. Most of the recent efforts on the topic have shifted to the design and implementation issues. We will briefly discuss some issues that may have significance in future applications.

Historically, the completeness or perfect reconstruction was the most desired feature in subband transforms. But there are infinitely many subband transform bases in the solution space. Therefore, the question is now how to pick the best solution for the application of interest.

Optimal filter bank designs are computationally costly. Therefore, the hierarchical filter banks using the optimal prototype filters repetitively have been very attractive. The critical point here is to define the most useful tree structure for the given input and application. It has been noted that performing a decomposition blindly might degrade the performance or not yield any improvement. Any signal or spectral decomposition should be justified first before it is performed.

The concept of input dependent tree structuring in signal processing has been discussed in [41, 10, 439]. It was successfully implemented in an adaptive time-frequency exciser for spread spectrum communications and in discrete multitone transceivers [397, 49]. A similar approach to tree structuring was also studied independently as the fashionable ***wavelet packets*** in [95]. It is expected that the smart subband structuring will find its use in many other signal processing applications. More recently, ***time-varying*** subband transforms were proposed. This approach switches from one subband transform basis to another based on the variations of the signal properties and still satisfies the ***completeness*** property [384]. The evaluations of performance and complexity for this new approach are the topics of future study.

Image coding is one of the most popular applications of subband transforms. Subjective performance studies in subband image coding have been very limited in the literature. More effort in this area is expected.

The joint time-frequency optimization approach is now popular in the signal processing field. We expect that a more relaxed approach like a domain-switchable time-frequency processing technique will be studied for certain classes of signals in the future. In this approach, the time and frequency domain properties of signals would still be monitored simultaneously. The difference is that the domain dominant features of a signal will be treated completely in a single domain. Therefore, the domains of processing will be switchable. A successful use of this concept was made in the smart time-frequency exciser proposed for spread spectrum communications [397]. This approach might also find its use in other applications.

More sophisticated software tools for the design and tailoring of subband transforms are expected to be available in the market. This will lead to meritful use of subband transforms in many other applications.

3

MATRIX THEORY FOR FILTER BANK ANALYSIS AND DESIGN

Kambiz Nayebi*, Thomas P. Barnwell, and Mark J. T. Smith

Digital Signal Processing Laboratory, School of ECE, Georgia Institute of Technology, Atlanta, Georgia, 30332

**Department of Electrical Engineering, Sharif University of Technology, Tehran, 11365, Iran*

1 INTRODUCTION

Recent developments in both multirate systems theory and microelectronic technology have led to widespread interest in using analysis/synthesis systems based on filter banks for an ever increasing number of applications. Typically the performance of such application-specific systems is affected significantly by both the properties of the individual analysis and synthesis filters as well as by the structure and quality of the overall analysis/synthesis system. The properties of interest associated with the individual filters include the passband deviations, the transition bandwidths and behavior, the stopband deviations, the phase characteristics, the filter type (i.e., finite impulse response, FIR, or infinite impulse response, IIR), the order of the filters, and the computational efficiency of the implementation. Structural properties include the number of frequency bands, the bandwidths of the filters, and the overall frequency coverage. The quality of the overall analysis/synthesis system is mainly a function of its ability to reconstruct the input signal at the output. Sometimes a small loss in reconstruction accuracy can be traded for gains elsewhere, such as in computational complexity. The design phase for application-specific systems based on filter banks should address all of these separate issues simultaneously in a single design context.

Matrix-based formulations provide an effective way to analyze and design filter banks of different types. In Chapter 2, transform-domain matrix representations were introduced, in particular, the AC-matrix and polyphase matrix formulations. In the next section we continue in this line, beginning with a brief summary of the AC-matrix formulation, as a precursor to lattice filter banks. Filter banks based on lattice formulations are constrained structurally to yield exact reconstruction, as we shall see. However, such formulations are limited in their ability to address simultaneously the gamut of applications-specific systems tradeoffs that might be attractive in practice. Hence, the remainder and major portion of the chapter is devoted to developing the time-domain matrix formulation of the filter bank problem.

2 TRANSFORM DOMAIN FORMULATIONS

Most of the early theoretical developments and design procedures were based on frequency domain analysis, where aliasing, magnitude, and phase distortions were isolated explicitly and controlled in the design. These distortions can then be minimized or completely eliminated by designing proper analysis and synthesis filters. The problem of reconstructing an M-band maximally decimated subband decomposition is a relatively complex process. This process can be formulated in the frequency domain or equivalently in the z-domain, in the sense that $e^{-j\omega} \Longleftrightarrow z^{-1}$.

The first analysis method introduced for exact reconstruction filter banks was the *aliasing component* or *AC-matrix* formulation, [371], [373]. It provides the unique perfect reconstruction synthesis filters for any given set of analysis filters. The AC-matrix theory is particularly useful for analysis and design of two-band filter banks, as was shown in Chapter 2. A limitation, however, is that it provides no mechanism for conveniently designing the analysis and synthesis filters to meet a general set of pre-specified requirements. This is most evident for M-band filter banks where M is greater than two. Quite often the synthesis filters obtained using the AC-matrix approach from a given set of analysis filters are IIR and often also do not have causal stable realizations. The problem is not in the AC-matrix, as it provides the one and only perfect reconstruction solution. Rather, the problem is that the AC-matrix formulation does not provide a mechanism for choosing the analysis filters so that the synthesis filters will be acceptable.

The polyphase component matrix formulation, which is also discussed in Chapter 2, does not solve any of the aforementioned problems per se. Rather it is an alternate but equivalent representation. Polyphase filters were introduced by Bellanger in [43], generalized in [426], and further developed in [416] and [417]. Given a two-band system with an arbitrary lowpass/highpass filter pair, $H_0(z)$ and $H_1(z)$, four polyphase filters can be constructed, $e_{0,0}[n], e_{0,1}[n], e_{1,0}[n], e_{1,1}[n]$, where

$$e_{0,\ell}[n] = h_0[2n+\ell] \tag{3.1}$$
$$e_{1,\ell}[n] = h_1[2n+\ell] \tag{3.2}$$

and $\ell = 0, 1$. Extending this to an M-band system with analysis filters $h_k(n)$ ($k = 0, 1, \ldots, M-1$), we obtain M^2 analysis polyphase filters given by

$$e_{k,\ell}[n] = h_k[Mn+\ell],$$

where $\ell = 0, 1, \ldots, M-1$.

With some simple algebra, it can be shown that the analysis filters are expressible in terms of their polyphase filters, and may be written as

$$H_k(z) = \sum_{\ell=0}^{M-1} z^{-\ell} E_{k\ell}(z^M). \tag{3.3}$$

This result was also given in Section 3 of Chapter 2. Since the synthesis filter section is the dual of the analysis, we can represent the synthesis filters in a form similar to that above, i.e.,

$$G_k(z) = \sum_{\ell=0}^{M-1} z^{-(M-1-\ell)} R_{\ell k}(z^M), \tag{3.4}$$

where $R_{\ell k}(z)$ is the set of polyphase synthesis filters, analogous to (3.3). The analysis filter bank, $\mathbf{h}(z)$, and the synthesis filter bank, $\mathbf{g}(z)$, can be represented by the polyphase component matrices $\mathbf{E}(z)$ and $\mathbf{R}(z)$ as

$$\mathbf{h}(z) = \begin{bmatrix} H_0(z) \\ H_1(z) \\ \vdots \\ H_{M-1}(z) \end{bmatrix} \tag{3.5}$$

$$= \underbrace{\begin{bmatrix} E_{00}(z^M) & E_{01}(z^M) & \ldots & E_{0,M-1}(z^M) \\ E_{10}(z^M) & E_{11}(z^M) & \ldots & E_{1,M-1}(z^M) \\ \vdots & \vdots & \ddots & \vdots \\ E_{M-1,0}(z^M) & E_{M-1,1}(z^M) & \ldots & E_{M-1,M-1}(z^M) \end{bmatrix}}_{\mathbf{E}(z^M)} \begin{bmatrix} 1 \\ z^{-1} \\ \vdots \\ z^{-(M-1)} \end{bmatrix}$$

and

$$\mathbf{g}^T(z) = \left[z^{-(M-1)}, \ldots, 1\right] \times \underbrace{\begin{bmatrix} R_{00}(z^M) & \ldots & R_{0,M-1}(z^M) \\ R_{10}(z^M) & \ldots & R_{1,M-1}(z^M) \\ \vdots & \ddots & \vdots \\ R_{M-1,0}(z^M) & \ldots & R_{M-1,M-1}(z^M) \end{bmatrix}}_{\mathbf{R}(z^M)}. \tag{3.6}$$

Using the polyphase component matrix, it can be shown that the necessary and sufficient condition for exact reconstruction is

$$\mathbf{C}(z) = \mathbf{R}(z)\mathbf{E}(z) = z^{-\Delta}\mathbf{I}, \tag{3.7}$$

where Δ is the system delay.

For a given set of analysis filters, the synthesis filters can be obtained by a direct inversion of $\mathbf{E}(z)$. But this inversion poses the same kind of problems that were encountered in inverting the AC-matrix, $\mathbf{H}(z)$, i.e., the potential of having poles in the synthesis filters and the potential of having long numerator polynomials, and thus long synthesis filters, (see Section 3 of Chapter 2).

2.1 Factored Matrix Formulations

To circumvent the synthesis filter expansion problem the polyphase matrix can be structured to be a cascade of coefficient and delay matrices, each of which is invertible. This structure guarantees that no poles will appear and that the synthesis filters will have the same length as the analysis filters.

Two-Band Lattice Filter Banks

To examine this point further, we start by considering two-band FIR paraunitary filter banks. Paraunitary (or unitary) filter banks satisfy the condition that

$$\mathbf{E}(z) = \mathbf{R}^{\dagger}(z) \tag{3.8}$$

where $\dagger$ denotes the conjugate transpose. These filter banks have the property that the resulting synthesis filters are time-reversed versions of the analysis filters, i.e.,

$$G_k(z) = z^{-(N-1)}H_k(z^{-1}) \qquad k = 0, 1, \ldots, M-1. \tag{3.9}$$

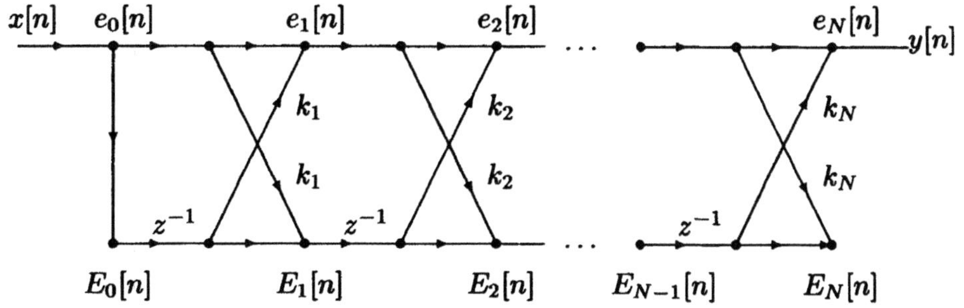

Figure 1 Flow graph of a two-band lattice structure.

In addition, paraunitary filter banks have the property that they reconstruct exactly, the filters are orthogonal, and the analysis and synthesis filters are power complementary, which means that

$$\sum_k |H_k(e^{j\omega})|^2 = \text{constant} \qquad -\pi \leq \omega \leq \pi. \tag{3.10}$$

Paraunitary two-band structures can be built by cascading 2×2 orthogonal matrices, i.e.,

$$\mathbf{E}(z) = \ \Theta_{N-1}D(z)\Theta_{N-2}D(z)\ldots\Theta_0 \tag{3.11}$$

where $D(z) = diag(1, z^{-1})$ and Θ_k is the 2×2 rotation matrix

$$\Theta_k = \begin{bmatrix} \cos\theta_k & \sin\theta_k \\ -\sin\theta_k & \cos\theta_k \end{bmatrix} = \cos\theta_k \begin{bmatrix} 1 & \alpha_k \\ -\alpha_k & 1 \end{bmatrix}, \tag{3.12}$$

where the rotation angles θ_k's (or α_k's) are the parameters for optimization. The system characterized by $\mathbf{E}(z)$ in Equation (3.11) can be realized using a lattice structure, like the one shown in Figure 1. Since the paraunitary condition guarantees perfect reconstruction, an unconstrained optimization may be used to obtain the θ_k parameters. A fairly significant part of the optimization problem is to find a good starting point for the process. This problem is addressed by choosing the starting lattice parameters to approximate a pre-designed frequency selective lattice filter [418]. Filters obtained using this method are orthogonal and satisfy the relationship

$$G_k(z) = z^{-(N-1)}H_k(z^{-1}) \qquad k = 0, 1. \tag{3.13}$$

Multiband Lattices

A simple approach to the design of multiband filter banks is to extend the aforementioned two-band approach to multiple bands. In effect, the design consists of building M-input, M-output paraunitary systems. This can be achieved by cascading $M \times M$ unitary matrices. Such orthogonal matrices can be a combination of rotations and delays such that

$$\mathbf{E}(z) = \Theta_{N-1} D_{N-1}(z) \Theta_{N-2} D_{N-2}(z) \ldots \Theta_1 D_1(z) \Theta_0 \tag{3.14}$$

where Θ_k is the combination of M rotation matrices (with M rotation angles) and $D(z) = diag(1, 1, \ldots, 1, z^{-1})$. For instance, for a three band system, $\Theta_k =$

$$\begin{bmatrix} \cos\theta_{1k} & \sin\theta_{1k} & 0 \\ -\sin\theta_{1k} & \cos\theta_{1k} & 0 \\ 0 & 0 & 1 \end{bmatrix} \begin{bmatrix} 1 & 0 & 0 \\ 0 & \cos\theta_{2k} & \sin\theta_{2k} \\ 0 & -\sin\theta_{2k} & \cos\theta_{2k} \end{bmatrix} \begin{bmatrix} \cos\theta_{3k} & 0 & \sin\theta_{3k} \\ 0 & 1 & 0 \\ -\sin\theta_{3k} & 0 & \cos\theta_{3k} \end{bmatrix}$$

and $D(z) = diag(1, 1, z^{-1})$. The rotation angles are the parameters that must be optimized to obtain the desired frequency response. Although analytically very attractive, there are a couple of pragmatic drawbacks. The task of finding good starting points is nontrivial, and relatively long optimization times make such structures less attractive for design purposes.

A more effective paraunitary structure is shown to be of the form

$$\mathbf{E}(z) = \mathbf{V}_{N-1}(z)\mathbf{V}_{N-2}(z)\mathbf{V}_1(z)\mathbf{T}, \tag{3.15}$$

where $\mathbf{V}_k$ is an $M \times M$ unitary matrix of the form

$$\mathbf{V}_k = (\mathbf{I} - \mathbf{v}_k\mathbf{v}_k^T + z^{-1}\mathbf{v}_k\mathbf{v}_k^T), \tag{3.16}$$

$\mathbf{T}$ is a unitary matrix defined by

$$\mathbf{T} = \pi(\mathbf{I} - 2\mathbf{w}_n\mathbf{w}_n^T), \tag{3.17}$$

and $\mathbf{v}_k$ and $\mathbf{w}_k$ are vectors with unity norms. In this structure, the elements of the vectors $\mathbf{v}_k$ and $\mathbf{w}_k$ are the optimization parameters. Many filter banks with good filtering properties may be designed using this structure. A shortcoming of the paraunitary structure is that it cannot be used to design biorthogonal filter banks. For example, linear phase filters cannot be obtained using paraunitary structures. But other similar structures are available for this purpose. The need for special purpose structures to handle filter banks of different types may be considered a drawback of this approach in comparison to the approach we introduce next.

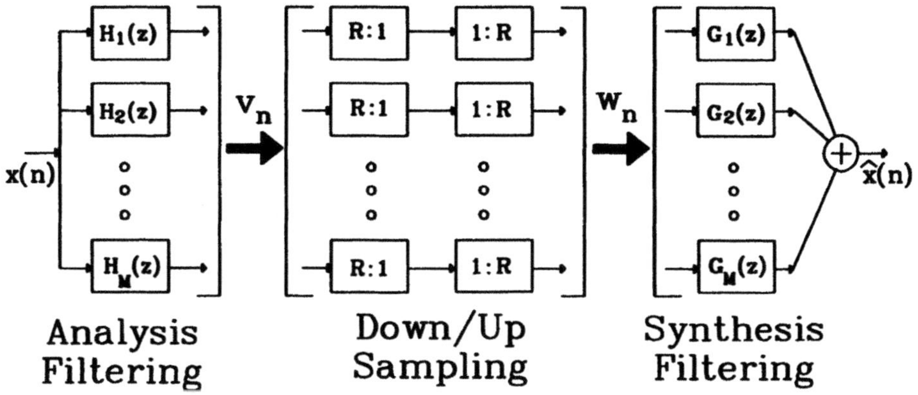

Figure 2 Block diagram of a uniform analysis/synthesis filter bank, partitioned into three different multi-input multi-output subsections.

3 TIME-DOMAIN FORMULATION

FIR analysis/synthesis filter banks can also be formulated in the time domain, which was originally proposed in [289] and later developed in [291], [290]. As it turns out, a powerful analysis and design framework can be constructed using a time domain representation [296], which has two major advantages. First, many different systems can be formulated in exactly the same form, resulting in the same design procedure for all. Among these systems are uniform band, nonuniform band, linear phase, modulated, low delay, and time-varying filter banks. The time domain formulation, as alluded to earlier, offers the flexibility to balance the tradeoffs among competing system features, such as reconstruction error and filter quality. The essence of the formulation [296] is a set of matrix equations that relates a block of output samples to blocks of input samples. As illustrated in Figure 2, the end-to-end system is partitioned into three parts: the analysis filtering section; the down/up sampling section; and the synthesis filtering section. For each section an input/output relationship is derived, and this set of relationships is used together to obtain the reconstruction conditions.

To obtain perfect reconstruction, the output, $\hat{x}(n)$, at time n, should be equal to the input at time $n-\Delta$, where Δ is the total reconstruction delay. For simplicity and without loss of generality we assume that all filters have the same length N, and that the filter lengths are assumed to be an integer multiple of the decimation factor, R, i.e., $N = LR$. Take note that these assumptions do not restrict generality, since N can always be viewed as the maximum length and thus any and all filters of shorter lengths can be expanded to this maximum length by appropriately adding zero coefficients.

Analysis Filtering Section

To begin, consider the ***analysis filtering*** section shown in Figure 2. At time n, the output of the mth analysis filter, $v_m(n)$, can be expressed as the inner product of the mth analysis filter coefficient vector, $\mathbf{h}_m$, and the input signal vector $\mathbf{x}_n$, i.e.,

$$v_m(n) = \langle \mathbf{h}_m, \mathbf{x}_n \rangle, \tag{3.18}$$

where

$$\mathbf{h}_m = [h_m(0), h_m(1), \ldots, h_m(N-1)]^T$$

and

$$\mathbf{x}_n = [x(n), x(n-1), x(n-2), \ldots, x(n-N+1)]^T.$$

The output sequences of the M analysis filters at time n, are represented by the vector sequence $\mathbf{v}(n)$ defined by

$$\mathbf{v}_n = [v_1(n), v_2(n), \ldots, v_M(n)]^T. \tag{3.19}$$

Using Equation (3.18), $\mathbf{v}_n$ can be expressed in terms of the input signal vector given by

$$\mathbf{v}_n = [\langle \mathbf{h}_1, \mathbf{x}_n \rangle; \langle \mathbf{h}_2, \mathbf{x}_n \rangle; \ldots; \langle \mathbf{h}_M, \mathbf{x}_n \rangle]^T = \mathbf{P}\mathbf{x}_n, \tag{3.20}$$

where $\mathbf{P}$ is an $M \times N$ transform matrix whose mth row is comprised of the mth analysis filter coefficients, i.e.

$$\mathbf{P} = \begin{bmatrix} h_1(0) & h_1(1) & h_1(2) & \ldots & h_1(N-2) & h_1(N-1) \\ h_2(0) & h_2(1) & h_2(2) & \ldots & h_2(N-2) & h_2(N-1) \\ h_3(0) & h_3(1) & h_3(2) & \ldots & h_3(N-2) & h_3(N-1) \\ \vdots & \vdots & \vdots & \vdots & \vdots & \vdots \\ h_M(0) & h_M(1) & h_M(2) & \ldots & h_M(N-2) & h_M(N-1) \end{bmatrix}. \tag{3.21}$$

These equations describe the input/output relationships for the ***analysis filtering*** section shown in Figure 2.

Down-up Sampling

The next section is the ***down/up sampling*** section. Here it is assumed that the decimation factor in all channels is R and $R \leq M$. Maximally decimated (or critically sampled) analysis/synthesis systems are obtained when $R = M$.

The output of this section in band m is $w_m(n)$ and can be expressed in terms of its input, $v_m(n)$, as

$$\mathbf{w}_n = \Lambda_n \mathbf{v}_n \tag{3.22}$$

where Λ_n is an $M \times M$ matrix, which is the identity matrix for $n = \ell R$, $\ell = 0, \pm 1, \pm 2, \ldots$, and zero otherwise. The vector $\mathbf{w}_n$ can be represented as

$$\mathbf{w}_n = [w_1(n), w_2(n), \ldots, w_M(n)]^T, \tag{3.23}$$

where $w_k(n)$ is the output of the kth upsampler. Thus Equation (3.22) is the input/output relationship for the down-up sampling section in Figure 2.

Synthesis Filtering Section

The last section is the ***synthesis filtering*** section, where the subband signals are filtered by the corresponding synthesis filters and then added together to produce the output $\hat{x}[n]$. Thus, at time n, the output signal can be determined by

$$\hat{x}(n) = \sum_{m=0}^{M-1} w_m(n) * g_m(n) = \sum_{m=1}^{M} \sum_{i=0}^{N-1} g_m(i) w_m(n-i). \tag{3.24}$$

To represent the synthesis filtering process in terms of inner products, an $M \times N$ matrix $\mathbf{Q}$ of synthesis filter coefficients is defined. As in matrix $\mathbf{P}$, each row of $\mathbf{Q}$ contains the coefficients of the corresponding synthesis filter coefficients, resulting in

$$\mathbf{Q} = \begin{bmatrix} g_1(0) & g_1(1) & g_1(2) & \ldots & g_1(N-2) & g_1(N-1) \\ g_2(0) & g_2(1) & g_2(2) & \ldots & g_2(N-2) & g_2(N-1) \\ g_3(0) & g_3(1) & g_3(2) & \ldots & g_3(N-2) & g_3(N-1) \\ \vdots & \vdots & \vdots & \vdots & \vdots & \vdots \\ g_M(0) & g_M(1) & g_M(2) & \ldots & g_M(N-2) & g_M(N-1) \end{bmatrix}. \tag{3.25}$$

For convenience, $\mathbf{Q}$ can be represented as

$$\mathbf{Q} = [\mathbf{q}_0 | \mathbf{q}_1 | \cdots | \mathbf{q}_{N-1}],$$

where the $\mathbf{q}_i$'s are the column vectors of $\mathbf{Q}$. Using the column representation of the synthesis filter matrix, Equation (3.24) can be written as

$$\hat{x}(n) = \sum_{i=0}^{N-1} \langle \mathbf{q}_i, \mathbf{w}_{n-i} \rangle. \tag{3.26}$$

Equations (3.20), (3.22) and (3.26) are the three basic equations that are to be combined to obtain the total input/output relationship for the system. A direct combination of Equations (3.20), (3.22) and (3.26) results in an input/output function of the form

$$\hat{x}(n+i) = \sum_{k=0}^{L-1} \langle \mathbf{P}^T \mathbf{q}_{kR+i}, \mathbf{x}_{n-kR} \rangle \qquad i = 0, 1, \ldots, R-1, \tag{3.27}$$

which relates the system output to the input vector $\mathbf{x}_n$. For the reconstruction to be exact, it is necessary to satisfy the equation

$$\hat{x}(n+i) = x(n+i-\Delta), \quad \text{for} \quad i = 0, 1, 2, \ldots, R-1. \tag{3.28}$$

To satisfy these, the necessary conditions to extract a specific sample of the input signal from the right hand side of Equation (3.27) should be identified.

Perfect Reconstruction Conditions

Equation (3.27) can be rewritten in a vector form as,

$$\hat{x}(n+i) = \mathbf{r}_i^T \hat{\mathbf{u}}(n), \tag{3.29}$$

where $\mathbf{r}_i$ and $\hat{\mathbf{u}}(n)$ are vectors of size $N = LM$ defined by

$$\mathbf{r}_i = [\mathbf{q}_i^T, \mathbf{q}_{i+R}^T, \mathbf{q}_{i+2R}^T, \ldots, \mathbf{q}_{i+LR-R}^T]^T, \tag{3.30}$$

and $\hat{\mathbf{u}}(n)$ is

$$\hat{\mathbf{u}}(n) = [\mathbf{P}x_n, \mathbf{P}x_{n-R}, \ldots, \mathbf{P}x_{n-LR+R}]. \tag{3.31}$$

The vector $\hat{\mathbf{u}}(n)$ can be expressed in a compact form as a matrix equation in terms of $M \times R$ submatrices $\mathbf{P}_j$ $(j = 0, 1, \ldots, L-1)$ where $\mathbf{P}_j$ is the jth $M \times R$ submatrix of the $M \times N$ matrix $\mathbf{P}$. With this notation, $\hat{\mathbf{u}}(n)$ can be expressed as

$$\hat{\mathbf{u}}(n) = \begin{bmatrix} \mathbf{P}_0^T & \mathbf{P}_1^T & \mathbf{P}_2^T & \cdots & \mathbf{P}_{L-1}^T & \mathbf{O} & \cdots & \mathbf{O} \\ \mathbf{O} & \mathbf{P}_0^T & \mathbf{P}_1^T & \cdots & \mathbf{P}_{L-2}^T & \mathbf{P}_{L-1}^T & \cdots & \mathbf{O} \\ \mathbf{O} & 0 & \mathbf{P}_0^T & \cdots & \mathbf{P}_{L-3}^T & \mathbf{P}_{L-2}^T & \cdots & \mathbf{O} \\ \vdots & \vdots & \vdots & \vdots & \vdots & \vdots & \ddots & \mathbf{O} \\ \mathbf{O} & \mathbf{O} & \mathbf{O} & \mathbf{O} & \mathbf{P}_0^T & \mathbf{P}_1^T & \cdots & \mathbf{P}_{L-1}^T \end{bmatrix} \mathbf{x}_I(n), \tag{3.32}$$

where $\mathbf{x}_I(n)$ is given as

$$\mathbf{x}_I(n) = [x(n), x(n-1), \ldots, x(n-2N+R)] \tag{3.33}$$

and $\mathbf{O}$ is the $R \times M$ zero matrix. Therefore the input/output function of the filter bank can be expressed in the form

$$\hat{x}(n+i) = \mathbf{r}_i^T \mathbf{A}^T \mathbf{x}_I(n). \tag{3.34}$$

Obviously, for exact reconstruction of the input at the output,

$$\mathbf{z}_i = \mathbf{A}\mathbf{r}_i = [1, 0, 0, \ldots, 0]^T. \tag{3.35}$$

For a perfect reconstruction filter bank with a delay of Δ,

$$\mathbf{z}_i = \mathbf{A}\mathbf{r}_i = [0, \ldots, 0, 1, 0, \ldots, 0]^T, \tag{3.36}$$

where 1 is the $\Delta + 1 - i$ th element of the vector.

By expressing the synthesis matrix, $\mathbf{Q}$, in terms of $M \times R$ submatrices $\mathbf{Q}_j$, (with $j = 0, 1, \ldots, L-1$), the conditions for perfect reconstruction can be expressed compactly as

$$\underbrace{\begin{bmatrix} \mathbf{P}_0^T & \mathbf{O} & \mathbf{O} & \ldots & \mathbf{O} \\ \mathbf{P}_1^T & \mathbf{P}_0^T & \mathbf{O} & \vdots & \mathbf{O} \\ \mathbf{P}_2^T & \mathbf{P}_1^T & \mathbf{P}_0^T & & \vdots \\ \mathbf{P}_3^T & \mathbf{P}_2^T & \mathbf{P}_1^T & \ddots & \\ \vdots & \vdots & \vdots & \ddots & \\ \mathbf{P}_{L-1}^T & \mathbf{P}_{L-2}^T & \mathbf{P}_{L-3}^T & \ldots & \mathbf{P}_0^T \\ \mathbf{O} & \mathbf{P}_{L-1}^T & \mathbf{P}_{L-2}^T & \ddots & \mathbf{P}_1^T \\ \vdots & \vdots & \vdots & \ddots & \vdots \\ \mathbf{O} & \mathbf{O} & \mathbf{O} & \mathbf{O} & \mathbf{P}_{L-1}^T \end{bmatrix}}_{\mathbf{A}} \underbrace{\begin{bmatrix} \mathbf{Q}_0 \\ --- \\ \mathbf{Q}_1 \\ --- \\ \vdots \\ --- \\ \mathbf{Q}_{L-2} \\ --- \\ \mathbf{Q}_{L-1} \end{bmatrix}}_{\mathbf{S}} = \underbrace{\begin{bmatrix} \mathbf{O}_R \\ -- \\ \vdots \\ -- \\ \mathbf{O}_R \\ -- \\ \mathbf{J}_R \\ -- \\ \mathbf{O}_R \\ -- \\ \vdots \\ -- \\ \mathbf{O}_R \end{bmatrix}}_{\mathbf{B}}, \tag{3.37}$$

where $\mathbf{O}$ is the $R \times M$ zero matrix, $\mathbf{O}_R$ is the $R \times R$ zero matrix, and $\mathbf{J}_R$ is the $R \times R$ anti-diagonal matrix,

$$\mathbf{J}_R = \frac{1}{R} \begin{bmatrix} 0 & 0 & \ldots & 0 & 1 \\ 0 & 0 & \ldots & 1 & 0 \\ \vdots & \vdots & & 0 & 0 \\ 1 & 0 & \ldots & 0 & 0 \end{bmatrix}.$$

The left side of Equation (3.37) is very similar to a convolution matrix equation while the columns of **B** are just the R impulse responses of the periodically time-varying analysis/synthesis filter bank. For perfect reconstruction, the **B** matrix has the form shown in Equation (3.37). Notice that for an aliasing-free system, the system is time-invariant and the columns of **B** are just shifted versions of each other.

To illustrate the formulation, a simple two-band example is presented next in which the filters have length 6 and the decimation factor is 2, i.e., $N = 6$ and $R = 2$. The **P** and **Q** matrices are then just

$$\mathbf{P} = \begin{bmatrix} h_0(0) & h_0(1) & h_0(2) & h_0(3) & h_0(4) & h_0(5) \\ h_1(0) & h_1(1) & h_1(2) & h_1(3) & h_1(4) & h_1(5) \end{bmatrix} \tag{3.38}$$

$$\mathbf{Q} = \begin{bmatrix} g_1(0) & g_1(1) & g_1(2) & g_1(3) & g_1(4) & g_1(5) \\ g_2(0) & g_2(1) & g_2(2) & g_2(3) & g_2(4) & g_2(5) \end{bmatrix}, \tag{3.39}$$

leading to the time domain matrix equations

$$\underbrace{\begin{bmatrix} h_0(0) & h_1(0) & 0 & 0 & 0 & 0 \\ h_0(1) & h_1(1) & 0 & 0 & 0 & 0 \\ h_0(2) & h_1(2) & h_0(0) & h_1(0) & 0 & 0 \\ h_0(3) & h_1(3) & h_0(1) & h_1(1) & 0 & 0 \\ h_0(4) & h_1(4) & h_0(2) & h_1(2) & h_0(0) & h_1(0) \\ h_0(5) & h_1(5) & h_0(3) & h_1(3) & h_0(1) & h_1(1) \\ 0 & 0 & h_0(4) & h_1(4) & h_0(2) & h_1(2) \\ 0 & 0 & h_0(5) & h_1(5) & h_0(3) & h_1(3) \\ 0 & 0 & 0 & 0 & h_0(4) & h_1(4) \\ 0 & 0 & 0 & 0 & h_0(5) & h_1(5) \end{bmatrix}}_{\mathbf{A}} \underbrace{\begin{bmatrix} g_0(0) & g_0(1) \\ g_1(0) & g_1(1) \\ g_0(2) & g_0(3) \\ g_1(2) & g_1(3) \\ g_0(4) & g_0(5) \\ g_1(4) & g_1(5) \end{bmatrix}}_{\mathbf{S}} = \underbrace{\begin{bmatrix} 0 & 0 \\ 0 & 0 \\ 0 & 0 \\ 0 & 0 \\ 0 & 1 \\ 1 & 0 \\ 0 & 0 \\ 0 & 0 \\ 0 & 0 \\ 0 & 0 \end{bmatrix}}_{\mathbf{B}}.$$

4 FILTER BANK DESIGN USING THE TIME DOMAIN EQUATIONS

In the time-domain design formulation, a constrained optimization procedure may be used to minimize a frequency error function that contains the passband and the stopband errors of the system filters, where the quadratic reconstruction constraints are given by

$$\mathbf{AS} = \mathbf{B}. \tag{3.40}$$

In the optimization process, the *reconstruction* error ϵ_r is reduced to zero, where

$$\epsilon_r = \|\mathbf{AS} - \mathbf{B}\|_F^2 \tag{3.41}$$

and $\|\cdot\|_F$ denotes the Frobenius matrix norm defined as

$$\|\mathbf{X}\|_F^2 = \sum_i \sum_j |\, x_{ij} \,|^2 . \tag{3.42}$$

Concurrently, the optimization also minimizes the *frequency* error defined by

$$\epsilon_f = \int_{\omega \in \text{stopband}} |H_m(\omega)|^2 d\omega + \int_{\omega \in \text{passband}} |1 - H_m(\omega)|^2 d\omega. \tag{3.43}$$

A more detailed discussion of the constrained optimization approach may be found in [435, 156].

Another design approach is to use an unconstrained optimization procedure to minimize a combination of ϵ_r and ϵ_f. Here the analysis filter coefficients are the optimization parameters and at each stage of the optimization, the synthesis filters are obtained using the least squares solution of Equation (3.40). This procedure, which was first reported in [296], makes a systematic tradeoff between design objectives possible. In general, an error term ϵ_j may represent any error term that might be required to impose a desired time or frequency domain characteristic on the system. Weighting factors, w_i's, can be used to combine the individual error terms. These factors play an important role in balancing the final result in terms of the filter responses, the reconstruction error, and/or any other characteristic that may be included in the optimization. For example, to force a near perfect reconstruction solution, the weight of the reconstruction error can be chosen to be much larger than the other weights. This can yield solutions that achieve better frequency response characteristics at the expense of a small reconstruction error.

The strength of the time domain formulation is that it has unprecedented flexibility in its design capability. It provides an environment in which it is very simple to impose constraints explicitly on the filter coefficients or to modify the matrix structure within the framework to accommodate many different cases. This is particularly true for two-band analysis/synthesis systems, because, compared to the general M-band case, they generally have fewer optimization parameters. This leads to faster design convergence and to better final solutions typically.

In the previous chapter, the design of two-band exact reconstruction filter banks was discussed using spectral factorization [373]. Two-band filter banks can also be designed using the time-domain formulation in a way that allows for much greater flexibility. Using Equation (3.37) with number of bands equal to two, and the iterative design procedure just discussed, a wide variety of different properties may be incorporated into the design process. In the sections that follow, we examine several special types of filter banks that can be designed conveniently within the time-domain framework.

4.1 Modulated and M-Band Filter Banks

The design of M-band uniform filter banks, which can be viewed as an extension of two-band systems, represents a challenging problem due to the greater number of aliasing components that must be cancelled. Equivalently, the number of time-domain equations that must be satisfied is large in comparison to the two-band filter bank. Hence in the context of the time-domain formulation, this can lead to large iterative optimizations. Thus M-band designs of this type are sometimes limited by their size.

In general, modulated filter banks have two major advantages over their non-modulated counterparts. First, they offer implementation structures that make their realizations very efficient; and second, their constrained nature results in a dramatic reduction in design parameters. Modulated filter banks are based on a single baseband filter. Cosine modulation is most common for filter banks. In cosine modulated systems, the analysis filters, $h_k(n)$, are constrained to be of the general form

$$h_k(n) = Ah(n)\cos\left(\frac{\pi}{M}(n+\alpha)(k+\gamma)+\phi\right) \qquad k = 0,\ldots,M-1, \tag{3.44}$$

where M is the number of frequency bands; α, γ, and ϕ are pre-defined offsets, $h(n)$ is the baseband prototype filter, and A is an amplitude scale factor. It is also assumed that $N = 2KM$ where N is the filter length and K is a positive integer. Similarly, the synthesis filters are also assumed to have a similar cosine modulated form.

For perfect reconstruction, the system filters need to satisfy the reconstruction conditions presented earlier for the general M-band case. In many cases, these conditions can be translated into a set of conditions on the baseband filter. For example, for the special case with

$$A = \sqrt{\frac{2}{M}}, \tag{3.45}$$

$$\alpha = \frac{M+1}{2}, \tag{3.46}$$

$$\gamma = \frac{1}{2}, \tag{3.47}$$

and $\phi = 0$, we obtain the analysis filters

$$h_k(n) = \sqrt{\frac{2}{M}} h(n) \cos[\frac{\pi}{M}(n + \frac{M+1}{2})(k + \frac{1}{2})]. \tag{3.48}$$

The reconstruction condition on the baseband filter, $h_k(n)$, can be expressed as

$$\sum_{i=0}^{2K-2\ell-1} h(n+iM)h(n+iM+2\ell M) = \delta(\ell) \tag{3.49}$$

for $n = 0, 1, \ldots, M/2 - 1$. The above conditions imply some orthogonality properties for the even symmetric baseband filter, $h(n)$. This specific type of cosine modulated system where the synthesis filters are assumed to be time-reversed versions of the analysis filters, i.e.,

$$g_k(n) = h_k(N - 1 - n), \tag{3.50}$$

was studied by Malvar [254].

The case of $N = 2M$, which was first reported in [324], was also presented in [252] under the name of lapped transforms. Designing modulated filter banks with $N = 2M$ can actually be done without using optimization. This is because the cosine kernel of Equation (3.48) automatically forces the conditions required for perfect reconstruction. An additional set of special cases for modulated filter banks was reported by Malvar in [253, 254]. In [292], a design approach using least squares is presented. There, paraunitary lattice structures were used successfully to design cosine modulated filter banks with the perfect reconstruction property [418]. A novel and flexible approach to cosine modulated filter banks is presented in [361], [360] and is based on an explicit cascade of matrices.

The time-domain approach is particularly well suited to the design of modulated filter banks because of the reduced number of parameters required in the optimization procedure. Examples of modulated filter banks designed using the constrained time-domain approach can be found in [288, 292]. In addition, a commercial filter bank design product that uses an unconstrained optimization approach [125] is also available.

4.2 Linear Phase Filter Banks

In some applications such as image processing, linear phase filter banks may be attractive. To design linear phase filter banks, a symmetry constraint is imposed on the filters. Imposing such symmetry can be done easily in the time-domain formulation. As an example, consider the two-band case given earlier in Equations (3.38) and (3.39). Imposing linear phase symmetry constraints results in

$$h_0(n) = h_0(N-1-n)$$
$$h_1(n) = -h_1(N-1-n).$$

For the example, this implies that

$$h_0(0) = h_0(5) \qquad h_1(0) = -h_1(5)$$
$$h_0(1) = h_0(4) \qquad h_1(1) = -h_1(4)$$
$$h_0(2) = h_0(3) \qquad h_1(2) = -h_1(3).$$

Therefore, we only need to optimize explicitly $h_0(0), h_0(1), h_0(2), h_1(0), h_1(1)$, and $h_1(2)$. Everything else in the design procedure remains the same. It is noteworthy that imposing linear phase actually makes the design process easier because the number of parameters is reduced by the symmetry.

It can be shown that the least squares solution used in the design results in linear phase synthesis filters. Thus choosing linear phase analysis filters results in linear phase synthesis filters. It is also appropriate to mention that linear phase can be imposed on cosine modulated filter banks. In such a case the baseband filter would be constrained to have the form

$$h(n) = h(N-1-n),$$

where N is the filter length.

4.3 Low Delay Filter Banks

Until fairly recently (1992), it was generally accepted that the system delay associated with a perfect reconstruction filter bank with filter lengths of N was $N-1$. That is, if $x(n)$ is the input to a perfect reconstruction analysis/synthesis system, then $x(n-N-1)$ is the output. It was through the time domain formulation [297] that it was shown that system delay could be treated as a variable parameter. The fact is that uniform M-band systems with a decimation factor of M in all bands can have a minimum delay of $M-1$ samples and a maximum delay of $(2L-1)M-1$ samples, where L is the number of submatrices (along a row) in the $\mathbf{A}$ matrix in Equation (3.37). By establishing the relationship between the structure of the time domain conditions and the reconstruction delay of the system, the design procedures just presented can be applied directly to the design of minimum delay and maximum delay analysis/synthesis systems, as well as to those with all other delays in between. The facilitating feature of the formulation is the $\mathbf{B}$ matrix, which embodies explicitly information about the delay through the filter bank. The position of the anti-diagonal matrix, $\mathbf{J}_M$, in $\mathbf{B}$ determines the system delay. For example, to design a minimum delay system with $M-1$ samples of delay, matrix $\mathbf{B}$ should be of the form

$$\mathbf{B}_{min} = [\mathbf{J}_M|\mathbf{0}^T|\ldots|\mathbf{0}^T|\mathbf{0}^T|\mathbf{0}^T]^T, \tag{3.51}$$

where $\mathbf{0}$ is the $R \times 1$ zero vector. A system delay of $2M-1$ samples requires a $\mathbf{B}$ matrix of the form

$$\mathbf{B}_1 = [\mathbf{0}^T|\mathbf{J}_M|\mathbf{0}^T|\ldots|\mathbf{0}^T|\mathbf{0}^T]^T. \tag{3.52}$$

Similarly for the maximum delay of $(2L-1)M-1$ samples

$$\mathbf{B}_{max} = [\mathbf{0}^T|\mathbf{0}^T|\ldots|\mathbf{0}^T|\mathbf{0}^T|\mathbf{J}_M]^T. \tag{3.53}$$

In some recent work, other approaches to low delay filter bank design were reported that employed a quadratically constrained optimization formulation [300], and a cascade matrix formulation [360]. All represent good alternative methods for the design of low delay filter banks.

The imposition of the low reconstruction delay on the analysis/synthesis systems will in general require analysis and synthesis filters with nonlinear phase. For example, most of the analysis and synthesis filters of a minimum delay system tend to be either minimum phase or very close to minimum phase. From a design point of view, this means that for satisfactory results and faster convergence, minimum phase starting filters are more suitable for designing low delay systems.

As one might expect, low delay represents a condition that degrades the quality of the analysis/synthesis filters, the severity of which is greatest for minimum and maximum delays. The best filter quality is typically achieved when the delay is $N-1$. Nonetheless, low delay filter banks can provide overall improvement in ***delay-sensitive*** applications. A notable example of this is the application to two-way digital subband speech coding. If delays greater than one-fourth second occur, the users find the delays annoying and disruptive. Although the speech quality itself may be perfect, the delays make the system objectionable. Low delay filter banks can be used in such cases to help confine the overall system delay to be within acceptable limits [293].

4.4 Nonuniform Filter Banks

Nonuniform band filter banks refer to systems where the bandwidths of the analysis filters and their corresponding downsampling rates are not equal. Tree structures composed of two-band, three-band, and M-band filter banks can be used to construct nonuniform band systems, as discussed in Chapter 2. Such an approach is straightforward conceptually. Trees of this type can be very useful, but can also have some drawbacks. They can, for instance, have excessive delays associated with them, which are directly related to the depth of the tree. Moreover, multilayer trees can make implementation difficult in some situations.

Less obvious, conceptually, is the construction of nonuniform band filter banks without the use of trees. Here there exists the potential to avoid some of the delay and implementation drawbacks that are sometimes present in tree-based approaches. Furthermore, there is the potential for greater flexibility and variety in the decompositions that are possible.

To investigate this point in further detail, we state the problem as follows. The filter bank structure under consideration has M-channels, each with an integer decimation factor, R_i, $i = 0, 1, \ldots, M-1$. The maximal decimation condition then implies that

$$\sum_{i=0}^{M-1} \frac{1}{R_i} = 1. \tag{3.54}$$

The question at hand is how do we design the analysis and synthesis filters such that we effectively achieve the desired frequency partitioning at the analysis section output and good reconstruction at the synthesis section output.

As in a uniform band filter bank, the same general three component time-domain formulation structure, illustrated in Figure 2, can be used. In fact the analysis filtering and the reconstruction subsystems are identical to those of the uniform band case, and their input/output functions are given as

$$\text{Analysis:} \quad \mathbf{v}_n = \mathbf{P}\mathbf{x}_n \tag{3.55}$$

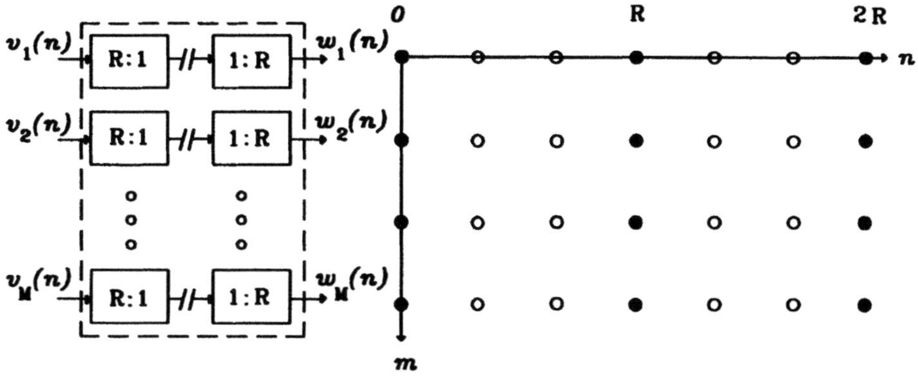

Figure 3 Time-frequency map for a uniformly decimated system. Heavy dots represent nonzero samples and empty circles represent zero samples. The notation $R:1$ and $1:R$ denotes downsampling and upsampling by a factor of R.

$$\text{Synthesis:} \quad \hat{x}(n) = \sum_{k=0}^{N-1} \langle \mathbf{q}_k, \mathbf{w}_{n-k} \rangle \tag{3.56}$$

where $\mathbf{x}_n$, $\mathbf{P}$, $\mathbf{v}_n$, $\hat{x}(n)$, $\mathbf{w}_n$, and $\mathbf{q}_k$ are all as defined in Equations (3.18)–(3.26). The difference between the uniform and nonuniform cases is in the down/up sampling section, represented by Equation (3.22).

To better visualize the difference, consider the output of the down/up sampling section of a uniform band system. The corresponding time-frequency map, is shown in Figure 3. In this map, m is the frequency index and n is the time index. The heavy dots represent nonzero samples and the open circles represent zero samples. The nth column in this map represents the vector $\mathbf{w}_n$. As it is clear from the output map of Figure 3, the map is periodic with a period of R, and the output vectors are nonzero at integer multiples of R. For the more general class of systems with different decimation rates in different bands, the output time-frequency map is also periodic but with period T, where T is the least common multiple of the decimation factors, R_i, $i = 0, \ldots, M-1$. This fact is evident from the three channel down/up sampling section example in Figure 4.

To obtain an input/output *function* for this down/up sampling subsystem, Λ_n becomes a sequence of T matrices, each of which is $M \times M$, and all of which have nondiagonal elements equal to zero. The diagonal elements are either 0 or 1 and their values at each time n depend on the output vector map at time n. Specifically, a heavy dot at the mth row and nth column of the map

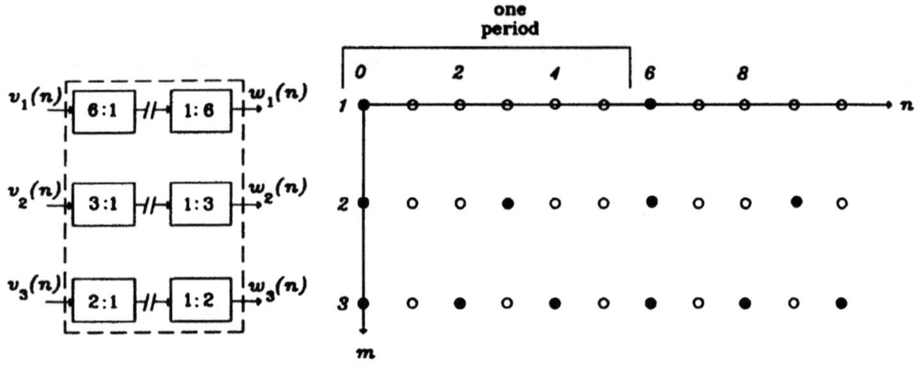

Figure 4 Down/up sampling section of a nonuniform system and its time-frequency map.

corresponds to a 1 at the mth diagonal element of Λ_n. Similarly, an empty circle in the same position results in a zero at the mth diagonal element of Λ_n.

As an example, consider a three-band system with decimation factors of 2, 3, and 6. The operator Λ_n is 3×3 and the period of the system is $T = 6$, which is the least common multiple of 2, 3, and 6. The Λ_n matrices for this case are

$$\Lambda_0 = \begin{bmatrix} 1 & 0 & 0 \\ 0 & 1 & 0 \\ 0 & 0 & 1 \end{bmatrix} \quad \Lambda_1 = \begin{bmatrix} 0 & 0 & 0 \\ 0 & 0 & 0 \\ 0 & 0 & 0 \end{bmatrix} \quad \Lambda_2 = \begin{bmatrix} 0 & 0 & 0 \\ 0 & 0 & 0 \\ 0 & 0 & 1 \end{bmatrix}$$

$$\Lambda_3 = \begin{bmatrix} 0 & 0 & 0 \\ 0 & 1 & 0 \\ 0 & 0 & 0 \end{bmatrix} \quad \Lambda_4 = \begin{bmatrix} 0 & 0 & 0 \\ 0 & 0 & 0 \\ 0 & 0 & 1 \end{bmatrix} \quad \Lambda_5 = \begin{bmatrix} 0 & 0 & 0 \\ 0 & 0 & 0 \\ 0 & 0 & 0 \end{bmatrix}. \tag{3.57}$$

It is easy to see that these matrices maps the input vector $\mathbf{v}_n$, to the proper output vector, $\mathbf{w}_n$.

Since the filter bank is a periodically time-varying system, there is only a finite number of such matrices, six in this case. The time domain conditions for nonuniform reconstruction can be expressed as a matrix equation similar to the one presented in Equation (3.37). From this point on the design process is very similar to that of the uniform band case [299]. The key modification presented here is matching the down/up subsystem to the implicit nonuniform subsampling implied by the nonuniform band system.

Much more can be said about nonuniform band filter banks, but such is beyond the scope of this chapter. To a large extent, we have only scratched the surface of this topic. A more complete discussion can be found in [298, 236]. It is worth pointing out, however, that it is also possible to construct more general nonuniform band filter banks via the time-domain formulation. An extended class of nonuniform banks consists of those in which each analysis channels involves upsampling the input, filtering, and then downsampling. Similarly, each channel of the synthesis also involves upsampling, filtering and downsampling. These systems allow for the design of filter banks that have an even greater variety of decompositions. For example, it is possible to design a two-band system with the first channel having a bandwidth of $2\pi/3$ and the second with a bandwidth of $\pi/3$.

4.5 Block Decimated Filter banks

Maximally decimated filter banks have traditionally been based on conventional downsampling and upsampling. In this way the number of samples in the input is equivalent to the aggregate number of samples in the analysis section output. This property of maximal decimation or critical subsampling is important in many applications (such as compression) where data expansion represents a problem. Thus conventional downsampling and upsampling provides an obvious solution. In principle, this data rate reduction can be achieved by other forms of downsampling that are not typically considered in multirate processing. In particular, we refer to the class of block downsamplers and upsamplers where samples are taken in contiguous blocks of R samples. The first R_1 of the R samples in a block are retained, while the remaining R_2 samples are discarded. (Note obviously that $R > R_1, R_2$.) Such an approach is an alternative to the convention and achieves a decimation rate of R to R_1. For example, if we let R be 4 and R_1 be 2, we achieve a block downsample that operates as a square wave. It retains the first two samples, discards the next two, keeps the next two, and so on. It is evident that one can even generalize further and achieve maximal decimation by allowing more arbitrary patterns of samples within the blocks to be discarded. Clearly this represents an alternate (and very different) way of formulating the filter bank problem. This approach also arose from the time domain formulation [295].

The construction of such systems follows the approach just presented for nonuniform band filter banks. The distinguishing aspect is the manner in which downsampling and upsampling are handled, which is addressable directly in the down/up subsystem in the time domain formulation. As before, the implication is that we have a set of matrices, Λ_n, that uniquely characterize the block (or pattern) decimation. As a simple example, consider a two-band system where in the first channel we keep the first and discard the next two samples in every contiguous block of three samples. In the second channel we keep the first two and discard the third in each block of three. This block decimation structure may be represented by the matrices

$$\Lambda_0 = \begin{bmatrix} 1 & 0 \\ 0 & 1 \end{bmatrix} \qquad \Lambda_1 = \begin{bmatrix} 0 & 0 \\ 0 & 1 \end{bmatrix} \qquad \Lambda_2 = \begin{bmatrix} 0 & 0 \\ 0 & 0 \end{bmatrix}. \tag{3.58}$$

Given that the down/up sampling is now represented in the formulation, the iterative optimization can be performed in exactly the same way as before.

Block decimation systems can be designed effectively using the time domain formulation. A more complete discussion of this topic can be found in [295]. Unlike some of the popular filter banks, however, block decimation filter banks have not yet found many areas of application for which they are attractive.

4.6 Time-Varying Filter Banks

One of the very novel and interesting concepts to emerge from the time domain formulation is the concept of time varying filter banks. This idea, which was introduced in 1992 [296], advanced the notion that the filter coefficients in the constituent analysis and synthesis filters could be changed during processing while still preserving the exact reconstruction property of the filter banks. More specifically, it demonstrated the feasibility of switching between two different sets of analysis/synthesis filter banks on the fly without incurring distortion in the process. In follow-on work, it was shown that, in a similar way, the number of frequency bands in the filter bank could be changed adaptively during processing while preserving exact reconstruction [384, 385, 381]. The obvious implication of this is that the potential exists to adapt the time and frequency resolution of the filter bank to match the input signal, which is rather interesting.

To elaborate upon this idea, envision it, for a moment, in the context of conventional filter banks. Consider first, a conventional uniform M-band decomposition in time and frequency, as depicted in Figure 5a. The frequency bands (shown on the ordinate) are plotted as a function of time, resulting in a classical time/frequency plot. Such a decomposition might be appropriate for a stationary input signal with wideband energy distribution. Similarly an octave-band decomposition, such as that shown in Figure 5b, might be well suited to sta-

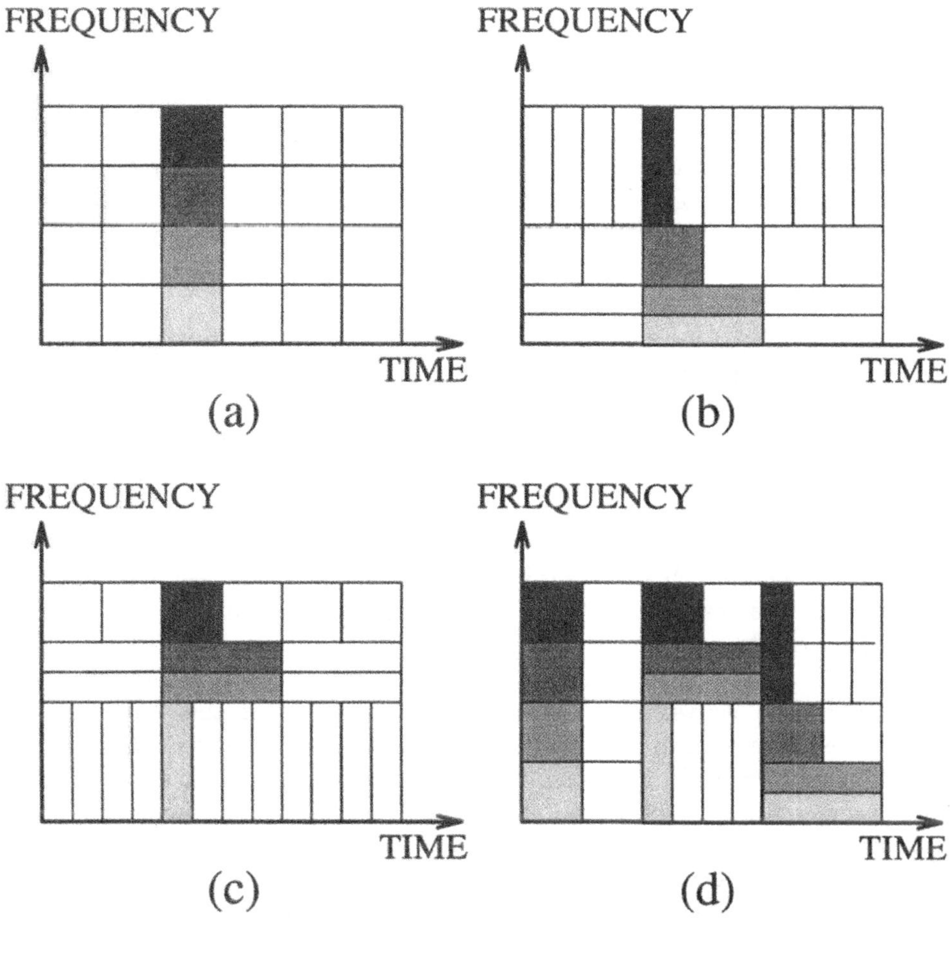

Figure 5 The time-frequency tilings for different time-frequency transforms: (a) a uniform 4-band filter bank; (b) an octave-band filter bank; (c) an arbitrary filter bank, illustrating an optimized decomposition; (d) an adaptive resolution-variant filter bank.

tionary inputs with strong lowpass frequency characteristics. Along these same lines and even better is a decomposition that is optimized specifically for a particular input. Such is illustrated in Figure 5c where the time/frequency tiling is seen to be irregular, hypothetically reflecting the particular characteristics of the signal. Assuming the signal is stationary (i.e. that these characteristics remain constant for all time), such a decomposition is expected to perform well.

In practice, however, inputs are rarely ever stationary. Their frequency content can vary radically from being lowpass to wideband, to highpass, and all points in between. Thus the notion of having a time/frequency decomposition that adapts to the dynamics of the input is attractive. This is illustrated graphically in Figure 5d, where over time we see the changing evolution of the decomposition.

Although one might envision using this newfound capability to facilitate a more accurate signal representation, the manner of implementation and adaptation control is a topic in need of further investigation. At the end of this chapter we will revisit this point and also summarize some of the other time-varying approaches that have been discussed recently. But the immediate and main focus here is on a particular method for addressing time-varying systems called the ***post filter*** method. It has a number of important advantages that make it attractive, among which are that it is easy to implement, it is easy to design, and it accommodates both time-varying M-band as well as adaptive time-frequency filter banks of the type shown in Figure 5d.

For convenience, we will develop the post filter method in the context of the specific case where we switch from a two-band system to a three-band system. Initially the filter bank is an exact reconstruction system with a lowpass and highpass output. At time n_0 we then switch ***directly*** to another exact reconstruction filter bank with three outputs, a lowpass, a mid-frequency, and a highpass. In the synthesis section, the two-band system is replaced with the three-band system at time n_0 or at time n_0 plus a delay, if desired. This is illustrated in Figure 6 and is called ***direct switching***.

Naturally during the steady-state operation of the two-band filter bank, there is no problem with exact reconstruction. Similarly the same is true during steady-state operation of the three-band system. However, during the transition from one filter bank to the other, exact reconstruction is not guaranteed, and hence transition distortion will result.

To correct this distortion, the corrupted output samples during the transition are processed with a correction filter (called the post filter). The post filter can

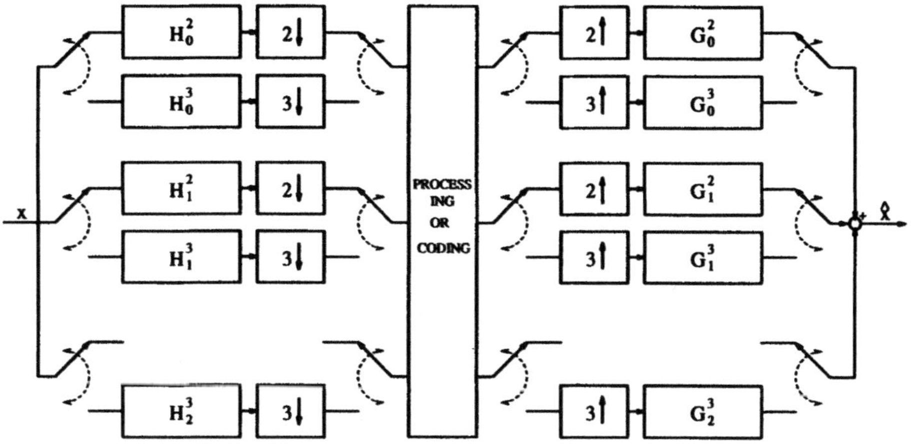

Figure 6 Time-varying analysis/synthesis filter bank that switches from a two-band to a three-band decomposition.

be viewed as an inverse filter that is active only during the transition period when it is needed.

The key element of the post filter approach is the notion of the time-varying impulse response of the analysis/synthesis system, which is a two-dimensional function given by $z(n, m)$. This impulse response, $z(n, m)$, is the response of the system at time n to a unit impulse at time m. For an exact reconstruction system, $z(n, m)$ is a line impulse along the diagonal and zero everywhere else in the (n, m)-plane.

During the switch from one exact reconstruction filter bank to another, $z(n, m)$ displays a deviation from its diagonal impulsive characteristic as shown in Figure 7. The post filter, which we will design, will correct this deviation.

In developing the post filter method, we find it convenient to represent the time-varying impulse response as a vector,

$$\mathbf{z}(n) = [z(n, n), z(n, n-1), \ldots, z(n, n - T_n - 1)],$$

where T_n denotes the width of the transition. The analysis/synthesis system can be expressed in terms of the analysis and synthesis filters in matrix form, resulting in

$$\mathbf{z}(n) = \mathbf{A}(n)\mathbf{s}(n), \tag{3.59}$$

where $\mathbf{A}(n)$ is the matrix of analysis filters and $\mathbf{s}(n)$ consists of the synthesis filter coefficients at time n. More specifically, the matrix $\mathbf{A}(n)$ can be written

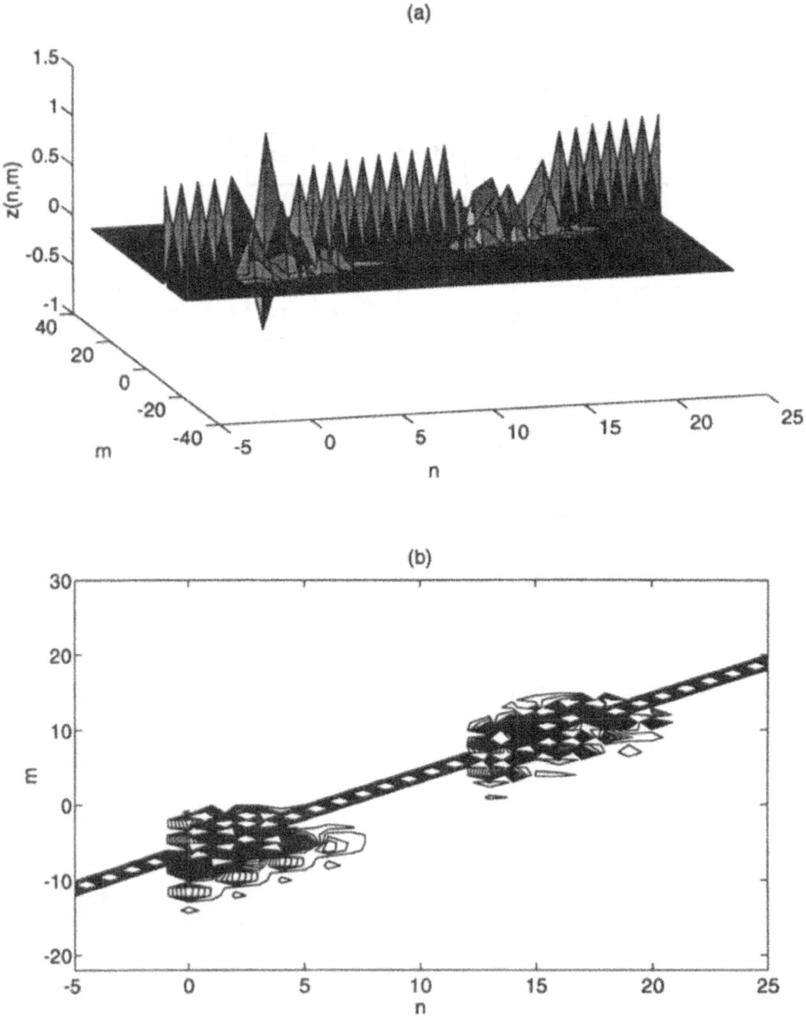

Figure 7 The time-varying impulse response for direct switching between a two-band and three-band analysis/synthesis filter bank. The filter bank is switched from the two-band to the three-band system at time $n = 0$ and switched back at time $n = 13$. Picture (a) shows the surface plot while (b) shows the contour plot.

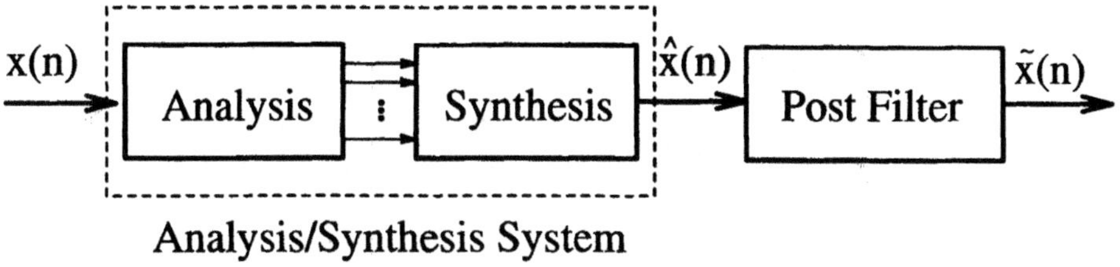

Figure 8 The block diagram of the time-varying filter bank and post filter.

as

$$\begin{bmatrix} \begin{bmatrix} \mathbf{P}(n)^T\mathbf{\Lambda}(n) \end{bmatrix} & \mathbf{O}^T & & \mathbf{O}^T \\ \mathbf{O}^T & \begin{bmatrix} \mathbf{P}(n-1)^T\mathbf{\Lambda}(n-1) \end{bmatrix} & & \vdots \\ \mathbf{O}^T & \mathbf{O}^T & \ddots & \mathbf{O}^T \\ \vdots & \vdots & & \begin{bmatrix} \mathbf{P}(n-T_n)^T\mathbf{\Lambda}(n-T_n) \end{bmatrix} \\ \mathbf{O}^T & \mathbf{O}^T & & \end{bmatrix}, \tag{3.60}$$

where $\mathbf{P}(n)$ and $\mathbf{\Lambda}(n)$ represent the time varying analysis coefficients after decimation, following Equation (3.37). For an exact reconstruction filter bank with a delay of Δ, it is necessary and sufficient that all but the $(\Delta+1)^{th}$ element in $\mathbf{z}(n)$ be zero at all times with the $(\Delta+1)^{th}$ entry equal to unity. Similar to Equation (3.37), which is the basic equation for the time-domain formulation, the time-varying filter bank is exactly reconstructing if

$$\mathbf{A}(n)\ \mathbf{s}(n) = \mathbf{b}(n) \qquad \text{for all } n, \tag{3.61}$$

where $\mathbf{b}(n)$ is the ideal impulse response as before. The post-filter approach consists of a cascade of a time-varying filter bank and a time-varying post filter. This overall system is depicted in Figure 8, where the post filter shown can be viewed as an inverse correction filter that operates during the transition period. For instance, if the filter banks are switched at time n_0 then distortion will occur in the interval $[n_0, n_0 + L]$, where L is the length of the transition. The goal here is to design the post filter, shown in Figure 8, to remove this distortion.

The time-varying impulse response associated with the time-varying analysis/synthesis system and the time-varying post filter are $\mathbf{z}(n)$ and $\mathbf{y}(n)$ respectively. As such, for generality, we can allow each to have a system delay, Δ for $\mathbf{z}(n)$ and Θ for $\mathbf{y}(n)$, in which case

$$\hat{x}(n) = \begin{cases} \text{Distorted} & \text{if } n_0 \le n < n_0 + L \\ x(n-\Delta) & \text{otherwise}, \end{cases} \tag{3.62}$$

and

$$\tilde{x}(n) = x(n - \Theta - \Delta). \tag{3.63}$$

Following the approach taken earlier, we again couch the formulation in terms of vectors and matrices, beginning by representing $\hat{x}(n)$ as the vector product

$$\hat{x}(n) = \mathbf{z}^T(n)\ \mathbf{x}_I(n). \tag{3.64}$$

Here $\mathbf{x}_I(n)$ denotes the input vector at time n, where

$$\mathbf{x}_I(n) = [x(n), x(n-1), x(n-2), \ldots, x(n-I+1)]^T,$$

and $\mathbf{z}(n)$ is a vector of length I that represents the time-varying impulse response of the filter bank at time n as defined in Equation (3.59). We can now define a ***transition impulse response matrix*** $\mathbf{Z}$ given by

$$\mathbf{Z} = \begin{bmatrix} [\ \mathbf{z}(n_0+L-1)\ ] & O & O \\ O & [\ \mathbf{z}(n_0+L-2)\ ] & O \\ O & O & \ddots & \vdots \\ \vdots & \vdots & & O \\ O & O & & [\ \mathbf{z}(n_0)\ ] \end{bmatrix}, \tag{3.65}$$

for a more compact description. Using the matrix $\mathbf{Z}$, the input/output relation of the filter bank in the transition period can be written as

$$\hat{\mathbf{x}}_L(n_0+L-1) = \mathbf{Z}^T \mathbf{x}_K(n_0+L-1) \tag{3.66}$$

where $\mathbf{Z}$ is a $K \times L$ matrix and $K = I + L - 1$. Notice that this matrix product includes samples before, during, and after the transition. Specifically, the $I - \Delta - 1$ samples preceding the transition period and the Δ samples trailing the transition period are involved in the evaluation of the output. Hence $\mathbf{x}_K(n_0+L-1)$ can be written as the concatenation of three vectors:

$$\mathbf{x}_K(n_0+L-1) = \begin{bmatrix} \mathbf{x}_a \\ \mathbf{x}_t \\ \mathbf{x}_b \end{bmatrix}, \tag{3.67}$$

where $\mathbf{x}_a$ and $\mathbf{x}_b$ are the input signals in the trailing and preceding regions, and $\mathbf{x}_t$ denotes the distorted input samples in the transition region. Using this partitioning, Equation (3.66) can be re-written as

$$\hat{\mathbf{x}}_L(n_0+L-1) = \mathbf{Z}_a^T \mathbf{x}_a + \mathbf{Z}_t^T \mathbf{x}_t + \mathbf{Z}_b^T \mathbf{x}_b \tag{3.68}$$

where

$$\mathbf{Z} = \begin{bmatrix} \mathbf{Z}_a \\ \mathbf{Z}_t \\ \mathbf{Z}_b \end{bmatrix}. \tag{3.69}$$

Replacing vectors $\mathbf{x}_b$ and $\mathbf{x}_a$ with their corresponding output vectors $\hat{\mathbf{x}}_a$ and $\hat{\mathbf{x}}_b$, $\mathbf{x}_t$ of Equation (3.68) can be written as

$$\begin{aligned} \mathbf{x}_t &= (\mathbf{Z}_t^T)^{-1}(\hat{\mathbf{x}}_t - \mathbf{Z}_a^T\hat{\mathbf{x}}_a - \mathbf{Z}_b^T\hat{\mathbf{x}}_b) \\ &= \mathbf{Y}^T\hat{\mathbf{x}}_K, \end{aligned} \tag{3.70}$$

which defines the post filter during the transition region. Thus $\mathbf{Y}$ is, in fact, the impulse response of the time-varying post filter and can be expressed as

$$\mathbf{Y} = \begin{bmatrix} -\mathbf{Z}_a\mathbf{Z}_t^{-1} \\ \mathbf{Z}_t^{-1} \\ -\mathbf{Z}_b\mathbf{Z}_t^{-1} \end{bmatrix}. \tag{3.71}$$

It is evident from Equation (3.71) that $\mathbf{Z}_t$ must have an inverse for the post filter to exist. To obtain the invertibility conditions associated with the transition response matrix $\mathbf{Z}_t$, we express the matrix as

$$\mathbf{Z}_t = \mathbf{\Psi}\,\mathbf{S} \tag{3.72}$$

where $\mathbf{\Psi}$ is the analysis transform applied to those input samples that are distorted during the transition period and $\mathbf{S}$ contains the synthesis filters during the transition period. A necessary condition for invertibility is that $\mathbf{\Psi}$ and $\mathbf{S}$ be full rank matrices. Note that the analysis filters for both exact reconstruction systems are embedded in $\mathbf{\Psi}$. Consequently if $\mathbf{\Psi}$ is full rank, then it is a simple matter to design the synthesis filters such that $\mathbf{S}$ and $\mathbf{Z}$ will also be full rank [382, 383].

As we bring this discussion to a close, it is appropriate to say a few words about some of the other approaches to time-varying filter banks that have surfaced in the literature [33, 375, 376, 197, 88, 34, 120, 170]. A common thread shared by many of these approaches is the use of intermediate analysis and synthesis filters to transit from one set of analysis/synthesis filter banks to another. Simple examples of this are approaches that use a constrained environment where exact reconstruction is guaranteed structurally. For instance, in [33] the required filter banks are implemented in the form of a lattice structure, as shown in Figure 1, where each lattice stage can be represented by a matrix. The lattice stages can be altered adaptively. Reconstruction is preserved simply by inverting adaptively the corresponding synthesis stage in the same order.

Similarly filter banks based on adaptive IIR filter banks have also appeared in the literature, introduced independently in [375] and [197]. An interesting aspect of this work is that the authors developed the adaptive filter banks in the context of subband image coding, where they show that some of the subjective distortions can be reduced by using time-varying IIR filter banks.

Both the IIR and FIR filter banks mentioned above rely on the structure of the analysis/synthesis filter bank to preserve exact reconstruction during switching. Implicitly these methods involve the use of intermediate filters during switching. Although the two individual analysis/synthesis banks used in the switching may have good properties, the intermediate filters used during the transition generally have poor characteristics. This could be a negative when practical applications are considered. In contrast, the post filter method allows for direct switching, which avoids intermediate analysis filters. Thus we can guarantee that the analysis filters have good properties before, during, and after the transition.

As a final note on this subject, we point out that tree structured filter banks provide an alternative approach to adaptive filter banks with time-varying frequency response. By adding and deleting layers in the subband tree dynamically, one can achieve time varying time-frequency decompositions of the form shown in Figure 5d. One approach based on using subband trees is given in [182, 181]. Thus we see there are several options for adaptive filter bank implementations. The selection of the particular method should, of course, be dictated by the application under consideration. It is expected that future work will investigate dynamic filter banks for use in a variety of applications.

Truly there are many interesting types of filter banks. The time-domain formulation discussed in this chapter is a very powerful tool in this regard and is useful in the analysis and design of FIR filter banks of all kinds. Time-varying filter banks, block decimation analysis/synthesis systems, and low-delay and variable delay filter banks are but three novel types of decompositions that have emerged as a result. Linear phase, mixed phase, cosine modulated, and a host of other FIR filter banks may also be designed within this same framework.

4

IIR FILTER BANKS AND WAVELETS

Rashid Ansari

Department of Electrical Engineering and Computer Science
University of Illinois, Chicago, Illinois 60680

1 INTRODUCTION

In digital filtering applications, it is known that Infinite-duration Impulse Response (IIR) filters [108], [372], [418], [26] provide a computational advantage over Finite-duration Impulse Response (FIR) filters that are designed to meet the same magnitude specifications. In the early work on multirate digital signal processing, this advantage of IIR filters did not carry over to their use in sampling rate conversions. The initial disadvantage arose from the fact that a conventional approach to IIR filter design and realization does not allow the exploitation of rate conversion in reducing the computational burden. However, a different approach, based on the use of a ***polyphase*** structure, provides a powerful framework that proves to be common to a low-complexity implementation of both IIR and FIR filters for rate conversion. In this chapter we examine the properties and design of IIR filter banks for one-dimensional and two-dimensional signals.

During the early work on multirate signal processing techniques, the focus was primarily on the use of FIR filters [356], due to the advantage offered by FIR filters over conventional IIR structures, especially when conversion factors for sampling rate are large. Consider for example the computational requirements in carrying out decimation by a factor of M. The samples in the output signal should be computed at a rate that is $1/M$ times the input sampling rate. This task is efficiently accomplished using an FIR filter by computing only those output samples that are ultimately retained in the final output signal. On the other hand, in a conventional implementation of an IIR filter of order N, one needs the $(N-1)$ most recent output samples in a recursive computation of any required output sample. This means that intermediate output samples that are eventually discarded still need to be computed [26], [356]. As a result the computational burden in implementing IIR filters, relative to FIR filters, increases with increasing values of M. However, by departing from conventional structures to a polyphase network proposed by Bellanger [43], one can arrive at a unified framework that leads to a computationally efficient implementation of both IIR and FIR filters in sampling rate conversion. A large part of this chapter deals with the construction of filter banks using polyphase networks, with special emphasis on networks composed of allpass building blocks.

As described in Chapter 2, a polyphase network is a parallel connection of component filters each of which provides a phase shift to the signal in the different parallel paths. The phase shifts can be suitably chosen to yield a required overall filter characteristic. If it is desired to use IIR filters in a polyphase configuration for the task of rate conversion by a factor M then one needs to express their transfer functions in a form in which the denominator polynomials contain only powers of z^M. Martinez and Parks [257] proposed a procedure in which the filter transfer function used in decimation by a factor M is constrained to have a denominator polynomial containing only powers of z^M, making the filter amenable to an efficient polyphase implementation. An approach for sampling rate conversion in which the polyphase filters are constrained to be allpass was proposed in [23], [99], [25]. For $M = 2$, special cases of classical Butterworth and elliptic filters can be cast in such a configuration. The choice of allpass filters is motivated by the fact that the ideal components in the different paths of a polyphase network turn out to be allpass filters. Allpass filters can be implemented with reduced computational requirements and low quantization noise. Design of filter banks in transmultiplexers using allpass filters was described in [394], [99], [24], [100]. The use of IIR filters in filter bank design was described in [41], [330], but the solution did not provide exact reconstruction. However, it was later found that polyphase networks composed of allpass filters led to exact reconstruction filter banks suitable for use in image subband coding [380], [20], [332].

This chapter mainly deals with the construction and design of IIR filter banks for one and two-dimensional signals, with considerable emphasis on filter banks composed of allpass sections. In Section 4.2 we review some background material on polyphase networks, allpass filters, and IIR halfband filters, and this material is relevant to the development of the content later in this chapter. Two-channel IIR filter banks and wavelets are examined in Section 4.3, and methods of designing and implementing IIR filters for two-channel filter banks are described in Section 4.4. Next we consider the problem of signal reconstruction in M-channel filter banks in Section 4.5, and present design methods for this case in Section 4.6. Two-dimensional filter banks are examined in Section

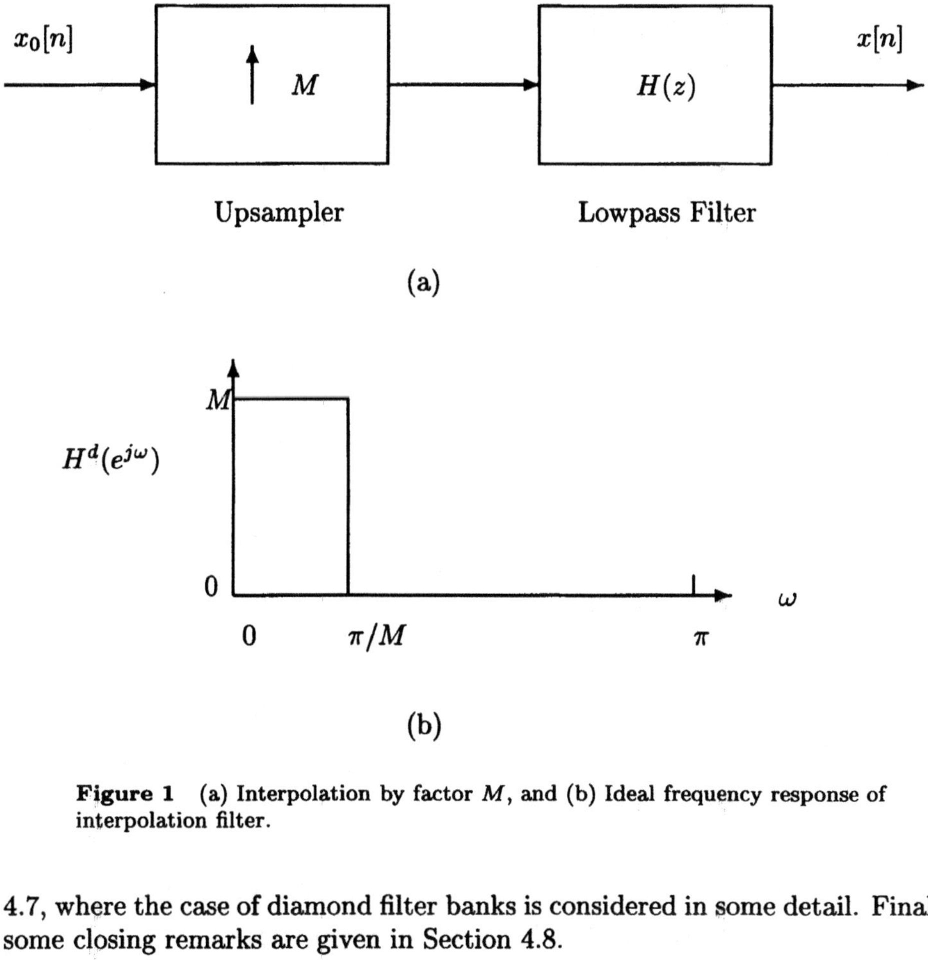

Figure 1 (a) Interpolation by factor M, and (b) Ideal frequency response of interpolation filter.

4.7, where the case of diamond filter banks is considered in some detail. Finally some closing remarks are given in Section 4.8.

2 IIR POLYPHASE NETWORKS

Polyphase networks are described in some detail in Chapter 2. In this section the structure of an IIR polyphase network for the task of interpolation by an integer factor is briefly reviewed. We consider the task of increasing the sampling rate of a signal by an integer factor M, which requires the computation of a signal $x[n]$ by the *interpolation* of a given signal $x_0[n]$. The interpolation operation is shown in Figure 1a, where the input is upsampled by a factor M and applied to a lowpass filter whose transfer function $H(z)$ is expressed in terms of its polyphase components as

$$H(z) = \sum_{p=0}^{M-1} z^{-p} P_p(z^M). \tag{4.1}$$

The polyphase component filters $P_p(z)$ approximate the ideal frequency response $P_p^d(e^{j\omega})$ given by

$$P_p^d(e^{j\omega}) = e^{j\omega p/M}, \quad |\omega| < \pi, p = 1, 2, \ldots, M-1. \tag{4.2}$$

If the filters $P_p(z)$ have the ideal frequency response $P_p^d(e^{j\omega})$ in Equation (4.2), then the filter $H(z)$ has an ideal lowpass frequency response, $H^d(e^{j\omega})$, with gain M and a cutoff at $\omega = \pi/M$ [43], as shown in Figure 1b.

It is seen in Chapter 2 that the task of *decimation* of a sequence requires lowpass filtering followed by downsampling. The same lowpass filter $H(z)$ used in interpolation can also be used in decimation by a factor M, but with filter gain equal to unity. If $H(z)$ in Equation (4.1) is multiplied by a factor $1/M$, then we get the transfer function required in decimation.

We next examine the form and properties of the IIR transfer function with a special choice of polyphase component filters. This will set the stage for the study of IIR filter banks.

2.1 Transfer Function of Polyphase Network Composed of Allpass Functions

The filter required in rate conversion by a factor M has an ideal lowpass response as shown in Figure 1b. In the case of interpolation, this response can be approximated by the transfer function $H(z)$ given by Equation (4.1), where each of the polyphase filters, $P_p(z)$, approximates the ideal frequency response given by Equation (4.2). Ideally we require the first polyphase component $P_0(z)$ to be unity since it just retains the samples in the input signal. This requirement ensures that with a well-approximated lowpass response, the phase response of $H(z)$ is approximately linear. In some applications, however, phase linearity may not be essential. In such cases the constraint of $P_0(z) = 1$ is relaxed, leading to a very significant reduction in computation.

In this chapter we will be largely concerned with polyphase networks in which the polyphase components $P_p(z)$ are allpass transfer functions with unit gain, i.e.

$$|P_p(e^{j\omega})| = 1. \tag{4.3}$$

Henceforth it will be assumed that all allpass filters have unit gain unless otherwise specified. A gain factor will be explicitly added if non-unity gain is desired.

An allpass rational transfer function, $H_{AP}(z)$, with unit gain in general has the form

$$H_{AP}(z) = z^{k_0} \prod_{k=1}^{N} \frac{a_k^* z + 1}{z + a_k}, \tag{4.4}$$

where a_k^* is the conjugate of a_k, and k_0 is an integer. Let $\tilde{H}_{AP}(z)$ be defined as the transfer function obtained by conjugating the coefficients in $H_{AP}(z^{-1})$. It can be verified that the transfer function $H_{AP}(z)$ has the following property

$$H_{AP}(z)\tilde{H}_{AP}(z) = 1. \tag{4.5}$$

In all cases of interest to us here, the coefficients of $H_{AP}(z)$ are real. In this case $\tilde{H}_{AP}(z) = H_{AP}(z^{-1})$, and $H_{AP}(z^{-1})H_{AP}(z) = 1$.

We will assume that regardless of the pole locations, the allpass filters $H_{AP}(z)$ and $\tilde{H}_{AP}(z)$ are both stable. This means that if the poles of $H_{AP}(z)$ are (a) all inside, (b) all outside, or (c) at least one on either side, of the unit circle then the filter has (a) right-sided, (b) left-sided, or (c) two-sided, infinite-duration impulse response respectively. And corresponding to the conditions (a), (b), or (c) on $H_{AP}(z)$, we note that $\tilde{H}_{AP}(z)$ has (a) left-sided, (b) right-sided, or (c) two-sided, infinite-duration impulse response. It turns out that for the polyphase networks used in rate conversion, the choice of real poles in the allpass transfer function $P_p(z)$ in all the polyphase paths is found to yield the best results [23].

In the important case of $M = 2$, the rate conversion filter with allpass polyphase components belongs to the class of filters called halfband filters. The properties of these halfband filter are discussed next.

2.2 IIR Halfband Filters

The term ***halfband filters*** is used to describe a special class of filters that is suitable for rate conversion by a factor of two. Here the term is used to describe stable, possibly non-causal, filters with transfer functions of the form

$$H_{HB}(z) = 1 + z^{-1}T(z^2), \tag{4.6}$$

where we have chosen to omit a scaling factor in Equation (4.6).

The term halfband was originally used to describe a zero-phase FIR lowpass filter whose frequency response exhibited certain anti-symmetry conditions about $\omega = \pi/2$ [45], [44]. It is used here in a more generalized sense [14], by allowing $H_{HB}(z)$ to be an IIR transfer function with an impulse response that is not necessarily symmetric.

If $H_{HB}(z)$ in Equation (4.6) has a lowpass frequency response, then $H_{HB}(-z)$ is the transfer function of a highpass filter with a complementary frequency response, since

$$H_{HB}(z) + H_{HB}(-z) = 2. \tag{4.7}$$

It should be noted that the expression for the transfer function $H_{HB}(z)$ in Equation (4.6) is in polyphase form for $M = 2$. The magnitude response $|H_{HB}(e^{j\omega})|$ is chosen to approximate an ideal lowpass filter response with a cutoff at $\omega = \pi/2$ and magnitude of 2. Another property of the halfband filter that we observe from Equation (4.6) is that the impulse response of the halfband filter satisfies

$$h_{HB}[n] = 0 \quad \text{for } n = \pm 2, \pm 4, \ldots \tag{4.8}$$

The discussion here is confined to halfband filters that have a real impulse response of infinite duration, with a frequency response that is complex-valued in general. From Equation (4.7) we can verify that the frequency response of the filter with transfer function $H_{HB}(z)$ defined in Equation (4.6) satisfies the condition

$$H_{HB}(e^{j\omega}) - 2 = -\{H^*_{HB}(e^{j(\pi-\omega)})\}. \tag{4.9}$$

The condition in Equation (4.9) implies that the complex deviations of $H_{HB}(e^{j\omega})$ from the desired values of 2 and 0 in the passband and stopband, respectively, are related, having the same absolute value. We observe that in the case of symmetric FIR halfband filters, the frequency response is real and, therefore, the magnitude deviations in the passband $[0, \omega_p]$ and the stopband $[\pi - \omega_p, \pi]$ are equal. On the other hand, in the case of IIR filters the magnitude deviations can be different. This relation between the passband and stopband magnitude deviations is discussed later in the context of filter design for two-channel filter banks.

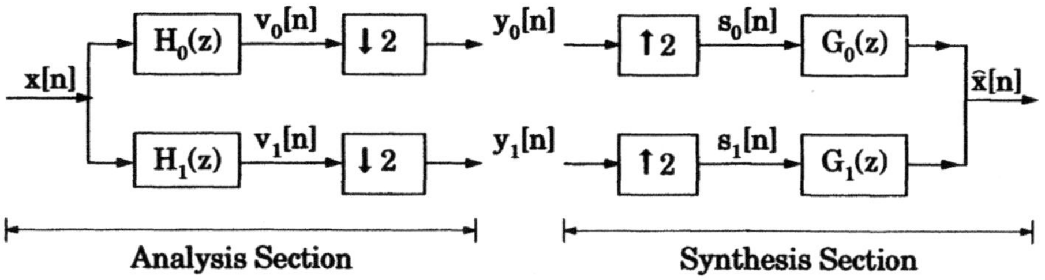

Figure 2 Two-channel IIR Filter Bank.

3 TWO-CHANNEL IIR FILTER BANKS AND WAVELETS

We revisit some of the material covered in Chapter 1 in order to derive conditions for constructing two-channel filter banks under a variety of conditions. A two-channel analysis/synthesis filter bank is shown in Figure 2.

The input signal $x[n]$ is decomposed into two signals with the analysis filter bank by processing it with a lowpass filter $H_0(z)$ in one path and with a highpass filter $H_1(z)$ in the other. The filtered signals $v_0[n]$ and $v_1[n]$ are downsampled by a factor of two to obtain the subband signals $y_0[n]$ and $y_1[n]$. The subband signals, $y_0[n]$ and $y_1[n]$ are upsampled and processed by filters $G_0(z)$ and $G_1(z)$ as shown in Figure 2. The output signals from the two paths are then added to yield $\hat{x}[n]$, which approximates the original input signal $x[n]$. The z-transform of the output $\hat{x}[n]$ can be expressed as

$$\begin{aligned}\hat{X}(z) &= \frac{1}{2}[G_0(z)H_0(z) + G_1(z)H_1(z)]X(z)\\ &\quad+\frac{1}{2}[G_0(z)H_0(-z) + G_1(z)H_1(-z)]X(-z).\end{aligned} \tag{4.10}$$

The first term on the right-hand-side is proportional to $X(z)$, and the second term is proportional to the aliasing component $X(-z)$ that arises due to a frequency translation caused by downsampling and upsampling.

3.1 Aliasing Cancellation with No Magnitude Distortion

We first consider the design of filter banks that ensures complete cancellation of aliasing distortion in the absence of coding and quantization errors. The undesirable aliasing term vanishes if the filters are chosen to satisfy the condition:

$$G_0(z)\, H_0(-z) + G_1(z)\, H_1(-z) = 0. \tag{4.11}$$

There are many solutions to the aliasing cancellation condition. One observation we make at this point is that the filters $G_0(z)$ and $H_0(z)$ approximate an ideal lowpass response, while $G_1(z)$ and $H_1(z)$ approximate an ideal highpass response. Now $H_0(-z)$ has a highpass response as a result of frequency translation, while $H_1(-z)$ has a lowpass response. Therefore one way of meeting condition (4.11) is to choose $G_0(z)$ proportional to $H_1(-z)$, and $G_1(z)$ proportional to $H_0(-z)$. It can be verified that the selection of synthesis filters as

$$G_0(z) = C(-1)^m z^{-k_0} H_1(-z)F(z), \tag{4.12}$$

and

$$G_1(z) = C(-1)^{m-1} z^{-k_0} H_0(-z)F(z), \tag{4.13}$$

ensures that Equation (4.11) is satisfied.

With aliasing cancelled, the input-output relationship is

$$\begin{aligned} \hat{X}(z) &= \frac{1}{2}[G_0(z)H_0(z) + G_1(z)H_1(z)]X(z) \\ &= D(z)X(z), \end{aligned} \tag{4.14}$$

where $D(z)$ determines the signal distortion. Using Equations (4.12) and (4.13), $D(z)$ can be expressed as

$$D(z) = \frac{1}{2}C(-1)^m z^{-k_0} F(z)[H_0(z)H_1(-z) - H_0(-z)H_1(z)]. \tag{4.15}$$

Now the transfer functions, $H_0(z)$ and $H_1(z)$, of the analysis filters can be expressed in terms of their generalized polyphase components (which were discussed in Chapter 2),

$$\begin{aligned} H_0(z) &= E_{00}(z^2) + z^{-1}E_{01}(z^2) \\ H_1(z) &= E_{10}(z^2) - z^{-1}E_{11}(z^2). \end{aligned} \tag{4.16}$$

Substituting Equation (4.16) in Equation (4.15) we get

$$D(z) = C(-1)^m z^{-k_0-1} F(z)[E_{01}(z^2)E_{10}(z^2) + E_{00}(z^2)E_{11}(z^2)]. \tag{4.17}$$

Conditions in addition to aliasing cancellation are now imposed. The highpass analysis filter is required to be obtained by a frequency translation of the low-pass filter to get a pair of mirror image responses. The transfer functions of the lowpass and highpass analysis filters are related by

$$H_0(z) = H_1(-z). \tag{4.18}$$

This combined with Equation (4.16) implies that

$$E_{00}(z^2) = E_{10}(z^2), \tag{4.19}$$

and

$$E_{01}(z^2) = E_{11}(z^2). \tag{4.20}$$

In this solution, if the analysis filters are chosen to be causal, then the synthesis filters are also causal.

Using Equations (4.19) and (4.20), $D(z)$ in Equation (4.17) is given by

$$D(z) = 2C(-1)^m z^{-k_0-1} F(z) E_{00}(z^2) E_{01}(z^2). \tag{4.21}$$

We now assume that the polyphase components $E_{00}(z^2)$ and $E_{01}(z^2)$ are all-pass, and that $F(z) = 1/(2C)$. In this case the magnitude of the signal distortion is given by

$$|D(e^{j\omega})| = 1. \tag{4.22}$$

This implies that the output signal has no magnitude distortion but only phase distortion.

3.2 Two-Channel Exact Reconstruction Filter Bank

In the previous section, some solutions for two-channel IIR filter banks that permit signal reconstruction without aliasing distortions, in the absence of coding and quantization errors, were considered. Solutions that allow exact reconstruction of the input signal are now examined. As seen from Equation (4.10), exact reconstruction requires that in addition to the condition implied by Equation (4.11) we need to ensure that

$$D(z) = z^{-n_0}, \tag{4.23}$$

where z^{-n_0} represents an integer delay of n_0. The matrix formulation of the problem has been considered in Chapters 2 and 3. From there it follows that if the analysis filters are assumed to satisfy the condition that the quantity $H_0(z)H_1(-z) - H_0(-z)H_1(z)$ is not zero for $|z| = 1$, then there exists a set of synthesis filters that yield exact signal reconstruction. In the following subsections, we consider methods of obtaining stable, possibly non-causal, IIR filters that meet the requirements for exact reconstruction. The filters are allowed to be non-causal with infinite-duration left-sided response, or are required to have strictly right-sided infinite-duration responses.

Non-Causal Solutions: Mirror Image Response

We now consider exact reconstruction solutions in which the analysis and synthesis filters satisfy the conditions in Equations (4.12) and (4.13) for cancelling aliasing. We begin by examining the case in which either the analysis or the synthesis filters are non-causal where, for brevity, the term non-causal filters is used to imply filters with an infinite-duration left-sided but not two-sided response.

The derivation of the solution will proceed along the lines of that in the case of aliasing cancellation discussed in the previous subsection. As in Equation (4.18) we first consider the situation in which the lowpass and highpass analysis filters have mirror-image magnitude responses about $\omega = \pi/2$, and are related by

$$H_0(z) = H_1(-z). \tag{4.24}$$

We assume that the polyphase components $E_{00}(z^2)$ and $E_{01}(z^2)$ are allpass. The overall transfer function given by Equation (4.21) is repeated here

$$D(z) = 2C(-1)^m z^{-k_0-1} F(z) E_{00}(z^2) E_{01}(z^2). \tag{4.25}$$

We choose $m = 0$ and $k_0 = 0$, and also set $n_0 = 0$. To satisfy Equation (4.23), we need to choose

$$F(z) = \frac{z}{2CE_{00}(z^2)E_{01}(z^2)}. \tag{4.26}$$

Since $E_{00}(z^2)$ and $E_{01}(z^2)$ are allpass with real coefficients,

$$F(z) = \frac{z}{2C} E_{00}(z^{-2}) E_{01}(z^{-2}). \tag{4.27}$$

From Equations (4.12) and (4.13) we therefore get

$$G_0(z) = \frac{1}{2}(E_{00}(z^{-2}) + zE_{01}(z^{-2})), \tag{4.28}$$

and

$$G_1(z) = \frac{1}{2}(E_{00}(z^{-2}) - zE_{01}(z^{-2})). \tag{4.29}$$

If the analysis filters $H_0(z)$ and $H_1(z)$ in Equation (4.24) are assumed to be causal then the synthesis filters $G_0(z)$ and $G_1(z)$ are non-causal. The situation can be reversed by choosing the synthesis filters to be causal, with the analysis filters non-causal. In real-time processing, an obvious drawback of non-causal filters is that a reversal of the signal in time or space is required to perform the filtering operations. However, in some applications such as image storage, it is permissible to use non-causal IIR filters in the analysis stage for performing non-causal processing. In accessing the information, causal synthesis filters are used. If a phase response that is approximately linear is desired, then a special class of filters [26], [337] can be used. For a design with strictly linear phase, symmetric two-sided IIR filters are needed. These are not very attractive for the purpose of implementation.

A special case of the above solution is one in which $E_{00}(z^2) = 1$ and $E_{01}(z^2)$ is allpass. The analysis filter transfer functions are given by

$$H_k(z) = 1 + (-1)^k z^{-1} E_{01}(z^2), \quad k = 0, 1. \tag{4.30}$$

It can be verified that the transfer functions of synthesis filters for exact reconstruction are given by

$$G_k(z) = \frac{1}{2}(1 + (-1)^k z E_{01}(z^{-2})), \quad k = 0, 1. \tag{4.31}$$

The solutions for exact reconstruction filter banks considered so far are characterized by a mirror image magnitude response of the highpass and lowpass filters. These solutions do not yield right-sided impulse responses for both analysis and synthesis filters.

Causal Solutions Obtained from a Halfband Filter: Non-mirror Image Response

We now consider solutions in which the condition of mirror image magnitude response is relaxed. This leads to solutions with a number of interesting properties [228], [230], [229], [17], [317], [316].

One approach to this problem is to consider the following factorization of a halfband filter, $H_{HB}(z)$, defined in Equation (4.6),

$$H_{HB}(z) = H_0(z)G_0(z). \tag{4.32}$$

Then, as a special case of Equations (4.12) and (4.13), we set

$$H_1(z) = 2(-1)^{m+k_0} z^{k_0} G_0(-z), \tag{4.33}$$

and

$$G_1(z) = \frac{1}{2}(-1)^{m-1} z^{-k_0} H_0(-z). \tag{4.34}$$

From Equations (4.32), (4.33), and (4.34), it follows that

$$\begin{aligned} H_1(z)G_1(z) &= (-1)^{k_0-1} H_0(-z)G_0(-z) \\ &= (-1)^{k_0-1} H_{HB}(-z). \end{aligned} \tag{4.35}$$

Given the conditions in Equations (4.33) and (4.34), we observe that Equation (4.11) is satisfied, and Equation (4.10) reduces to

$$\begin{aligned} \hat{X}(z) &= \frac{1}{2}[G_0(z)H_0(z) + G_1(z)H_1(z)]X(z) \\ &= \frac{1}{2}[H_{HB}(z) + (-1)^{k_0-1} H_{HB}(-z)]X(z), \end{aligned} \tag{4.36}$$

where we have used Equations (4.32) and (4.35). Choosing k_0 to be odd and noting that $H_{HB}(z)$ satisfies Equation (4.7), we get

$$\hat{X}(z) = X(z). \tag{4.37}$$

Therefore the choice of filters defined by Equations (4.32), (4.33), (4.34), satisfies the conditions of exact reconstruction.

Causal Solutions Obtained from a Pair of Halfband Filters

Partitioning any given halfband filter into factors $H_0(z)$ and $G_0(z)$ such that each of the filters $H_0(z)$ and $G_0(z)$ has a good lowpass response is neither straightforward nor guaranteed to yield satisfactory results. We need a procedure in which we have some control over obtaining a good response. With this objective let us examine an alternate approach [228] that leads to several interesting results. The work was motivated by a need to design an exact reconstruction filter bank in which the analysis lowpass filter $H_0(z)$ itself was

constrained to be a halfband filter [17]. The challenge was to come up with a solution in which the remaining filters in the filter bank yield good approximations to the desired response. The approach in [228] is outlined here, and related results can be found in [230], [229], [17], [227], [317], [316]. The result in [228] leads to a filter bank in which causal, stable IIR filter banks can be used. The filters can be designed to have sharp transition bands. The same filters lead to linear phase FIR filter banks in which similar sharp transition bands can be obtained.

We return to the approach described earlier, where we considered the factorization of a halfband filter in Equation (4.32). A halfband filter transfer function, $H_{HB1}(z)$, is factored into $H_0(z)$ and $F_0(z)$, where this time $F_0(z)$ is not the synthesis lowpass filter as in Equation (4.32). The factorization is given by

$$H_{HB1}(z) = H_0(z)F_0(z) = 1 + z^{-1}T_1(z^2), \tag{4.38}$$

where we have used a form similar to that given in Equation (4.6) for a halfband filter.

The idea is to choose $H_0(z)$ with a good lowpass response. However $F_0(z)$ may not provide a good approximation to the desired response. One could choose $H_0(z) = H_{HB1}(z)$ and $F_0(z) = 1$. Now we will construct $G_0(z)$ such that its product with $H_0(z)$ leads to a different halfband filter, $H_{HB2}(z)$,

$$H_{HB2}(z) = H_0(z)G_0(z) = 1 + z^{-1}T_2(z^2). \tag{4.39}$$

We shall attempt to find a suitable $G_0(z)$, which will be assumed to be of the form

$$G_0(z) = F_0(z) + F_0{}'(z). \tag{4.40}$$

The object is to determine the form of $F_0{}'(z)$, the second term on the right-hand side of Equation (4.40), that will allow us to obtain satisfactory solutions. Multiplying both sides of Equation (4.40) by $H_0(z)$ and rearranging terms, we get

$$\begin{aligned} F_0{}'(z)H_0(z) &= H_0(z)G_0(z) - H_0(z)F_0(z) \\ &= z^{-1}T_2(z^2) - z^{-1}T_1(z^2), \end{aligned} \tag{4.41}$$

where we have used Equations (4.38) and (4.39). A choice of $F_0{}'(z)$ that can satisfy the requirement that the product $F_0{}'(z)H_0(z)$ contains only odd powers of z as in Equation (4.41) is given by

$$F_0{}'(z) = z^{-1}F_1(z^2)H_0(-z). \tag{4.42}$$

With this choice, $G_0(z)$ can be expressed as

$$G_0(z) = F_0(z) + z^{-1}F_1(z^2)H_0(-z). \tag{4.43}$$

By choosing $H_1(z)$ and $G_1(z)$ according to Equations (4.33) and (4.34), with k_0 odd, the requirement for exact reconstruction is satisfied.

The choice of $H_0(z)$ and $G_0(z)$ to approximate a lowpass response with an ideal cutoff at $\pi/2$ is now examined. In some applications, one may need to choose $H_0(z)$ itself to be a halfband filter [228], [17]. The transfer function can be expressed in the form,

$$H_0(z) = 1 + z^{-1}A(z^2). \tag{4.44}$$

As before, it is assumed that the magnitude response of $H_0(z)$ approximates the value 2 in the passband and 0 in the stopband.

With $H_0(z)$ as defined in Equation (4.44) and the requirement that the product $H_0(z)F_0(z)$ in Equation (4.38) be allpass, one can choose $F_0(z) = 1$. Therefore

$$\begin{aligned} G_0(z) &= 1 + z^{-1}F_1(z^2)(1 - z^{-1}A(z^2)) \\ &= \frac{1}{2}(2 + z^{-1}B(z^2)(1 - z^{-1}A(z^2))), \end{aligned} \tag{4.45}$$

where $B(z^2) = 2F_1(z^2)$. Ideally, $H_0(-e^{j\omega}) = 0$ in the passband of $H_0(z)$, and $H_0(-e^{j\omega}) = 2$ in the stopband. Therefore the requirement on $z^{-1}B(z^2)$ is that it should ideally approximate the value -1 in the stopband. From the polyphase representation of a halfband filter, it follows that $B(e^{j\omega})$ should approximate

$$B(e^{j\omega}) \approx e^{j\omega/2}, \quad 0 \leq \omega \leq 2\omega_{pB}, \tag{4.46}$$

where $\omega_{pB} < \pi/2$ is the passband edge of a halfband filter with transfer function $1 + z^{-1}B(z^2)$.

The solution therefore requires the selection of a *pair* of halfband filters, from which we obtain the transfer functions $A(z)$ and $B(z)$. The resulting analysis and synthesis filters can be either FIR, IIR, or hybrid. One attractive feature here is that frequency responses with very narrow transition bands can be approximated by choosing a suitable order for the halfband filters. The FIR analysis and synthesis filters can all be chosen to have linear phase. The IIR filters can all be chosen to be causal. Also filters can be chosen and implemented in a way that the exact construction property is preserved even when filter coefficients are quantized.

We will briefly look at the frequency response of two exact reconstruction solutions from [228]. In the first solution $A(z)$ and $B(z)$ are both allpass and

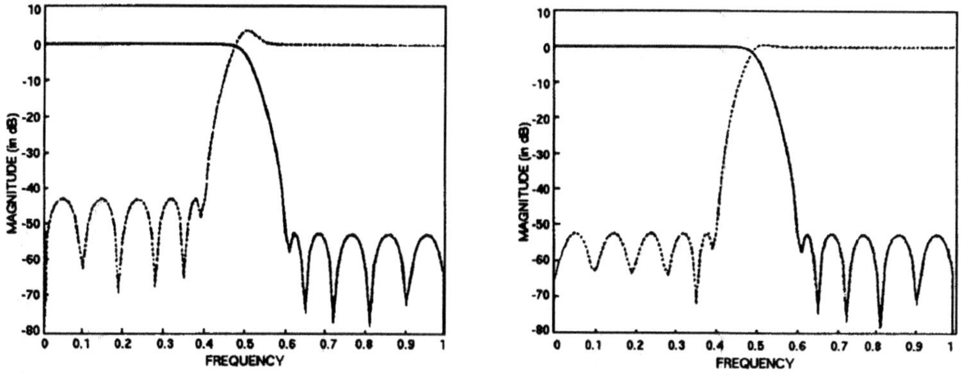

(a) (b)

Figure 3 (a) Magnitude responses of IIR analysis filters obtained from a pair of IIR halfband filters. (b) Magnitude responses of IIR/FIR hybrid analysis filters obtained from FIR and IIR halfband filters. [Adapted from [228] with permission from IEEE].

identical. In the second solution $A(z)$ is the same as in the first example, but $B(z)$ is FIR and it is obtained by the Parks-McClellan design. The magnitude responses for the examples are shown in Figure 3.

3.3 Wavelets Derived from IIR Filter Banks

The relation between filter banks and wavelets has been examined in [342], [429], to mention a few, and is treated in Chapters 2 and 5. Here we point to some results on wavelets that are obtained from IIR analysis and synthesis filters. The wavelets generated by IIR filters have infinite support. Both orthogonal and biorthogonal wavelets can be derived from IIR filter banks, [230], [404], [183].

Orthogonal wavelet and scaling function derived from Butterworth filters are described in some detail in [183]. Using the exact reconstruction filter banks with non-mirror frequency responses for the lowpass and highpass filters, one can arrive at biorthogonal wavelets. An interesting case is that of wavelets derived from symmetric IIR filter banks with two-sided support. Figure 4 shows the wavelet and scaling function for a biorthogonal case described in [230].

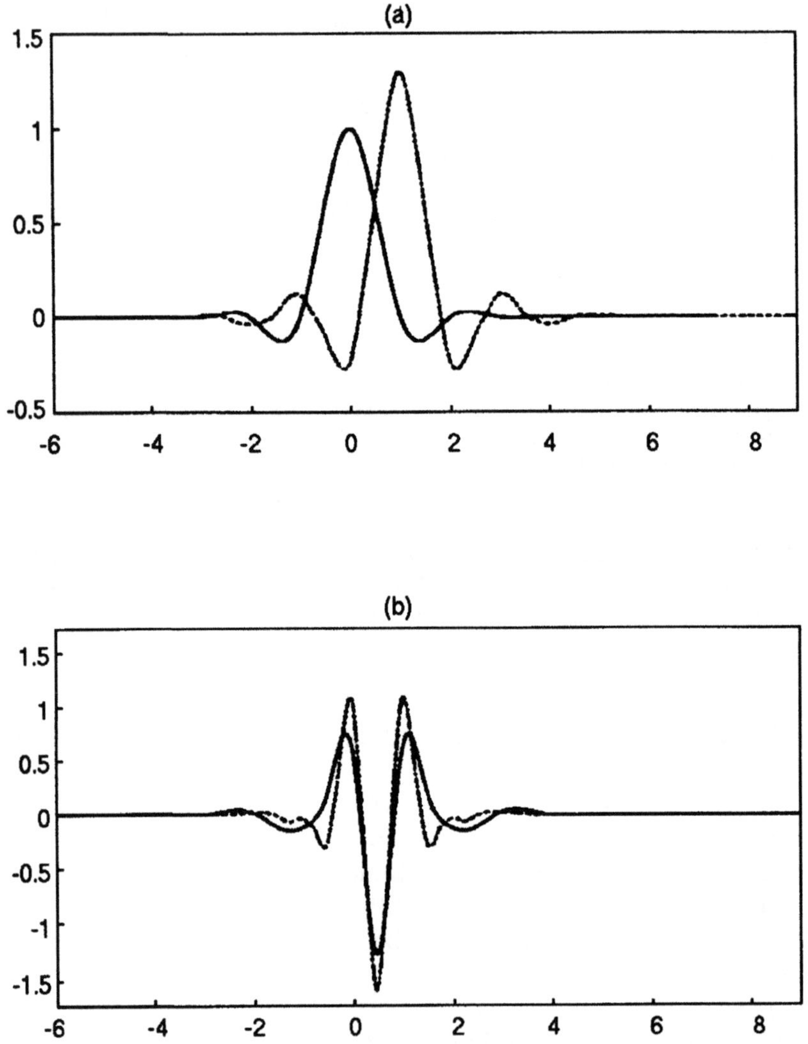

Figure 4 Biorthogonal (a) analysis and synthesis scaling functions and (b) analysis and synthesis wavelets obtained from Butterworth filters. Adapted from [230] with permission from IEEE.

4 IIR FILTER DESIGN AND IMPLEMENTATION OF TWO-CHANNEL FILTER BANKS

Analytic solutions to the design of IIR filters suitable for use in two-channel filter banks are derived in this section. The case considered is one in which the analysis lowpass filter is a halfband filter with transfer function $H_0(z)$ given by the right hand side of Equation (4.6):

$$H_0(z) = 1 + z^{-1}T(z^2), \tag{4.47}$$

where $T(z^2)$ is allpass.

The system with transfer function defined in Equation (4.47) may be non-causal and stable with a possibly two-sided infinite duration impulse response. In other words, we allow the possibility that some of the poles of $T(z^2)$ may be outside the unit circle. Taking this into account, $T(z^2)$ is expressed as

$$T(z^2) = E_{01}(z^2)/E_{00}(z^2), \tag{4.48}$$

where $E_{00}(z^2)$ and $E_{01}(z^2)$ are transfer functions of allpass filters, obtained by partitioning $T(z^2)$ in such a way that the poles of both $E_{00}(z^2)$ and $E_{01}(z^2)$ are inside the unit circle. Using Equation (4.48) we define a stable filter $H_0'(z)$ that has the same magnitude response as $H_0(z)$ in Equation (4.47) but now has a right-sided impulse response. The filter transfer function is given by

$$H_0'(z) = E_{00}(z^2) + z^{-1}E_{01}(z^2). \tag{4.49}$$

Unlike $H(z)$ in Equation (4.47) the sum of $H_0'(z)$ and $H_0'(-z)$ is not constant but has constant magnitude.

The form of the transfer function $H_0(z)$ is now considered. The choice of the allpass transfer function $T(z^2)$ in Equation (4.48) that leads to a computationally efficient design is one in which the poles are located on the imaginary axis:

$$T(z^2) = \prod_{i=1}^{L} \frac{\alpha_i z^2 + 1}{z^2 + \alpha_i}, \tag{4.50}$$

where the α_i's are real and positive.

As discussed earlier in Section 4.2, the form of the transfer function $H_0(z)$ forces some constraints on the nature of the frequency response in the passband and the stopband regions. It is assumed that

$$\omega_p = \pi - \omega_s, \quad 0 < \omega_p < \pi/2. \tag{4.51}$$

The dependence of the stopband and the passband deviation is now briefly examined. Looking at Equations (4.6) and (4.9), we note that the frequency response of the filter with transfer function $H_0(z)$ can be expressed as

$$H_0(e^{j\omega}) = 1 + e^{j(\omega+\phi(2\omega))}, \tag{4.52}$$

where $\phi(2\omega)$ is the phase response of $T(z^2)$. For any ω in the passband, let us evaluate the response at $\pi - \omega$:

$$H_0(e^{j(\pi-\omega)}) = 1 - e^{-j(\omega+\phi(2\omega))}. \tag{4.53}$$

It follows from Equations (4.52) and (4.53) that

$$\mid H_0(e^{j\omega}) \mid^2 + \mid H_0(e^{j(\pi-\omega)}) \mid^2 = 4. \tag{4.54}$$

Equation (4.54) shows that the passband magnitude deviations $2\delta'$ and the stopband magnitude deviation 2δ are related by

$$(1 - \delta')^2 + \delta^2 = 1. \tag{4.55}$$

When the stopband deviation δ is much less than 1, then

$$\delta' \approx \frac{1}{2}\delta^2, \tag{4.56}$$

in which case the passband ripple is much smaller than the stopband ripple.

Suppose that we need to design a filter with δ_p and δ_s given as the maximum deviations in the passband and stopband, assuming that the desired passband magnitude is 1. In view of the relation between the actual passband and stopband magnitude deviations δ' and δ, the filter should be designed to meet the more stringent of the specifications δ_p and δ_s. Toward this end, define

$$\delta_{min} = min(\delta_s, (2\delta_p - \delta_p^2)^{1/2}). \tag{4.57}$$

The quantity δ_{min} is used as the required stopband magnitude deviation.

4.1 Butterworth Filter Design

A halfband filter with frequency response that is maximally flat at $\omega = 0$ can be obtained from a specialized Butterworth design, with possibly a two-sided

Table 1 Butterworth polyphase filter coefficients

E_{00}	L= 1	0.33333				
E_{01}	L= 1					
E_{00}	L= 2	0.10557				
E_{01}	L= 2	0.52786				
E_{00}	L= 3	0.05210	0.63596			
E_{01}	L= 3	0.23191				
E_{00}	L= 4	0.03109	0.33333			
E_{01}	L= 4	0.13247	0.70409			
E_{00}	L= 5	0.02067	0.20856	0.75083		
E_{01}	L= 5	0.08622	0.41301			
E_{00}	L= 6	0.01474	0.14383	0.47645		
E_{01}	L= 6	0.06075	0.27546	0.78486		
E_{00}	L= 7	0.01105	0.10557	0.33333	0.81073	
E_{01}	L= 7	0.04518	0.19823	0.52786		
E_{00}	L= 8	0.00859	0.08095	0.24795	0.57027	
E_{01}	L= 8	0.03494	0.15008	0.38338	0.83105	
E_{00}	L= 9	0.00687	0.06413	0.19241	0.42684	0.84744
E_{01}	L= 9	0.02785	0.11786	0.29287	0.60580	
E_{00}	L=10	0.00562	0.05210	0.15403	0.33333	0.63596
E_{01}	L=10	0.02272	0.09515	0.23191	0.46484	0.86093

infinite duration impulse response [25]. In this case the transfer function $T(z^2)$ that appears in Equation (4.47) is given by Equation (4.50) where the α_i's are equal to

$$\alpha_i = \cot^2(\frac{\pi i}{2L+1}). \tag{4.58}$$

In order to obtain the transfer function with a right-sided response as in Equation (4.50), we partition $T(z^2)$ and arrive at the following two allpass sections:

$$E_{00}(z^2) = \prod_{i=1}^{J} \frac{\alpha_{0i} z^2 + 1}{z^2 + \alpha_{0i}} \tag{4.59}$$

with $J = \lfloor L/2 \rfloor$, and

$$\alpha_{0i} = \cot^2(\frac{\pi i}{2L+1}); \tag{4.60}$$

and

$$E_{01}(z^2) = \prod_{i=J+1}^{L} \frac{\alpha_{1i}z^2 + 1}{z^2 + \alpha_{1i}}, \tag{4.61}$$

where α_{1i} is given by

$$\alpha_{1i} = \tan^2(\frac{\pi i}{2L+1}). \tag{4.62}$$

It should be noted that once the required filter order is known, then the coefficients are immediately determined according to Equations (4.58) and (4.60). The coefficients for $L = 1, 2, \ldots, 10$ are given in Table 1. The magnitude responses for $L = 1, 3, 10$ are plotted in Figure 5.

4.2 Elliptic Halfband Filter Design

After having considered the design of filters with maximally flat response, we now look at the design of halfband filters [23], [157], [14]. Using an approach similar to the Butterworth case, we arrive at a solution, which in this case, consists of a special class of elliptic filters. The design is briefly described, and the reader is referred to [14] for details.

The values of δ_s, δ_p, ω_p, and $\omega_s = \pi - \omega_p$ are assumed to be given. The procedure in [14] consists of the following steps:

(i) Compute δ_{min} using Equation (4.57).

(ii) Compute the values of elliptic function parameters, $k = \tan^2(\omega_p/2)$ and $k' = (1-k^2)^{1/2}$.

(iii) Compute p defined as

$$p = \frac{1}{2}\frac{1-\sqrt{k'}}{1+\sqrt{k'}}, \tag{4.63}$$

and, using an approximation, compute the value of q:

$$q \approx p + 2p^5 + 15p^9 + 150p^{13}. \tag{4.64}$$

(iv) Estimate the filter order $2L+1$ according to

$$2L+1 = \left[\frac{2\ln(4(1-\delta_{min}^2)/\delta_{min}^2)}{-\ln q}\right]_o, \tag{4.65}$$

where $[\cdot]_o$ denotes the smallest odd integer equal to or exceeding the argument.

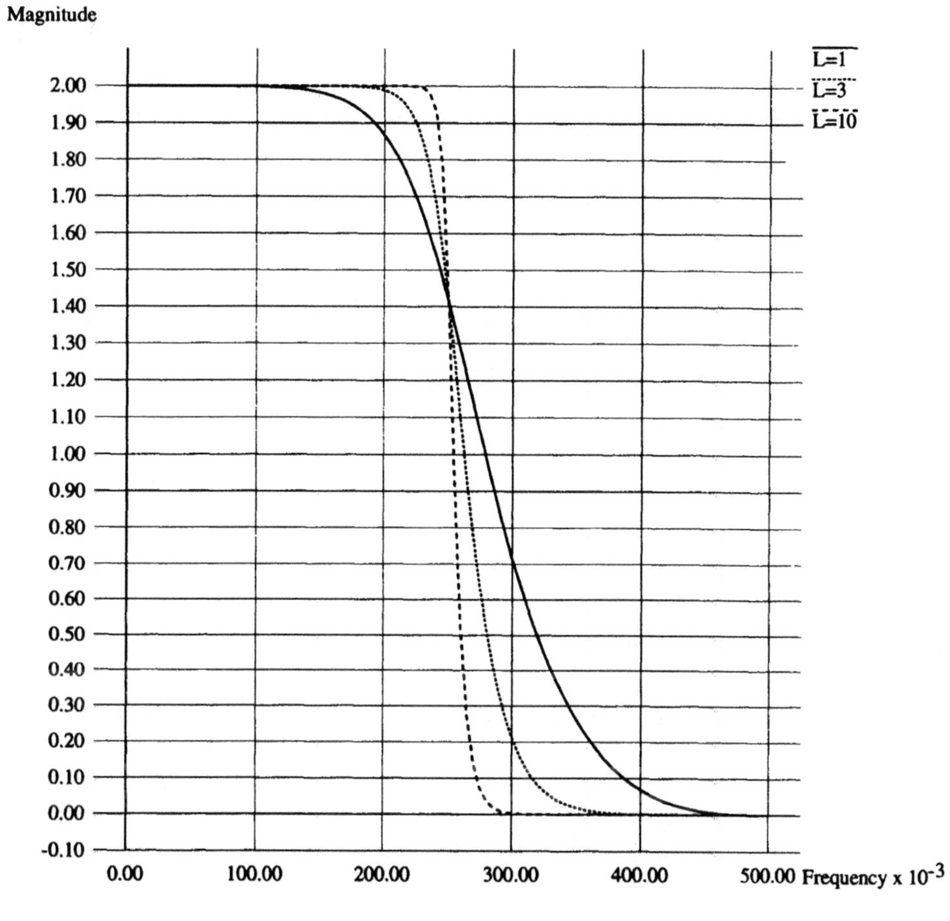

Figure 5 Magnitude responses of halfband Butterworth filters for $L = 1, 3, 10$.

(v) Compute the analog extremal frequencies Ω_i. Find

$$r_i = [(1 - k\Omega_i^2)(1 - \Omega_i^2/k)]^{1/2} \tag{4.66}$$

and determine

$$\cos\theta_i = \frac{(-1)^{i+1} r_i}{1 + \Omega_i^2}. \tag{4.67}$$

(vi) Compute $\alpha_i = (1 - \cos\theta_i)/(1 + \cos\theta_i), i = 1, 2, ..., L$.

The values α_i computed in (vi) are then substituted in Equation (4.52) to get the transfer function containing poles both inside and outside the unit circle. A right-sided impulse response is obtained using the manipulations described before.

Consider an example of an elliptic filter design for rate conversion. In order to see the frequency response more clearly, we choose somewhat large allowed deviations. Suppose $\delta_p = 0.1$, $\delta_s = 0.2$, $\omega_p = 0.49\pi$, $\omega_s = \pi - \omega_p$. The design solution requires a value of $L = 2$ with $N = 5$. The transfer function, with gain 2 at $\omega = 0$, is of the form:

$$H_0(z) = \frac{0.56110z^2 + 1}{z^2 + 0.56110} + z^{-1}\frac{0.93154z^2 + 1}{z^2 + 0.93154}. \tag{4.68}$$

The normalized filter magnitude response is plotted in Figure 6. The actual passband deviation is 0.0121 and the stopband deviation is 0.1550. In this case the filter order is determined by the stopband deviation requirement.

In this section we considered analytic solutions for the case in which the version of the solution with a right-sided impulse response does not provide a good approximation to a linear-phase frequency response. If an approximately linear-phase causal filter is desired, then in general an analytic solution is not known and a numerical method described later for the M-channel case can be used.

4.3 Implementation Issues

Having looked at the design of two-channel filter banks composed of allpass polyphase, we now consider some aspects of their implementation. The basic operation required is essentially the implementation of a first order allpass section. There are many ways of realizing first-order and second-order allpass sections [275]. The choice is decided by the arithmetic precision used in the implementation and the tolerance bounds on the resulting quantization noise.

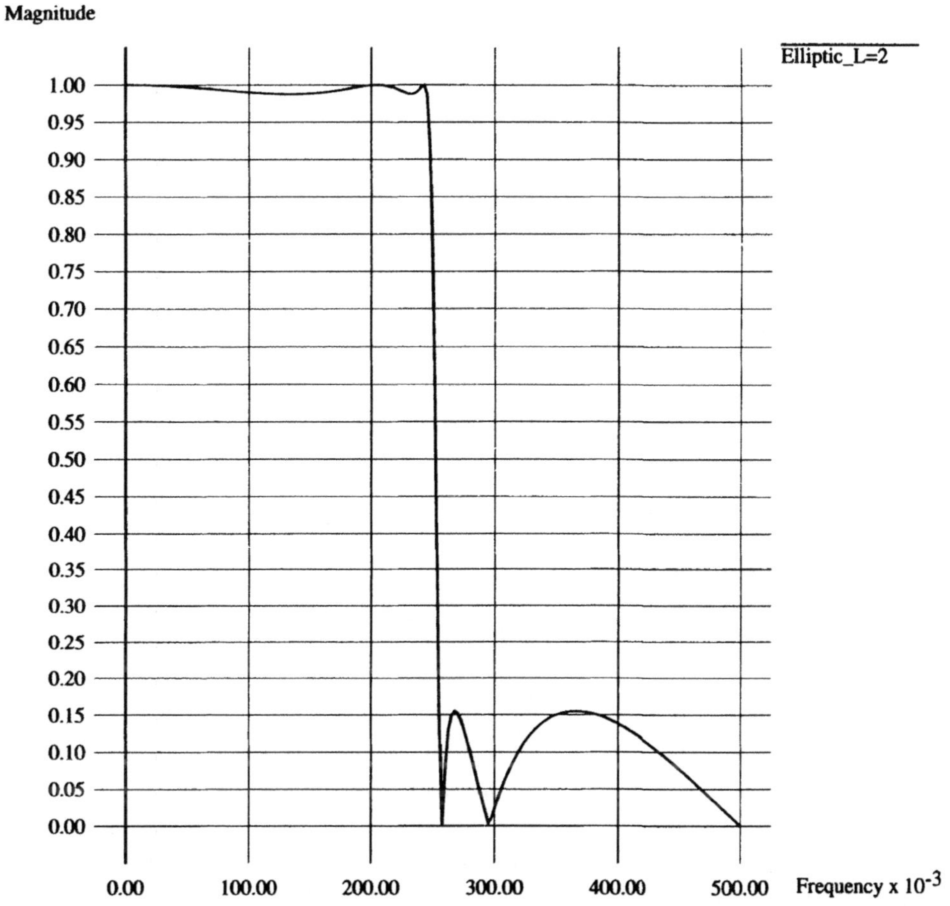

Figure 6 Magnitude response of elliptic filter described in example.

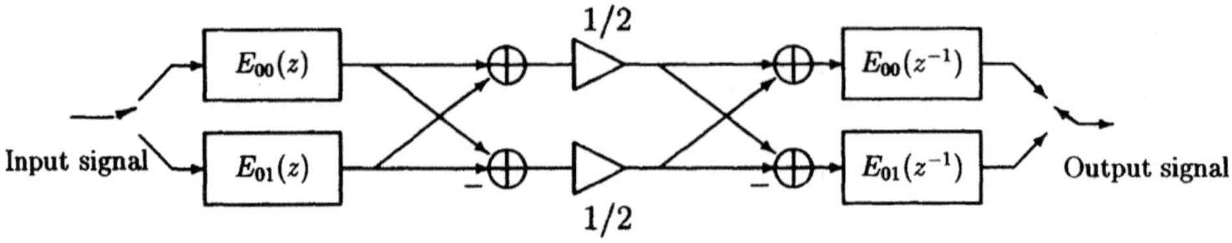

Figure 7 Reconfigured two-channel analysis filter bank for implementation.

If floating point arithmetic is used, then the quantization noise is negligible in most practical applications. In this case, an implementation requiring the smallest number of arithmetic operations can be used. Let us examine the implementation of a two-channel filter bank in which the analysis filters are given by

$$H_k(z) = E_{00}(z^2) + (-1)^k z^{-1} E_{01}(z^2), \quad k = 0, 1. \tag{4.69}$$

Let us examine the two-channel analysis filter bank shown in Figure 2, and its reconfiguration for implementation shown in Figure 7.

The input signal is partitioned into two streams by moving the downsampler ahead of the polyphase filters. The two streams correspond to the downsampled signals obtained from the original input and its version that is shifted with unit delay. The operations performed by the polyphase filters are shared by the lowpass and highpass filters. Assume that the filters E_{0p}, $p = 0, 1$ are allpass with real poles. The filters can be implemented as a cascade of first order sections, and the fact that the coefficients in the numerator and denominator are equal can be exploited to reduce the multiplication requirements. The simplification where the polyphase filter operations are shared by the channel filters in the filter bank can be extended to the M-channel case.

In performing the filtering, it is clear that IIR filters would yield infinite-extent output signals even when the input is of finite extent, as in the case of a typical image. In practice, one would attempt to limit the amount of data that one needs to compute while ensuring that the signal can be reconstructed exactly. This can be accomplished by making sure that the necessary boundary conditions are available at the synthesis stage [379], [36].

5 M-CHANNEL IIR FILTER BANKS

Let us now consider the decomposition of the input signal into $M \geq 2$ component signals, using an M-channel filter bank. Frequently, the case $M = 2^L$, is encountered, where L is a positive integer. In this case a tree-structured filter bank [380], [336] with L stages of two-channel filter banks can be used. We now consider the more general case where M is not necessarily equal to 2^L. This requires an extension of the framework developed earlier for the two-channel case. Attention is restricted to the case of uniform critical sampling where each channel signal is downsampled by a factor, M, equal to the number of channels. The use of IIR filters for perfect reconstruction is examined, and it is seen that filters with allpass polyphase sections lead to a solution for perfect reconstruction.

We begin by examining the expression for the output signal in terms of the input signal and its modulated versions [378], [427], [390]. Figure 1 in Chapter 1 shows a back-to-back connection of M-channel analysis and synthesis filter banks. As shown in Chapter 1, the relation between the z-transforms of the input signal $x[n]$ and the output signal $\hat{x}[n]$ can be expressed as

$$\hat{X}(z) = \frac{1}{M} \sum_{k=0}^{M-1} \left(G_k(z) \sum_{i=0}^{M-1} H_k(zW_M^i)\, X(zW_M^i) \right), \tag{4.70}$$

where W_M is defined as

$$W_M = e^{-j2\pi/M}, \quad k = 0, 1, \ldots, M - 1. \tag{4.71}$$

Equation (4.70) is compactly expressed as

$$\hat{X}(z) = \frac{1}{M} \mathbf{g}^T(z)\, \mathbf{H}(z)\, \mathbf{x}(z), \tag{4.72}$$

where the element in the i-th row and k-th row of matrix $\mathbf{H}(z)$ is $H_k(zW_M^i)$, $k, i = 0, 1, ..., M - 1$, the k-th element of M-vector $\mathbf{g}(z)$ is $G_k(z)$ and the i-th element of M-vector $\mathbf{x}(z)$ is $X(zW_M^i)$. Note that the rows and columns of matrices and elements of vectors are labeled beginning $0, 1, ..., M - 1$.

If $\det[\mathbf{H}(z)]$ is not zero on the unit circle, then given a set of stable analysis filters, one can find stable, possibly non-causal synthesis filters, that provide exact reconstruction as follows

$$\mathbf{g}(z) = M z^{-n_0}\, \mathbf{H}^{-1}(z)\, \mathbf{e}, \tag{4.73}$$

where $\mathbf{e}^T = [1\ 0\ 0\ \ldots\ 0]$.

The derivation of filters for exact reconstruction is facilitated by a polyphase decomposition of the analysis filters [416]. In particular, the transfer function $H_k(z)$ of the analysis filter in the k-th channel can be expressed as:

$$H_k(z) = \sum_{p=0}^{M-1} z^{-p} E_{kp}(z^M). \tag{4.74}$$

The analysis filters are assumed to be IIR, so that $H_k(z)$ are rational transfer functions, and therefore so are $E_{kp}(z)$. We note that in the time domain, the polyphase representation implies

$$e_{kp}[n] = h_k[Mn + p]. \tag{4.75}$$

Let the polyphase decomposition of the input signal $x[n]$ be described by the relation

$$x_p[n] = x[Mn - p], \tag{4.76}$$

where we note the direction of polyphase partition is the reverse of that in Equation (4.75). We define the vector $\mathbf{x}$ (not the same as $\mathbf{x}(z)$ in Equation (4.72)) in terms of the z-transforms of the polyphase components $x_p[n]$ as follows:

$$\mathbf{x}^T = [X_0(z)\ X_1(z)\ \ldots\ X_{M-1}(z)] \tag{4.77}$$

The subband signal $y_k[n]$ in the k-th channel can be expressed as

$$\begin{aligned} y_k[n] &= \sum_{q=-\infty}^{\infty} h_k[Mn - q]x[q] \\ &= \sum_{m=-\infty}^{\infty} \sum_{p=0}^{M-1} h_k[M(n-m) + p]x[Mm - p] \\ &= \sum_{m=-\infty}^{\infty} \sum_{p=0}^{M-1} e_{kp}[n-m]x_p[m] \\ &= \sum_{p=0}^{M-1} e_{kp}[n] * x_p[n]. \end{aligned} \tag{4.78}$$

If $\mathbf{y}$ is defined as

$$\mathbf{y}^T = [Y_0(z)\ \ Y_1(z)\ \ldots\ Y_{M-1}(z)], \tag{4.79}$$

then from Equation (4.78) it follows that

$$\mathbf{y} \;=\; \mathbf{E}(z)\;\mathbf{x}, \tag{4.80}$$

where $\mathbf{E}(z)$ denotes a matrix whose (k,p)-th element is $E_{kp}(z)$.

Our objective is to recover the input signal, with a possible delay, given the set of subband signals defined by the vector $\mathbf{y}$ in (4.79). In order to arrive at the solution to the signal recovery problem, we define $\tilde{\mathbf{E}}(z)$ to be the matrix obtained from the transpose of the matrix $\mathbf{E}(z^{-1})$ with a conjugation of coefficients.

We will now require that $\mathbf{E}(z)$ be chosen to give exact reconstruction under the condition that the M polyphase components of the output signal $\hat{x}[n]$ be obtained from $\mathbf{y}$ in Equation (4.80) according to

$$\hat{\mathbf{x}} \;=\; z^{-n_0}\;\tilde{\mathbf{E}}(z)\;\mathbf{y}. \tag{4.81}$$

where $\hat{\mathbf{x}}^T \;=\; [\hat{X}_0(z)\;\;\hat{X}_1(z)\;\;\ldots\;\;\hat{X}_{M-1}(z)]$. $\hat{X}_p(z)$ denotes the z-transform of the p-th polyphase component of the signal $\hat{x}[n]$, analogous to the definition in (4.77). The reconstructed signal can therefore be expressed as

$$\hat{\mathbf{x}} \;=\; z^{-n_0}\tilde{\mathbf{E}}(z)\;\mathbf{E}(z)\;\mathbf{x}. \tag{4.82}$$

The requirement of exact reconstruction reduces to choosing $\mathbf{E}(z)$ such that

$$\tilde{\mathbf{E}}(z)\;\mathbf{E}(z) \;=\; \mathbf{I} \tag{4.83}$$

in a region $r_1 < |z| < r_2$, where $r_1 < 1 < r_2$ and $\mathbf{I}$ is an $M \times M$ identity matrix. If we choose $\mathbf{E}(z)$ to be unitary on the unit circle, i.e.

$$[\mathbf{E}^*(e^{j\omega})]^T\;\mathbf{E}(e^{j\omega}) \;=\; \mathbf{I}, \tag{4.84}$$

then we note that by analytic continuation of Equation (4.84) the condition in Equation (4.83) holds in a region $r_1 \;<\; |z| \;<\; r_2$ containing the unit circle, and $\mathbf{E}(z)$ is paraunitary.

Here we are interested in IIR solutions to the problem. The choice of $\mathbf{E}(z)$ obtained from IIR analysis filters is now considered. Having seen the usefulness of polyphase filters composed of allpass sections in the case of two-channel filter banks, let us now consider $H_k(z)$ to be composed of allpass sections, all derived by a frequency translation of a (lowpass) generating filter with transfer function $F(z)$:

$$H_k(z) \;=\; F(W_M^k z) \;=\; \sum_{k=0}^{p} W_M^{-kp}\;z^{-p}\;F_p(z^M). \tag{4.85}$$

It can be shown that with the choice of $H_k(z)$ in Equation (4.85), $\mathbf{E}(z)$ can be expressed as

$$\mathbf{E(z)} = \boldsymbol{\Omega}\, \mathbf{F}(z), \tag{4.86}$$

where $\boldsymbol{\Omega}$ is the matrix whose (k,p)-th element is $\Omega_{kp} = W_M^{-kp}$ and $\mathbf{F}$ is a diagonal matrix with (p,p)-th diagonal element $F_p(z)$.

The problem of filter bank design thus reduces to a proper choice of $F(z)$. If we require that $F_p(z)$ is stable, causal, and allpass, then $F_p(z^{-1})$ would be stable but non-causal allpass. $\mathbf{E}(z)$ satisfies Equation (4.84) and its implementation as well as that of $\tilde{\mathbf{E}}(z)$ would involve DFT computations.

6 IIR FILTER DESIGN FOR M-CHANNEL FILTER BANK

The preceding section showed how an exact reconstruction M-channel filter bank can be derived using a single generating filter $F(z)$. The filter with transfer function $F(z)$ is assumed to be in a polyphase form with M allpass components whose coefficients are determined by a suitably formulated design. If one of the polyphase paths is constrained to be a pure delay, then the design solution has an approximately linear-phase response [25]. This approach can be used in the case of $M = 2$ to obtain causal filters that have approximately linear phase. Efficient methods of designing filters with this structure are described in [129], [337], [338]. The frequency response of a "lowpass" filter designed with this allpass polyphase structure, produces spikes in the magnitude response at certain frequencies. This may make it unsuitable in some applications. However in other applications, such as a transmultiplexer design, the response can be cascaded with a suitable lowpass filter to suppress the spikes. The filter bank used in a transmultiplexer does not require exact reconstruction of the signal.

In this section a method of designing a polyphase network with allpass polyphase components, based on [337], is described. As noted earlier, the choice of polyphase filters with real poles is found to yield the best approximation in the design. The transfer function $F(z)$ of the generating filter is assumed to be of the form

$$F(z) = \sum_{p=0}^{M-1} z^{-p} F_p(z^M), \tag{4.87}$$

where

$$F_p(z^M) = z^{-k_pM} \prod_{k=1}^{J_p} \frac{\alpha_{pk} z^M + 1}{z^M + \alpha_{pk}}. \tag{4.88}$$

Appropriate choice of integers J_p and k_p in Equation (4.88) is discussed later. The design problem then consists of choosing coefficients α_{pk}. There are restrictions on the magnitude response that come about due to the choice of allpass sections in the polyphase paths. Let us consider the case $k_p = 0$ and $J_p = K$, both independent of p. In this case the magnitude response at the frequencies $\omega = m\pi/M$, $1 \leq m \leq M$, is given by [24]:

$$|F(e^{jm\pi/M})| = \begin{cases} 0 & \text{for } m \text{ even}, \\ \csc \frac{m\pi}{2M} & \text{for } m \text{ odd}. \end{cases} \tag{4.89}$$

We note that the response at $\omega = m\pi/M$, for M odd and $1 \leq m \leq M$, is not under the designer's control.

Properties of the filters given by Equations (4.87) and (4.88) and efficient techniques for designing the filters are described in [129], [337] and [338]. It turns out that as in the case of two-channel filters, in the M-channel case too there is a relation among the magnitude deviations in the passband and the stopbands. To explore this, we define a sequence $\{F_\omega[k]\}_{k=0}^{M-1}$ as follows

$$F_\omega[k] \quad = \quad F(e^{j(\omega+2\pi k)/M}), \tag{4.90}$$

where $0 \leq \omega \leq \pi$. From Equations (4.87) and (4.88) it follows that

$$F_\omega[k] = \sum_{p=0}^{M-1} e^{-j2\pi pk/M} F_p(e^{jM\omega}) e^{-jp\omega}, \quad k = 0, 1, ..., M-1. \tag{4.91}$$

Equation (4.91) shows that $F_\omega[k]$ is the DFT of the sequence $\{f_\omega[p]\}_0^{M-1}$ given by

$$f_\omega[p] = F_p(e^{jM\omega}) e^{-jp\omega}, \quad p = 0, 1, ..., M-1. \tag{4.92}$$

It follows from Parseval's relation that

$$\sum_{k=0}^{M-1} |F_\omega[k]|^2 = M \sum_{p=0}^{M-1} |f_\omega[p]|^2. \tag{4.93}$$

Noting that $F_p(z)$ is allpass, so that $|f_\omega[p]| = 1$, Equation (4.93) leads to

$$\sum_{k=0}^{M-1} |F(e^{j(\omega+2\pi k/M)})|^2 = M^2. \tag{4.94}$$

The relation (4.94) represents the constraint on the passband and stopband magnitudes. Specifically, for each frequency ω_0 in the range from 0 to π/M, the sum of the squared magnitude responses at frequencies in the range $[0,\pi]$ given by $2k\pi/M \pm \omega_0$ is equal to M^2. In view of the dependence of the response at these frequencies, one makes a provision for *don't-care* bands about the frequency $\omega_m = m\pi/M$ with m odd and $1 \leq m \leq M$. Also the constraint of Equation (4.94) implies that if the squared magnitude response is approximately equal to M^2 in a passband $[0, \alpha\pi/M]$, $0 < \alpha < 1$, then the stopband magnitude response is automatically close to zero, at frequencies in each of the stopband intervals $[(2k-\alpha)\pi/M, (2k+\alpha)\pi/M]$, $k = 1, 2, \ldots, \lfloor M/2 \rfloor$. With maximum passband magnitude normalized to 1, suppose that the maximum stopband deviation is $\delta_s \ll 1$. The passband deviation δ_p is then guaranteed to be bounded by

$$\delta_p \leq 1 - \sqrt{1-(M-1)\delta_s^2}. \tag{4.95}$$

In the design, one is required to specify the values of k_p and J_p in Equation (4.88). It is found [336] that if the approximation of phase linearity is ignored, the smallest stopband attenuation is obtained with $k_p = 0$ for all p and

$$J_p = \begin{cases} K+1, & 0 \leq p < P, \\ K, & P \leq p \leq M-1, \end{cases} \tag{4.96}$$

where $0 \leq P \leq M-1$. For approximately linear-phase design, $k_P = K$ and $J_P = 0$ is chosen in the P-th path, while in the other paths $k_P = 0$ and J_p is given by Equation (4.96), for $p \neq P$. With the parameters k_P and J_p selected, one proceeds to the selection of the unknown α_{pk} using a Remez-type algorithm in which an initial guess of attenuation zeros is iteratively adjusted. Design details and examples can be found in [337].

7 2-D IIR FILTER BANKS

We will now consider the decomposition and reconstruction of two-dimensional signals using analysis/synthesis filter banks. Two-dimensional filter banks are useful in image coding [425], [447], [163], [439], [448], as they allow efficient bit allocation to be made based on the amount of significant information in the different frequency bands. The topics of subband image and video coding are discussed in Chapters 7 and 8. Here a brief description of IIR filter banks is given, with some emphasis on IIR diamond filter banks, to point out aspects of 2-D filter banks that are different from a direct extension of 1-D filter banks.

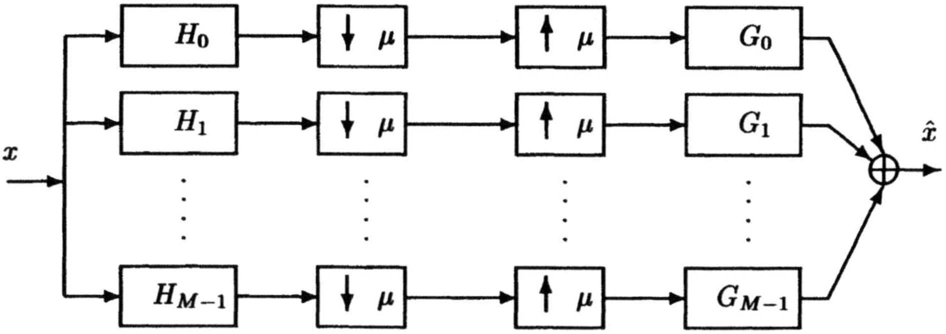

Figure 8 Nonseparable filter bank.

In this section we first discuss ***separable*** IIR filter banks where the results in the 1-D case that we obtained earlier can be extended to the case of rectangular resampling grids [380]. Next we consider the case of analysis/synthesis filter banks for non-rectangular periodic sampling of 2-D signals. Then the special case of a diamond filter bank is considered in some detail.

Signal partitioning using a separable filter bank is the most commonly used method of decomposing 2-D signals. In a separable filter bank the 2-D signal is processed horizontally across each row and downsampled to yield M_H horizontal subband signals. Each of these horizontal subband signals is processed vertically across each column to yield M_V subband signals per horizontal subband signal. Overall we get $M_H \times M_V$ subband signals. It is common to use one or more stages of a 2×2 partition, where one can use 1-D exact reconstruction IIR filter banks operating along the rows and columns [380], [20], [332], [237]. The procedure can be extended to the case of partitioning into $M \times M$ subband signals [201], [200], [202]. Similar methods can be developed for the more general case of filter banks that are not necessarily separable.

7.1 Non-Separable Filter Banks

Rate conversion for 2-D signals is defined by a matrix μ which is a 2×2 nonsingular matrix with integer entries [448]. Consider the filter bank shown in Figure 8a where $M = |\det[\mu]|$ and the rate conversion matrix is considered to be the same in all paths.

The input $x[\mathbf{n}]$ is applied to the set of M analysis filters with impulse response $h_i[\mathbf{n}]$ and frequency response $H_i(\boldsymbol{\omega})$, where $\mathbf{n} \in \mathcal{Z}^2$ and $\boldsymbol{\omega} \in \mathcal{R}^2$. The outputs $v_i[\mathbf{n}]$ are downsampled to yield the signals y_i, where

$$y_i[\mathbf{n}] = v_i[\mu\mathbf{n}]. \tag{4.97}$$

Only samples at locations defined by linear combinations (with integer coefficients) of the column vectors of μ are retained after downsampling. It is convenient to associate the following index set $P(\mu)$ with a matrix

$$P(\mu) = \{\mathbf{k} \in \mathcal{Z}^2 : \mu^{-1}\mathbf{k} \in [0,1)^2\}. \tag{4.98}$$

This set consists of sample locations in $\mathcal{Z}^2$ contained in the parallelogram defined by the two column vectors of μ except the samples on the two outer boundaries.

In the synthesis filter bank, the signals $y_i[\mathbf{n}]$ are upsampled to get the signals $s_i[\mathbf{n}]$, which are filtered by the synthesis filters with frequency response $G_i(\boldsymbol{\omega})$. Using notation analogous to the one-dimensional M-channel case used earlier, it can be shown [448] that the Fourier transform $\hat{X}(\boldsymbol{\omega})$ of the output $\hat{x}[\mathbf{n}]$ is related to the input by

$$\hat{X}(\boldsymbol{\omega}) = \frac{1}{M}\,\mathbf{g}^{\mathbf{T}}(\boldsymbol{\omega})\mathbf{H}(\boldsymbol{\omega})\mathbf{x}(\boldsymbol{\omega}). \tag{4.99}$$

This relation is similar to Equation (4.72). Proceeding as before in the 1-D case, we can derive the requirements on the filters for exact reconstruction.

7.2 IIR Diamond Filter Bank

A specific case of nonrectangular filter banks is now considered. Here $M = 2$, and the ideal lowpass filter has a diamond-shaped passband. The design and implementation of IIR diamond filters [19], [16] is described here. The ideal frequency response of a diamond lowpass filter is shown in Figure 9.

A procedure for approximating this response using a 1-D prototype filter will now be described. We note that the filter in Figure 9 can be obtained from the filter in Figure 10a by a suitable transformation of variables. Let $F_a(z_1, z_2)$ be the transfer function of the filter that approximates the ideal response in Figure 10a. The diamond filter transfer function $D(z_1, z_2)$ is obtained by the transformation

$$D(z_1, z_2) = F_a(z_1^{\frac{1}{2}} z_2^{\frac{1}{2}},\ z_1^{-\frac{1}{2}} z_2^{\frac{1}{2}}). \tag{4.100}$$

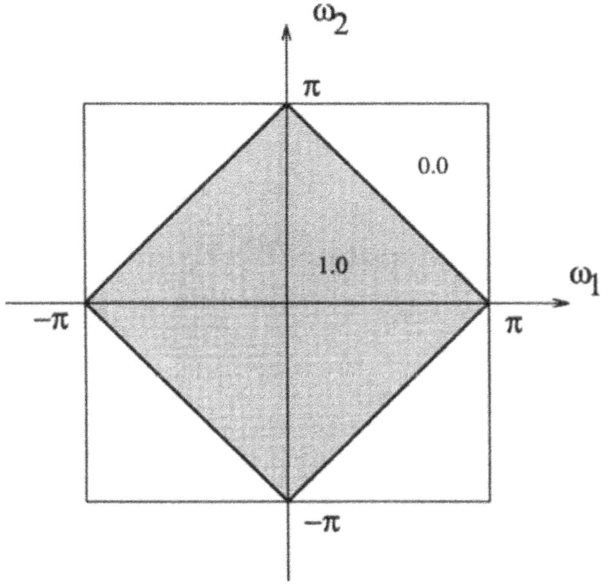

Figure 9 Ideal frequency response of a diamond lowpass filter.

The frequency response of the filter with transfer function $D(z_1, z_2)$ will approximate the ideal response in Figure 9.

We observe that the ideal frequency response in Figure 10a can be expressed as the sum of the two responses shown in Figures 10b and 10c. We also note that the ideal frequency responses in Figures 10b and 10c are frequency domain translates of each other. Therefore, if $F_b(z_1, z_2)$ is the transfer function of the filter approximating the response in Figure 10b then

$$F_c(z_1, z_2) = F_b(-z_1, -z_2) \tag{4.101}$$

will approximate the response in Figure 10c.

The ideal frequency response in Figure 10b can be expressed as the product of two ideal 1-D lowpass filters each having the response shown in Figure 10d, with one filter operating along the vertical direction of the 2-D array, and other along the horizontal. The ideal 1-D frequency response is a suitable candidate for approximation by a halfband filter. The transfer function, $H(z)$, of the 1-D halfband filter that approximates the response in Figure 10d can be expressed as

$$H(z) = 1 + z^{-1}T(z^2). \tag{4.102}$$

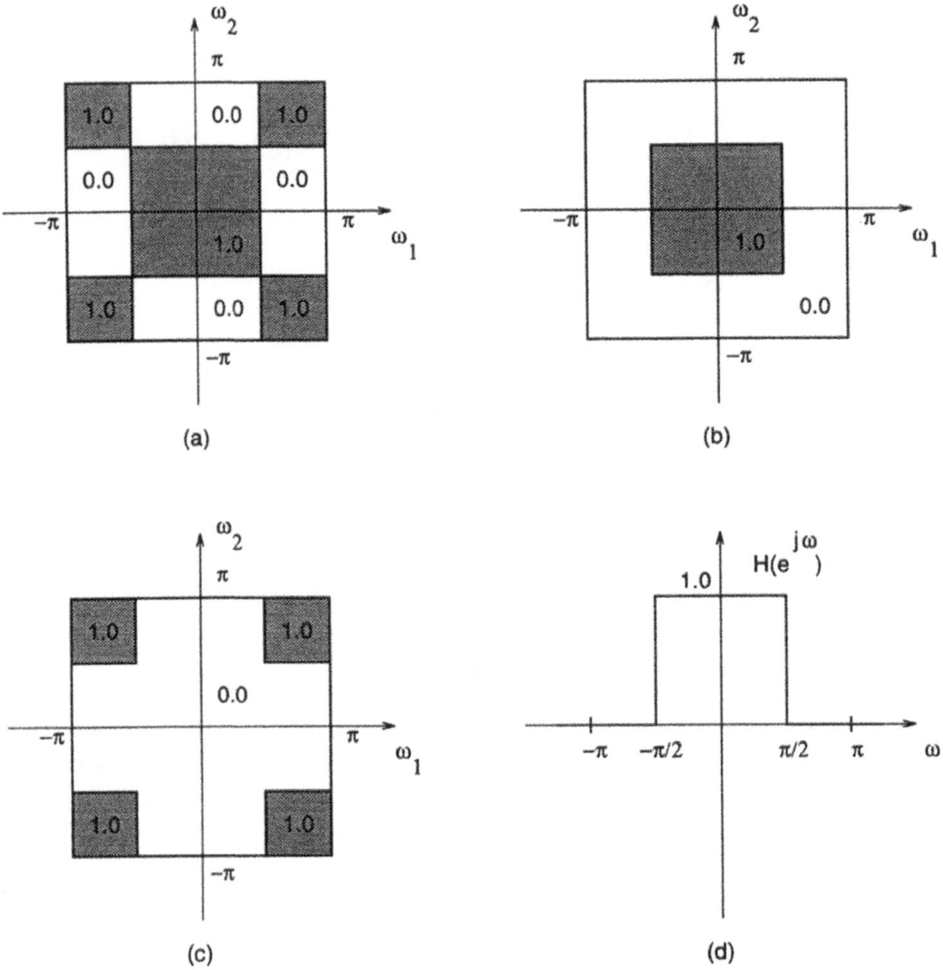

Figure 10 Ideal frequency response of (a) $F_a(z_1, z_2)$, (b) $F_b(z_1, z_2)$, (c) $F_c(z_1, z_2)$, and 1-D filter $H(z)$.

The transfer function F_a can be expressed as the sum of products:

$$F_a(z_1, z_2) = H(z_1)\, H(z_2) + H(-z_1)\, H(-z_2). \tag{4.103}$$

It follows from Equations (4.100)–(4.103) that

$$D(z_1, z_2) = 2 + 2\, z_2{}^{-1} T(z_1 z_2) T(z_1{}^{-1} z_2). \tag{4.104}$$

The halfband filter in Equation (4.102) can be an IIR filter, with $T(z^2)$ chosen to be an allpass filter. For this choice of analysis filters one can readily obtain synthesis filters that allow exact reconstruction in the absence of coding and quantization [19]. If the analysis filters are causal then the synthesis filters will be non-causal. Alternate methods of designing IIR diamond filter banks, including solutions with all causal filters, are described in [229].

We observe that the transfer function $D(z_1, z_2)$ given in Equation (4.104) can be viewed as being expressed as a ***halfband diamond*** filter with two polyphase components, one of which is a constant. The other component is a product of transfer functions with arguments $z_1 z_2$ and $z_1{}^{-1} z_2$. The filters can be implemented as 1-D filters, with processing carried out along appropriate diagonal directions. This processing can be considered as a generalization of the notion of separable processing using 1-D filters. This generalized separable processing can be extended to the case of non-rectangular filter banks more general than diamond filter banks. The directional processing is applicable in systems with multiband partitions based on some clever manipulations [38], [37].

8 CONCLUDING REMARKS

This chapter dealt with the construction and design of IIR filter banks for one- and two-dimensional signals. The main advantage of IIR filters is that they offer a computational advantage over FIR filters that are designed to meet the same magnitude specifications. This advantage is significant when phase linearity of the filters is not important. When approximately linear phase is desired, then either IIR filters with a two-sided impulse response response or those obtained using a special design should be used. IIR filters composed of allpass polyphase components lead to an implementation with reduced complexity.

A significant part of this chapter was devoted to a discussion of filter banks composed of allpass sections. Properties and design of filter banks for two-channel and M-channel cases were considered. Details of Butterworth and elliptic filter design for two-channel filter banks are included here the convenience of designers. A brief discussion of the two-dimensional case with emphasis on diamond filter banks was presented.

5

THE CONTINUOUS WAVELET TRANSFORM, FROM 1 TO 3 DIMENSIONS

Jean-Pierre Antoine and Romain Murenzi*

Institut de Physique Théorique,
Université Catholique de Louvain
B - 1348 Louvain-la-Neuve, Belgium

**Center for Theoretical Studies of Physical Systems,*
Clark Atlanta University, Atlanta, GA 30314, USA

1 MOTIVATION

Wavelet analysis is a particular time- or space-scale representation of signals which has found a wide range of applications in physics and mathematics in the last few years. In order to understand its success, let us consider first the case of one-dimensional signals. As a matter of fact, most real life signals are nonstationary. They often contain transient components, sometimes physically significant, and mostly cover a wide range of frequencies. In addition, there is frequently a direct correlation between the characteristic frequency of a given segment of the signal and the time duration of that segment. Low frequency pieces tend to last a long interval, whereas high frequencies occur in general for a short moment only. Human speech signals are typical in this respect. Vowels have a relatively low mean frequency and last quite long, whereas consonants contain a wide spectrum, especially in the attack, and are often very short.

The Fourier transform, which gives a purely frequency domain representation, does not provide explicit time localization information, which is important for such signals. For this reason, signal analysts turn to *time-frequency* (TF) representations. The idea is that one needs *two* parameters: one called a, which refers to frequency; the other called b, which indicates the position in the signal. Thus a general TF transform of a signal x will take the form

$$x(t) \leftrightarrow c(a,b) = \int_{\mathbb{R}} \overline{\psi_{ab}(t)}\, x(t)\, dt, \tag{5.1}$$

where ψ_{ab} is the analyzing function and $\overline{\psi_{ab}}$ is its complex conjugate. This concept of a TF representation is in fact quite old and familiar. The most obvious example is simply a musical score!

Among the possible TF transforms, two stand out as particularly simple and efficient: the windowed or short-time Fourier transform (STFT) and the wavelet transform (WT). They are both linear and may be derived from group representation theory (only the group differs). The essential difference between the two is in the way the frequency parameter a is introduced in the analyzing function. In both cases, b is simply a time translation.

The kernels of the two transforms can be written as follows:

(1) *For the short-time Fourier transform*

$$\psi_{ab}(t) = e^{jt/a}\,\psi(t-b). \tag{5.2}$$

Here ψ is a window function and the a-dependence is a modulation ($1/a \sim$ frequency); the window has constant width, but the lower a, the larger the number of oscillations in the window (see Figure 1a). (This transform was originally introduced by Gabor, with the window function ψ taken as a Gaussian. For this reason, it is sometimes called the Gabor transform, but this terminology is confusing since Gabor's was in fact a discretized version).

(2) *For the wavelet transform*

$$\psi_{ab}(t) = \frac{1}{\sqrt{a}}\,\psi(\frac{t-b}{a}). \tag{5.3}$$

The effect of a on the function ψ (which must be oscillating, see Section 2.1) is a dilation ($a > 1$) or a contraction ($a < 1$). The shape of the function is unchanged, it is simply spread out or squeezed (see Figure 1b).

In the remainder of this chapter, we will focus on the WT, and mostly on the continuous WT.

Combining Equations (5.1) and (5.3), one obtains the basic formula for the 1-D CWT:

$$W_x(a,b) = \frac{1}{\sqrt{a}} \int_{\mathbb{R}} \overline{\psi}\left(\frac{t-b}{a}\right) x(t)\, dt. \tag{5.4}$$

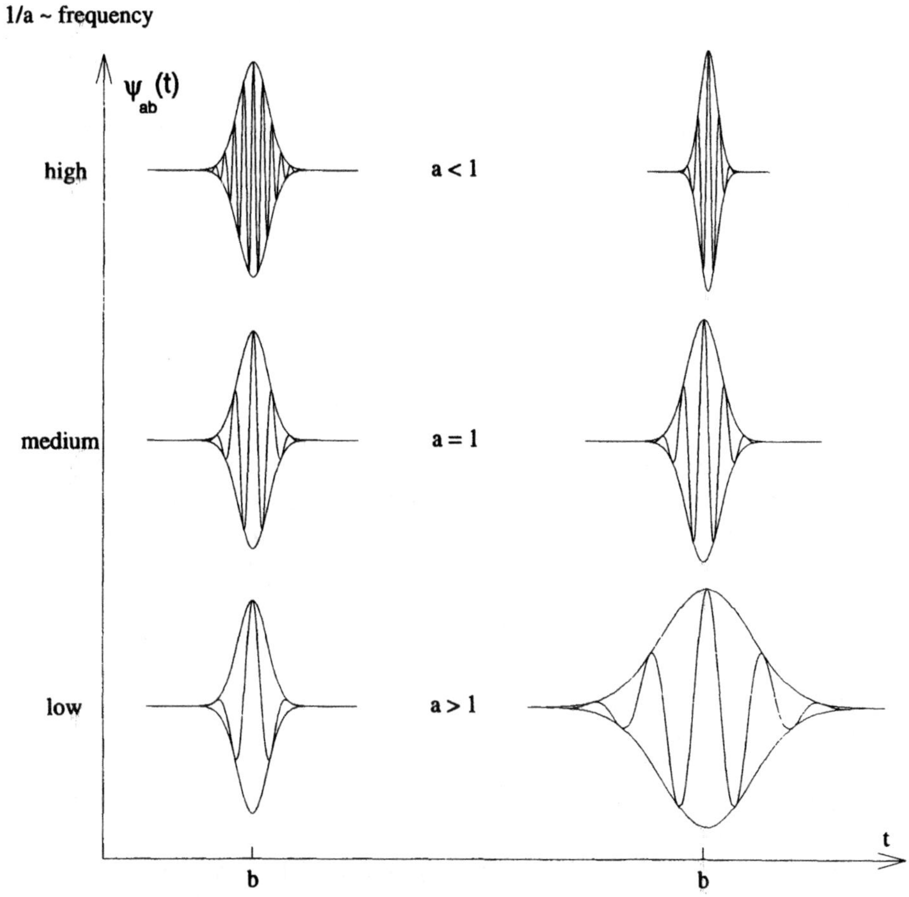

Figure 1 Varying the scale parameter a in the case of: (a) the short-time Fourier transform; (b) the wavelet transform.

The CWT may be better understood by recognizing it as a convolution of the (finite energy) signal x with the wavelet ψ, which is assumed to be well localized *both* in the time domain and in the frequency domain. In addition ψ must satisfy an admissibility condition, which in most cases may be reduced to the requirement that ψ has zero mean: $\int \psi(t)\, dt = 0$. Combining these two facts, one sees that the WT $x \mapsto W_x$ provides a *local (i.e. bandpass) filtering*, both in time b and in scale a, which works at constant relative bandwidth, $\Delta\Omega/\Omega =$ constant. Thus it is more efficient at high frequency, i.e. small scales, in particular for the detection of singularities in the signal. These properties of the 1-D CWT will be discussed at length in Sections 2–5 below, together with the main applications. The advantage of this presentation of the CWT (which is ultimately based on group representation theory) is that it is easily extendible to two or more dimensions. In 2-D, the analog of Equation (5.4) is

$$W_x(a, \theta, \vec{b}) = \frac{1}{a} \int_{\mathbb{R}^2} \bar{\psi}(\frac{1}{a} r_{-\theta}(\vec{t} - \vec{b}))x(\vec{t})\, d^2\vec{t}, \tag{5.5}$$

where, as before, $x(\vec{t})$ is the 2-D signal (the image to be analyzed) and $\psi(\vec{t})$ is the analyzing wavelet, which is translated by the vector $\vec{b} \in \mathbb{R}^2$, dilated by $a > 0$ and rotated by an angle θ ($r_{-\theta}$ is the rotation operator). Since the wavelet ψ is required to have zero mean, we have again a bandpass filtering effect, i.e. the analysis is *local* in all four parameters $a, \theta, \vec{b}$, and here too it is particularly efficient at detecting discontinuities or specific sharp features in images. It also allows good directional filtering, an important operation in image processing. We will discuss the 2-D case in detail in Section 6. Some three-dimensional physical phenomena also depend crucially on scale, typically in fluid dynamics. This suggests a 3-D wavelet analysis, which is equally easy to set up, by the same scheme (see Section 7).

Finally a word about references. The literature on wavelet analysis has grown exponentially, so that some guidance may be helpful. As a first contact, the introductory article [342] may be a good suggestion, followed by the elementary book of Meyer [265]. For a survey of the various applications, and a good glimpse of the chronological evolution, there is still no better place to look than the proceedings of the three large wavelet conferences, Marseille 1987 [98], Marseille 1989 [264] and Toulouse 1992 [266]. Finally a systematic study requires a textbook. Among the increasing number of books and special issues of journals appearing on the market, we recommend in particular the volume of I. Daubechies [115], the collection of review articles in [347] and the special issue of the IEEE Transactions on Information Theory [203].

2 THE MAIN FEATURES OF THE 1-D CWT

2.1 Definitions

As is clear from Equation (5.4), the CWT is a projection of the signal, in the L^2 sense, onto the family $\{\psi_{ab}, a > 0, b \in \mathbb{R}\}$ generated from the single function ψ by translation and dilation:

$$\begin{aligned} W_x(a,b) &= \langle \psi_{ab} | x \rangle \\ &= \frac{1}{\sqrt{a}} \int_{-\infty}^{\infty} \overline{\psi}\left(\frac{t-b}{a}\right) x(t)\, dt \\ &= \sqrt{a} \int_{-\infty}^{\infty} \overline{\Psi}(a\Omega) X(\Omega) e^{jb\Omega}\, d\Omega, \end{aligned} \tag{5.6}$$

where Ψ and X denote the Fourier transforms of ψ and x respectively. Thus the transform $W_x(a,b)$ lives in the time-scale half-plane $\mathbb{R}_+^* \times \mathbb{R} = \{a > 0, b \in \mathbb{R}\}$. The analyzing wavelet ψ satisfies a number of conditions:

(i) For the formalism to make sense, $\psi(t)$, hence also $\Psi(\Omega)$, should be square integrable ($\psi \in L^2(\mathbb{R})$).

(ii) ψ must be ***admissible***, that is, the integral

$$(2\pi)^{-1} c_\psi = \int_{-\infty}^{\infty} |\Psi(\Omega)|^2 \frac{d\Omega}{|\Omega|} < \infty \tag{5.7}$$

must converge. This condition implies (and for ψ regular enough, is equivalent to)

$$\Psi(0) = 0, \tag{5.8}$$

which in turn is equivalent to the *zero mean* condition,

$$\int_{-\infty}^{\infty} \psi(t)\, dt = 0, \tag{5.9}$$

which implies ψ must be oscillating.

(iii) In order to get an efficient transform (good bandpass filtering, both in time *and* in frequency), $\psi(t)$ and $\Psi(\Omega)$ should be both well localized. It suffices to require that ψ be also integrable ($\psi \in L^1 \cap L^2$), but in practice a better localization will be useful.

(iv) In addition to (ii), ψ *may* be required to have a certain number of *vanishing moments*, i.e.

$$\int_{-\infty}^{\infty} t^n \psi(t)\, dt = 0, \quad n = 0, 1, \ldots N. \tag{5.10}$$

This property improves the efficiency of ψ at detecting singularities in the signal.

(v) Finally, ψ is often required to be ***progressive***, that is, Ψ real and $\Psi(\Omega) = 0$ for $\Omega \leq 0$. Such a ψ is also called an analytic signal or a Hardy function.

We emphasize that (i) and (ii) are essential requirements for the CWT while, (iii) ensures a clear interpretation, and thus, in a practical sense, is necessary. Conditions (iv) and (v) make life easier, but are not compulsory.

As we shall see in the following sections, a wavelet ψ that satisfies these requirements generates, by Equation (5.6), a transform $x(t) \mapsto W_x(a, b)$ that yields a good analysis of the signal. It also allows an efficient reconstruction $W_x(a, b) \mapsto x(t)$ of the signal from its transform. In fact, $x \mapsto W_x$ will be a unitary map with respect to appropriate norms, as will be discussed in Section 3.

2.2 Common Wavelets for 1-D Analysis

The two analyzing wavelets most used in practice are the mexican hat and the Morlet wavelets.

The *mexican hat* is simply the second derivative of the Gaussian,

$$\psi_H(t) = (1 - t^2)\, e^{-t^2/2}, \quad \Psi_H(\Omega) = \Omega^2\, e^{-\Omega^2}. \tag{5.11}$$

This is an admissible, real wavelet, with two vanishing moments ($n = 0, 1$).

The *Morlet wavelet* is just a modulated Gaussian, given by

$$\psi_M(t) = e^{j\Omega_o t}\, e^{-t^2/2\sigma_o^2} + \text{corr.}, \quad \Psi_M(\Omega) = e^{-[(\Omega-\Omega_o)\sigma_o]^2/2} + \text{corr.} \tag{5.12}$$

In fact the first term alone does *not* satisfy the admissibility conditions (5.8)-(5.9), hence the necessity of a correction. However, for Ω_o large enough (typically $\Omega_o \geq 5.5$), this correction term is numerically negligible. Notice that, without the correction term, Equation (5.12) is just a Gabor function.

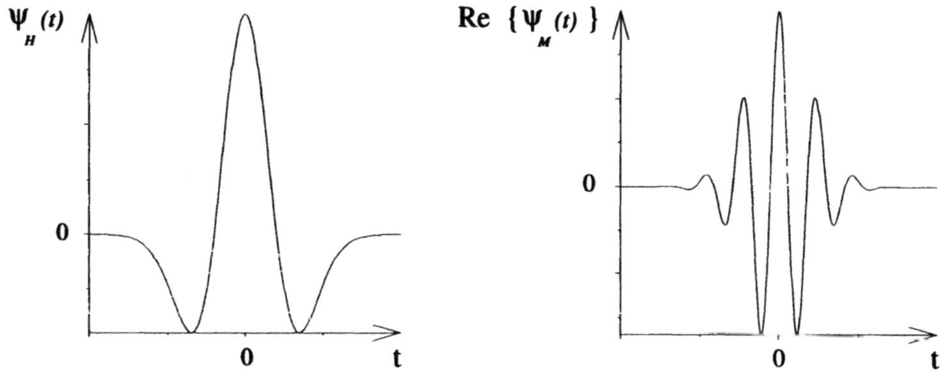

Figure 2 Two common one-dimensional wavelets: (a) the 1-D mexican hat; (b) the real part of 1-D Morlet wavelet, for k=5.6.

The Morlet wavelet is complex, hence the corresponding transform $W_x(a,b)$ is also, and one may treat separately its phase and its modulus. It turns out that the phase of the transform is a crucial ingredient for the algorithm of singularity detection in a signal, for instance the localization of spectral lines (Section 5).

2.3 Localization Properties and Interpretation

The main virtues of the CWT follow from the support properties of ψ. Assume ψ and Ψ to be as well localized as possible (compatible with the Fourier uncertainty principle). More specifically, assume that ψ has an 'essential' support of width T, centered around 0, while Ψ has an essential support of width F, centered around Ω_o. Then the transformed wavelets ψ_{ab} and Ψ_{ab} have, respectively, an essential support of width aT around b and an essential support of width F/a around Ω_o/a.

Notice that the product of the two widths is constant. We know it has to be bounded below by a fixed constant, by Fourier's theorem, also known as the uncertainty principle. We illustrate this vital fact in Figure 3, which is a time-frequency representation. Remember that $1/a$ behaves like a frequency.

Therefore, if $a \gg 1, \psi_{ab}$ is a wide window, whereas Ψ_{ab} is very peaked around a small frequency Ω_o/a. This transform will be most sensitive to *low frequencies*. If $a \ll 1$, then ψ_{ab} is a narrow window and Ψ_{ab} is wide and centered around

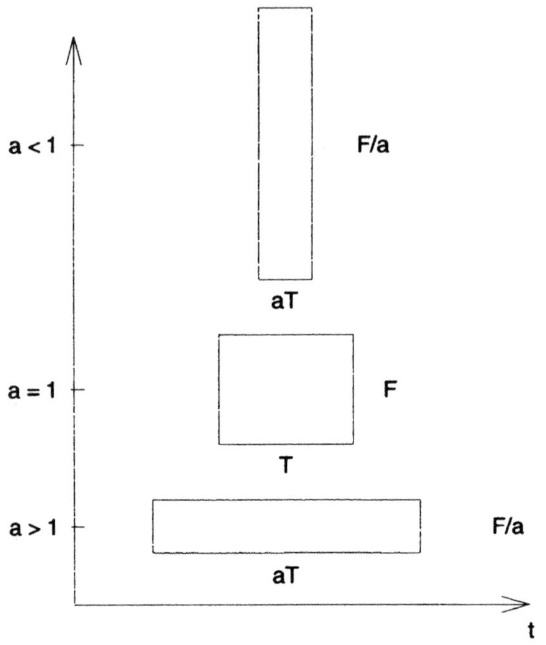

Figure 3 Support properties of ψ_{ab} and Ψ_{ab}. The time axis is horizontal. The frequency axis is vertical.

a high frequency Ω_o/a. This wavelet has a good localization capability in the time domain and is mostly sensitive to ***high frequencies.*** Thus we have obtained a tool that reproduces the correlation between duration and average frequency discussed in the introduction. Low frequency portions of the signal tend to be long, whereas high frequencies occur briefly in general.

Combining now these localization properties with the zero mean condition and the fact that ψ_{ab} acts like a filter (convolution), we see that the CWT performs a *local filtering*, both in time and in scale. The WT, $W_x(a,b)$, is nonnegligible only when the wavelet ψ_{ab} matches the signal, that is, it filters the part of the signal, if any, that lives around the time b and the scale a.

Taking all these properties together, one is naturally led to the interpretation of the CWT as a *mathematical microscope*, with optics ψ, position b and global magnification $1/a$ [32]. In addition, the analysis works at constant relative bandwidth ($\Delta\Omega/\Omega$ = constant), so that it has a better resolution at high frequency, i.e. small scales. This property makes it an ideal tool for detecting *discontinuities* in a signal, and also scale dependent features, in particular for analyzing *fractals* [32].

3 MATHEMATICAL ANALYSIS OF THE 1-D CWT

3.1 Main Properties

Given an admissible wavelet ψ, i.e. such that $c_\psi < \infty$ (see Equation (5.7)), the corresponding CWT $T_\psi : x(t) \mapsto W_x(a, b)$ has the following properties:

(1) $T_\psi : x(t) \mapsto W_x(a, b)$ is a linear map, covariant under time translation and under dilation (scale change).

(2) T_ψ conserves the energy of the signal,

$$\int |x(t)|^2 \, dt = c_\psi^{-1} \iint |W_x(a,b)|^2 \, \frac{da\, db}{a^2}. \tag{5.13}$$

This relation means that the map T_ψ is an isometry from the space of signals $L^2(\mathbb{R})$ onto a closed subspace of $L^2(\mathbb{R}^2_+, da\, db/a^2)$. An equivalent statement is that the wavelet ψ generates a *resolution of the identity,*

$$c_\psi^{-1} \iint |\psi_{ab}\rangle \langle \psi_{ab}| \frac{da\, db}{a^2} = I. \tag{5.14}$$

(3) As a consequence, the map T_ψ is invertible on its range, and the inverse transformation is simply the adjoint of T_ψ. Thus the signal $x(t)$ may be reconstructed from its wavelet transform by the formula,

$$x(t) = c_\psi^{-1} \iint \psi_{ab}(t)\, W_x(a,b) \frac{da\, db}{a^2}. \tag{5.15}$$

This means that the WT provides a decomposition of the signal as a linear superposition of the wavelets ψ_{ab} with coefficients $W_x(a, b)$.

(4) The projection from $L^2(\mathbb{R}^*_+ \times \mathbb{R}, da\, db/a^2)$ onto the range of T_ψ, that is, the space of transforms, is an integral operator, whose kernel

$$K(a', b'; a, b) = c_\psi^{-1} \langle \psi_{a'b'} | \psi_{ab} \rangle \tag{5.16}$$

is the autocorrelation function of ψ. It is also called a *reproducing kernel,* because the statement above means precisely that a function $f \in L^2(\mathbb{R}^*_+ \times \mathbb{R}, da\, db/a^2)$ is the WT of a certain signal iff it satisfies the reproduction property

$$f(a', b') = \iint K(a', b'; a, b)\, f(a, b) \frac{da\, db}{a^2}. \tag{5.17}$$

Remark: The relation (5.15) means that the *same* wavelet ψ has been used for the analysis *and* for the reconstruction. This is an unnecessary restriction, however. Indeed one may use two different (admissible) wavelets, provided they satisfy a mutual compatibility relation. In this way one may obtain simpler formulas, in particular for reconstruction, as shown originally by Morlet.

3.2 Mathematical Justification

From condition (5.13), one sees that $|W_x(a,b)|^2$ may be interpreted as an energy density in the (a,b)-half-plane $\mathbb{R}^2_+$, and that the natural geometry of the latter is not the usual Euclidean one. Indeed the measure $da\,db/a^2$ is invariant not only under time translation, but also under dilation. The reason behind these facts and the nice properties described above is to be found in group representation theory. The natural operations on a 1-D signal are precisely time translations and dilations, and these together constitute the '$ax+b$' group, that is, the connected affine group G_{aff} of the line. Then the relation

$$(U(a,b)f)(t) = f_{ab}(t) \equiv \frac{1}{\sqrt{a}} f\left(\frac{t-b}{a}\right) \tag{5.18}$$

defines a unitary irreducible representation of G_{aff} in the (Hardy) Hilbert space $\mathcal{H}_+ = \{f \in L^2(\mathbb{R}), F(\Omega) = 0 \text{ for } \Omega \le 0\}$. Furthermore, and this is the crucial feature, this representation is *square integrable*. That is, there exists at least one and in fact a dense set of admissible vectors, i.e. vectors ψ such that the matrix element $\langle U(a,b)\psi|\psi\rangle$ is square integrable over the group, with respect to the (left Haar) measure $da\,db/a^2$. Then a straightforward calculation shows that the squared norm of that matrix element equals precisely the constant c_ψ given in Equation (5.7). Thus the two notions of admissibility we have introduced indeed coincide.

It is worth emphasizing here that this pattern is by no means restricted to the affine group G_{aff}. Exactly the same situation prevails for many other transforms that are similarly based upon a square integrable representation of some Lie group. These are in fact particular instances of the construction of coherent state systems, a concept widely used in quantum physics [232]. A few interesting examples are

rotation group SO(2)	$\Longleftrightarrow$	Fourier series,
Weyl-Heisenberg group	$\Longleftrightarrow$	short time Fourier transform,
similitude group of $\mathbb{R}^n$	$\Longleftrightarrow$	multidimensional wavelet transform.

The Weyl-Heisenberg group consists of phase space translations, while the similitude group is made of translations, rotations and dilations. The last case is of special interest to us, for it yields the CWT in two and three dimensions, which we shall discuss at length in Section 6.

4 DISCRETIZATION OF THE 1-D CWT

The reproduction property (5.17) implies that the information content of the CWT, $W_x(a,b)$, is highly redundant. In fact the signal has been unfolded from one to two dimensions, and this explains the practical efficiency of the CWT for disentangling parts of the signal that live at the same time, but on different scales. This redundancy may be exploited in several ways. The principal one is that it must be possible to obtain the full information about the signal from a small subset of the values of the transform, $W_x(a,b)$. Two applications at least have been described in the literature, the theory of frames [116] and the ridge or skeleton representation [123].

4.1 Frames

Let $\Gamma = \{a_m, b_n, m, n \in \mathbb{Z}\}$ be a discrete lattice in the (a,b)-half-plane. We say that Γ yields a good discretization if the CWT of an arbitrary signal $x(t)$ may be represented as a superposition

$$x(t) = \sum_{m,n\in\mathbb{Z}} \widetilde{\psi_{mn}} \langle \psi_{mn} | x \rangle, \tag{5.19}$$

where $\psi_{mn} \equiv \psi_{a_m b_n}$ and $\widetilde{\psi_{mn}}$ may be explicitly constructed from ψ_{mn}. We emphasize that Equation (5.19) must be an *exact* representation, i.e. there is no loss of information as compared to the continuous reconstruction (5.15). Actually Equation (5.19) means that the signal $x(t)$ may be replaced by the set $\{\langle \psi_{mn} | x \rangle\}$ of its wavelet coefficients. In addition, one wants the reconstruction of $x(t)$ from its coefficients to be numerically stable (i.e. a small error in the coefficients implies a small error in the reconstructed signal). As shown by Daubechies et al. [115, 116, 114], this is achieved provided the following condition holds true, for some constants $A > 0, B < \infty$:

$$A \, \|x\|^2 \;\le\; \sum_{m,n\in\mathbb{Z}} |\langle \psi_{mn} | x \rangle|^2 \;\le\; B \, \|x\|^2. \tag{5.20}$$

The lower bound guarantees the numerical stability. In that case, one says that the set $\{\psi_{mn}\}$ constitutes a ***frame***, with ***frame bounds*** A and B. This notion was originally introduced in the 50's, in the context of nonharmonic analysis. If $A = B > 1$, the frame is said to be ***tight***. Of course, if $A = B = 1$, the set $\{\psi_{mn}\}$ is simply an orthonormal basis. The important point here is that, for all practical purposes, a good frame is almost as good as an orthonormal basis. By 'good frame', we mean that the expansion (5.19) converges sufficiently fast. A detailed analysis [115, 116, 114] shows this to be the case if $|B/A - 1| \ll 1$, i.e. if the frame is tight.

Precisely at this point arises the basic difference between the discretized CWT and the DWT. In the former case, the wavelet ψ is chosen a priori (with very few constraints, see Section 2), and the question is whether one can find a lattice Γ such that $\{\psi_{mn}\}$ is a frame with decent frame bounds A, B. In the other approach, one usually imposes that the set $\{\psi_{mn}\}$ be an orthonormal basis and tries to construct a function ψ to that effect. The construction is rather indirect. It proceeds through the determination of suitable QMF filters and the resulting function is usually very complicated. Often it has a fractal behavior.

As was to be expected, the solution to the frame problem for a given wavelet ψ must take into account the non-Euclidean geometry of the time-scale half-plane, i.e. the lattice Γ must be invariant under discrete dilations and translations. For the scales, one chooses naturally $a_m = a_o^m, m \in \mathbb{Z}$, for some $a_o > 1$. For the times, one takes $b_n = n\, b_o\, a_o^m, m, n \in \mathbb{Z}$. Thus

$$\psi_{mn}(t) = a_o^{-m/2}\, \psi(a_o^{-m} t - n), \quad m, n \in \mathbb{Z}. \tag{5.21}$$

The most common choice is $a_o = 2$ (octaves) and $b_o = 1$, which results in

$$\psi_{mn}(t) = 2^{-m/2}\, \psi(2^{-m} t - n), \quad m, n \in \mathbb{Z}. \tag{5.22}$$

It is worth noticing that this so-called ***dyadic*** lattice, shown in Figure 4, is exactly the same that indexes the DWT, although the two approaches are totally different. For this choice indeed, the relevant theorems may be proven [115, 114]. Both the mexican hat and the Morlet wavelet yield good frames. Actually the same question may be asked for the STFT. There the geometry is Euclidean. Typically Γ is a square lattice and the frame theorem simply says that a frame is obtained if the density of Γ is larger than a critical value. This is the well-known result of von Neumann concerning the density of canonical coherent states [232]. It is in fact a reincarnation of a standard theorem known in different circles under the names of Fourier, Heisenberg (uncertainty relations), Nyquist or Shannon.

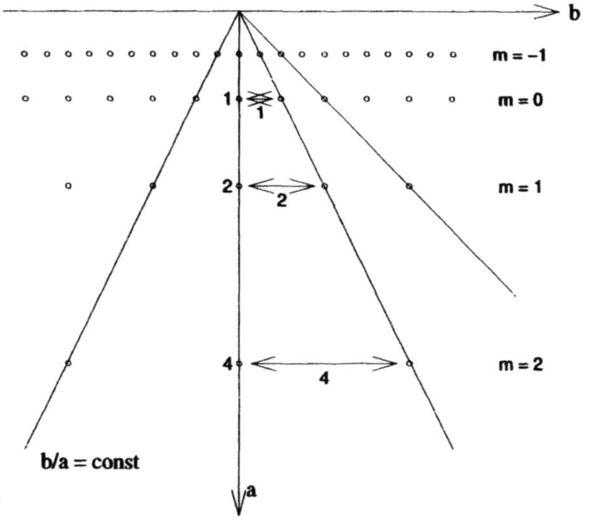

Figure 4 The dyadic lattice.

In conclusion, the two standard wavelets yield very good frames (nontight, however). This explains their efficiency in signal analysis, and thus their popularity for applications. We refer to [115, 114] for a detailed analytical and numerical discussion of this problem.

4.2 Ridges and Skeletons

Many signals are well approximated by a superposition of spectral lines,

$$x(t) = \sum_j A_j(t) \exp(i\phi_j(t)), \tag{5.23}$$

where the amplitude $A_j(t)$ varies slowly with respect to the phase $\phi_j(t)$. Such signals are sometimes called ***asymptotic***. Typical examples are spectra in NMR spectroscopy [123]. For a signal of this kind, the CWT (5.6) in the time domain is a rapidly oscillating integral, the essential contribution to which is given by the stationary points of the phase of the integrand. Assume for simplicity there is only one such point $t_s = t_s(a,b)$. Then the ***ridge*** of the WT is defined as the set of points (a,b) for which $t_s(a,b) = b$. These constitute a curve in the (a,b)-half-plane and a detailed analysis shows that, on this curve, the WT, $W_x(a,b)$, coincides, up to a small correction, with the analytic signal $Z(b)$ associated with $x(t)$ [123]. It follows that the restriction of the WT, $W_x(a, b)$, to its ridge, the so-called *skeleton*, contains the whole information. In particular, the so-called frequency modulation law $t^{-1} \arg\{x(t)\}$ of $x(t)$ is easily recovered from the skeleton. Thus it is not necessary to compute the whole CWT, but only its ridge. This is of course much less costly computationally, because there are fast algorithms available.

Here again, a clever exploitation of the intrinsic redundancy of the CWT improves spectacularly the efficiency of the wavelet analysis method.

5 IMPLEMENTATION OF THE 1-D CWT

5.1 Academic Signals

Faced with this new tool, one must begin by learning the rules of the trade, that is, one must learn how to read and understand a CWT [173]. The simplest way is to get some practice on very simple academic signals, such as a simple discontinuity in time or a monochromatic signal (pure sinusoid). Two remarks are in order here. First, the L^2 normalization used in Equation (5.6) was chosen mainly for mathematical reasons. It is the one for which the wavelet transform $x(t) \mapsto W_x(a, b)$ is a unitary map. However, in practice, this may be changed. For instance, it is common to use instead the so-called L^1 normalization, which amounts to replacing the factor $a^{-1/2}$ in Equation (5.6) by a factor a^{-1}. The effect is to enhance the small scales, that is, to make more conspicuous the discontinuities in the signal, which is precisely one of the aims of the WT. Second, it is natural to use a logarithmic scale for the scale parameter a. The visual effect is that the lines, $b/a =$ constant, are no longer straight lines, as in Figure 4, but they become hyperbolic curves. At the same time, the horizon $a = 0$ recedes to minus infinity.

We will now analyze the two academic signals mentioned above. The analyzing wavelet ψ is supposed to be complex, so that we may treat separately the modulus and the phase of the transform. The scale axis, in units of $\log a$, points downward, so that high frequencies (small a) correspond to the top of the plots, and low frequencies (large a) to the bottom. The results are presented by coding the height of the function by density of points (12 levels of grays, from white to black). The phase is 2π-periodic. When it reaches 2π, it is wrapped around to the value 0. Thus the lines of constant phase with value $2k\pi$ are lines of discontinuity, where the density of points drops abruptly from

1 (black) to 0 (white). In addition, the functions plotted are thresholded at 1% of the maximum value of the modulus of $W_x(a, b)$.

A Simple Discontinuity

The simplest signal is a simple discontinuity in time, at $t = t_o$, modeled by $x(t) = \delta(t - t_o)$. The WT is obtained immediately and reads

$$W_x(a, b) = \frac{1}{\sqrt{a}} \overline{\psi}\left(\frac{t_o - b}{a}\right). \tag{5.24}$$

The following features may be read off Equation (5.24):

- The phase of $W_x(a, b)$ is constant on the lines, $b/a =$ constant, originating from the point $b = t_o$ on the horizon. These lines point towards the position of the singularity like a finger.
- On the same lines of constant phase, the modulus of $W_x(a, b)$ increases as $a^{-1/2}$ when $a \to 0$, so that the singularity is enhanced. The effect is even more pronounced if one uses the L^1 normalization.

These two properties are illustrated on Figure 5, which shows the modulus and phase of the WT of a delta function, using a standard Morlet wavelet. The result is independent of the choice of ψ. Notice that here and in the following pictures, we have used a rather rough discretization in a for reasons of computational economy, but this does not prevent the relevant features from standing out clearly. These properties explain why the CWT is primarily a tool for detecting singularities in the signal. Indeed they follow only from the fact that the delta function is homogeneous of order -1,

$$\delta(\lambda(t - t_o)) = \frac{1}{\lambda}\, \delta(t - t_o), \quad \lambda > 0, \tag{5.25}$$

and they remain true for any signal which is homogeneous of order α around a given point t_o. Thus

$$x(\lambda(t - t_o)) = \lambda^{\alpha} x(t - t_o)) \ \text{ for } \ t \sim t_o, \lambda > 0, \tag{5.26}$$

implies (using the covariance property of the CWT under dilation)

$$W_x(a, b - t_o) = a^{\alpha + \frac{1}{2}}\, W_x(1, \frac{b - t_o}{a}). \tag{5.27}$$

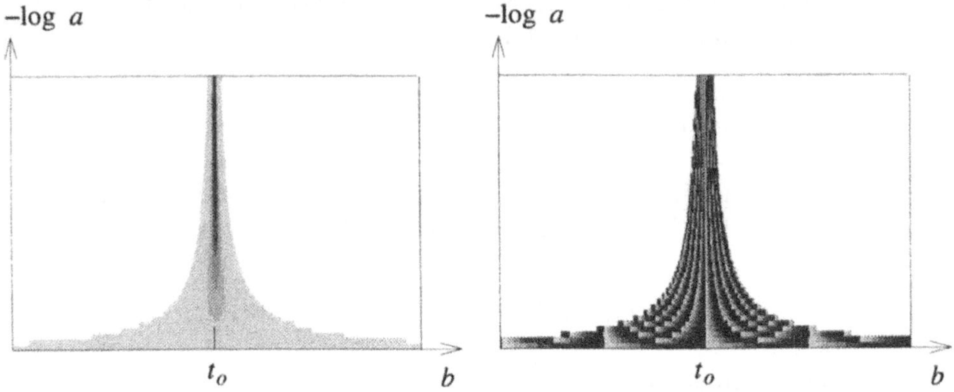

Figure 5 WT of a delta function with the Morlet wavelet: (a) modulus; (b) phase.

Such a homogeneous behavior, with different values of α on the left and on the right of t_o, models almost any type of discontinuity in t_o. In fact, the CWT allows the strength of singularities to be estimated. Indeed, for such a homogeneous function (5.26), the exponent α characterizing the singularity on the right hand side of t_o may be read off a plot of $\log|W_x(a,b)|$ vs. $\log a$, as b tends to t_o from the right and similarly from the left.

The interesting point is that this behavior is extremely robust. For instance, the 'finger' pointing to a delta-singularity remains clearly visible when the latter is superposed on a continuous signal (even if the amplitude of the delta function is too small to be visible on the signal itself), or even in the presence of substantial background noise. Similarly, the discontinuity corresponding to the abrupt onset of a signal is readily identified with the CWT, a situation common in seismology. We refer to [173] for several spectacular examples.

A Single Monochromatic Wave

Equally simple is a single harmonic signal (monochromatic wave),

$$x(t) = e^{j\Omega_s t} \quad \Longleftrightarrow \quad X(\Omega) = \frac{1}{\sqrt{2\pi}}\,\delta(\Omega - \Omega_s), \tag{5.28}$$

which gives

$$W_x(a,b) \;=\; \sqrt{\frac{a}{2\pi}}\;\Psi(a\Omega_s)\,e^{j\Omega_s b}$$

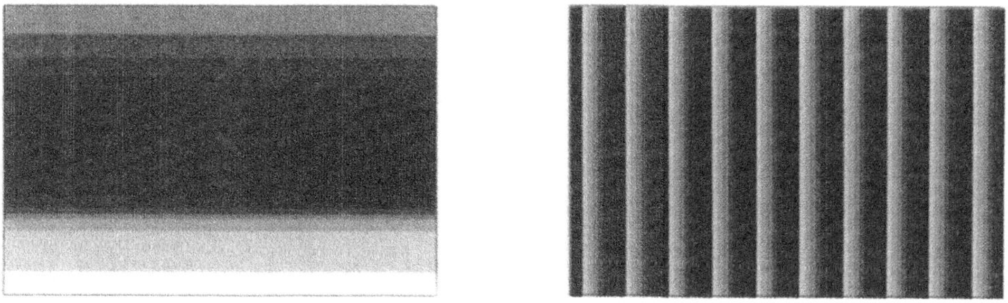

Figure 6 WT of a single sinusoid with the Morlet wavelet: (a) modulus; (b) phase.

$$= W_x(a,0)\, e^{j\Omega_s b}. \tag{5.29}$$

The same relations remain true for a real monochromatic signal, $x(t) = \sin \Omega_s t$ or $x(t) = \cos \Omega_s t$, if the wavelet ψ is progressive ($\Psi(\Omega) = 0$ for $\Omega \le 0$). Again two important properties may be read off immediately from Equation (5.29). First, the modulus of $W_x(a,b)$ is independent of b, i.e.

$$|W_x(a,b)| = \sqrt{\frac{a}{2\pi}}\, \Psi(a\Omega_s). \tag{5.30}$$

The graph of $|W_x(a,b)|$ consists of horizontal bands and the profile for a fixed time b essentially reproduces the profile of Ψ. Second, the phase of $W_x(a,b)$ is linear in b. Since the phase is 2π-periodic, the graph of $\Phi(a,b) \equiv \arg W_x(a,b)$ is a linear sawtooth function,

$$\Phi(a,b) = \Omega_s b \pmod{2\pi}. \tag{5.31}$$

These properties are illustrated in Figure 6 for a single sine function analyzed with a Morlet wavelet.

Both the modulus and the phase allow for the recovery of the frequency Ω_s of the signal. If the modulus of the wavelet $\Psi(\Omega)$ has a single maximum for $\Omega = \Omega_o$, Equation (5.30) gives immediately $\Omega_s = \Omega_o / a_M$, where a_M is the scale corresponding to the maximum in the profile of $|W_x(a,b)|$ for fixed b. For instance, the Morlet wavelet $\psi(t) = \exp(j\Omega_o t)\exp(-t^2/2)$ yields

$$W_x(a,b) = \sqrt{a}\, e^{-\frac{1}{2}(a\Omega_s - \Omega_o)^2}\, e^{j\Omega_s b}, \tag{5.32}$$

and the result is obvious. As for the phase, Equation (5.31) gives, at least locally,

$$\frac{\partial \Phi(a,b)}{\partial b} = \Omega_s = \frac{\Omega_o}{a_M}. \tag{5.33}$$

For an asymptotic, locally monochromatic signal (see Equation (5.23))

$$x(t) = A(t)\exp(j\phi(t)), \tag{5.34}$$

the result given in Equation (5.33) may be recovered if one introduces the notion of ***instantaneous frequency***,

$$\Omega_{\text{inst}} = \frac{\partial \Phi(a,b)}{\partial b}. \tag{5.35}$$

Then it can be shown that Equation (5.33) generalizes to

$$\frac{\partial \Phi(a,b)}{\partial b} = \frac{\Omega_o}{a}. \tag{5.36}$$

The solution of Equation (5.36) corresponds to the points where the instantaneous frequency of the signal coincides with the frequency of the wavelet, i.e. the values of a corresponding to the peaks of the signal (center of the spectral lines). This relation may then be further extended to a general asymptotic signal analyzed by a general asymptotic wavelet, and this leads precisely to the notion of a ridge, which was discussed in Section 4.2 [123].

5.2 Physical Applications

The CWT has found a wide variety of applications in various branches of physics and/or signal processing. We will list here a representative selection. All of them, and many more, may be found in the proceedings volumes [98, 264, 266]. In all cases, the CWT is primarily used for analyzing transient phenomena, detecting abrupt changes in a signal or comparing it to a given pattern.

- *Sound and acoustics.*
  The first applications of the CWT were in the field of acoustics. A few examples are musical synthesis, speech analysis, and modeling of the sonar system of bats and dolphins. Other examples include various problems in underwater acoustics, such as the disentangling of the different components of an underwater refracted wave and the identification of an obstacle, such as a submarine.

- *Geophysics.*
  This is the origin of the method, which was designed in an empirical fashion by J. Morlet for analyzing the recordings of microseisms used in oil prospecting. More recently, the CWT has been applied to the analysis of various long time series of geophysical origin, e.g. in gravimetry (fluctuations of the local gravitational field) or in astronomy (fluctuations of the length of the day).

- *Fractals, turbulence.*
  As mentioned above, the CWT is an ideal tool for studying fractals, or more generally phenomena with particular properties under scale changes. Thus it is quite natural that the CWT has found many applications in the analysis of 1-D and 2-D fractals, artificial diffusion limited aggregates or natural arborescent growth phenomena. Related to these is the use of the CWT in the analysis of developed turbulence, e.g. identification of coherent structures, and uncovering of hierarchical structure.

- *Spectroscopy.*
  This was one of the earliest and most successful applications, in particular for NMR spectroscopy, where the method proved extremely efficient in subtracting unwanted spectral lines or filtering out background noise. We will discuss this example in more detail below.

- *Medical applications.*
  The CWT has been used for analyzing or monitoring various electrical or mechanical phenomena in the brain (EEG) or the heart (ECG).

- *Industrial applications.*
  Here again the important aspect is monitoring, for instance in detecting anomalies in the functioning of nuclear, electrical or mechanical installations.

Before concluding this section, let us illustrate the use of the CWT by the example of NMR spectroscopy. The physical phenomenon may be described as follows. A sample is irradiated by a magnetic field, which is then switched off. The protons contained in the sample tend to align on the field, then relax to their equilibrium state with frequencies characteristic of their chemical environment. The signal to be analyzed is the Fourier transform (spectrum) of the damped response curve of the protons. It contains a large number of narrow peaks, the spectral lines, but many among them are useless, coming for instance from the protons of the solvent. These peaks, which may be quite big, must be subtracted, and the position of the relevant ones measured with precision.

A typical analysis of real NMR spectra is given in Figure 7. On the left is an example of peak subtraction. The original spectrum (a) exhibits a huge parasite peak, that completely masks the interesting structures. The analysis consists of isolating, on the CWT, this peak and then subtracting it from the spectrum. The result (b) is a spectrum where all the fine details are now clearly visible, and have not been perturbed by the removal of the large peak. Indeed, the prominent structures appear at exactly the same place on the frequency (horizontal) axis in both pictures. Note that, for better visibility, the vertical scales are different in the two cases. The reason for the remarkable efficiency of the method in this case is, that the huge line and the rest of the spectrum live at different scales, hence they are decoupled in the CWT ('unfolding') and can be readily separated with very little distortion.

The pictures on the right hand side in Figure 7 show an example of noise filtering. The original signal (c) is very noisy. The ten dominant peaks are localized with the help of the so-called CWT ridge algorithm. The remnant of the spectrum is subtracted and the filtered spectrum is reconstructed, using one of the reconstruction formulas. The result (d) is visibly enhanced.

6 THE CWT IN TWO DIMENSIONS

As mentioned before, the formulation of the 1-D CWT given in the previous sections extends readily to the 2-D case. The key is to start from a suitable representation of the similitude group of the plane, which consists of 2-D translations, (global) dilations and plane rotations.

6.1 Mathematical Properties

We begin by listing the basic properties of the CWT in two dimensions, which are completely analogous to those discussed in Section 3 for the 1-D case. By an image, we mean a 2-D signal of finite energy, represented by a complex-valued, square integrable function $x \in L^2(\mathbb{R}^2, d^2\vec{t})$. This condition may be relaxed, to allow, for instance, a plane wave or a δ function. In practice, a black and white image will be represented by a bounded non-negative function: $0 \leq x(\vec{t}) \leq M, \forall \vec{t} \in \mathbb{R}^2, M < \infty$, the discrete values of $x(\vec{t})$ corresponding to the level of gray of each pixel.

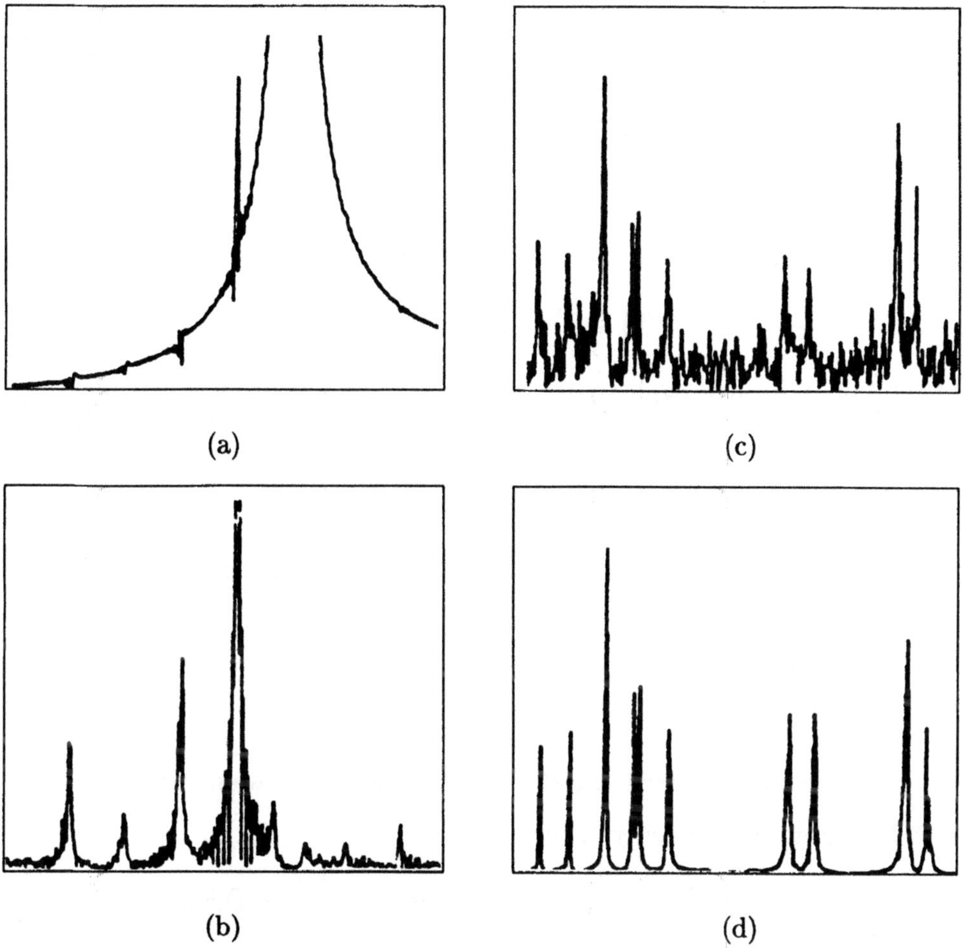

Figure 7 Application of the CWT in NMR spectroscopy. Subtraction of an unwanted peak (a, b) and noise filtering (c, d). Figures (a) and (c) show the original spectrum. Figures (b) and (d) show the spectrum reconstructed after subtraction and filtering, respectively.

Given an image $x \in L^2(\mathbb{R}^2)$, its WT with respect to the wavelet ψ is

$$W_x(a,\theta,\vec{b}) = \langle \psi_{a,\theta,\vec{b}} | x \rangle \equiv a^{-1} \int \bar{\psi}(a^{-1} r_{-\theta}(\vec{t} - \vec{b}))\, x(\vec{t})\, d^2\vec{t}. \tag{5.37}$$

In this relation, $\vec{b} \in \mathbb{R}^2$ is the translation parameter, $a > 0$ the dilation, and $r_{-\theta}(0 \le \theta < 2\pi)$ denotes the usual 2×2 rotation matrix. The wavelet $\psi \in L^2(\mathbb{R}^2)$ and its Fourier transform Ψ are both supposed to be well localized. In addition, ψ must be ***admissible***. That is,

$$(2\pi)^{-2} c_\psi \equiv \int |\Psi(\vec{k})|^2 \frac{d^2\vec{k}}{|\vec{k}|^2} < \infty. \tag{5.38}$$

If ψ is regular enough ($\psi \in L^1(\mathbb{R}^2) \cap L^2(\mathbb{R}^2)$ suffices), the admissibility condition simply means that the wavelet has zero mean, i.e.

$$\Psi(\vec{0}) = 0 \iff \int d^2\vec{t}\, \psi(\vec{t}) = 0. \tag{5.39}$$

In addition, ψ is often required to have a certain number of vanishing moments. As a matter of fact, the WT as well as the admissibility condition originate from the natural representation of the similitude group of $\mathbb{R}^2$, $G \equiv SIM(2) = (\mathbb{R}^+_* \times SO(2)) \wedge \mathbb{R}^2$, in the Hilbert space $L^2(\mathbb{R}^2)$:

$$(U(a,\theta,\vec{b})\psi)(\vec{t}) = \psi_{a,\theta,\vec{b}}(\vec{t}) \equiv a^{-1}\, \psi(a^{-1} r_{-\theta}(\vec{t} - \vec{b})), \tag{5.40}$$

which is unitary, irreducible, and square integrable.

The main properties of the wavelet transform $T_\psi : x \mapsto W_x$ may be summarized as follows:

- T_ψ is *covariant* under translations, dilations and rotations [282, 281].
- T_ψ conserves the energy of the signal, i.e.

$$c_\psi^{-1} \iiint |W_x(a,\theta,\vec{b})|^2 \frac{da}{a^3} d\theta d^2\vec{b} = \int |x(\vec{t})|^2\, d^2\vec{t}. \tag{5.41}$$

  It is an isometry from the space of signals into the space of transforms, which is a closed subspace of $L^2(G, dg)$, where $dg \equiv a^{-3} da d\theta d^2\vec{b}$ is the natural (Haar) measure.

- As a consequence, T_ψ is invertible on its range and the inverse transformation is simply the adjoint of T_ψ. Thus one has an exact ***reconstruction formula***,

$$x(\vec{t}) = c_\psi^{-1} \iiint \psi_{a,\theta,\vec{b}}(\vec{t})\, W_x(a,\theta,\vec{b})\, \frac{da}{a^3} d\theta d^2\vec{b}\,. \tag{5.42}$$

In other words, the 2-D wavelet transform, like its 1-D counterpart, provides a decomposition of the signal in terms of the analyzing wavelets $\psi_{a,\theta,\vec{b}}$, with coefficients $W_x(a,\theta,\vec{b})$.

- The projection from $L^2(G,dg)$ onto the range of T_ψ is an integral operator, whose kernel $K(a',\theta',\vec{b}'|a,\theta,\vec{b})$ is the autocorrelation function of ψ (also called the *reproducing kernel*):

$$K(a',\theta',\vec{b}'|a,\theta,\vec{b}) = c_\psi^{-1}\,\langle\psi_{a',\theta',\vec{b}'}|\psi_{a,\theta,\vec{b}}\rangle. \tag{5.43}$$

Therefore, transforms satisfy the *reproduction property*

$$W_x(a',\theta',\vec{b}') = \iiint K(a',\theta',\vec{b}'|a,\theta,\vec{b})\,W_x(a,\theta,\vec{b})\,\frac{da}{a^3}d\theta d^2\vec{b}. \tag{5.44}$$

6.2 The Two Basic Representations

The first problem one faces in practice is one of visualization. Indeed $W_x(a,\theta,\vec{b})$ is a function of four variables: two position variables $\vec{b} \in \mathbb{R}^2$, and the pair $(a,\theta) \in \mathbb{R}_*^+ \times [0,2\pi) \simeq \mathbb{R}_*^2$.

As a consequence [28], there are two natural ways of presenting the CWT, using two-dimensional sections of the parameter space. The first is the *position representation*, where a and θ are fixed and the CWT is considered as a function of position $\vec{b}$ alone. The second is the *scale-angle representation*, for fixed $\vec{b}$. The CWT is considered as a function of scale and angle (a,θ), i.e. of spatial frequency. In other words, one looks at the full CWT through a keyhole located at $\vec{b}$, and observes all scales and all directions at once. The position representation is the standard one, and it is useful for the general purpose of image processing such as detection of position, shape and contours of objects, pattern recognition and image enhancement by resynthesis after elimination of unwanted features (such as noise). The scale-angle representation will be particularly interesting whenever scaling behavior (as in fractals) or angular selection is important, in particular when directional wavelets are used. In fact, both representations are needed for a full understanding of the properties of the CWT in all four variables, and the reproducing kernel K should be studied in both.

For numerical evaluation, the WT is sampled and FFT algorithms are used. This leads to numerical complexity of $3N_1N_2\log(N_1N_2)$, where N_1, N_2 denote the number of sampling points in the variables (b_{t_1}, b_{t_2}) or (a,θ).

Whichever representation we use, we end up with a function of two variables, either in Cartesian coordinates $\vec{b}$, or in polar coordinates (a, θ). In both cases, the function may be real or complex (see Figures 8 and 9). In the latter case, it will be often represented through its modulus and phase. It turns out that the phase is particularly instructive, as was already the case in 1-D. We refer to [28] for a more detailed discussion.

6.3 Choice of the 2-D Analyzing Wavelet

The next step is to choose an analyzing wavelet ψ. The most popular ones are the familiar mexican hat and Morlet wavelets from the 1-D case.

The 2-D Mexican Hat or Marr Wavelet

In its isotropic version, this is simply the Laplacian of a Gaussian,

$$\psi_H(\vec{t}) = (2 - |\vec{t}|^2)\ \exp(-\frac{1}{2}|\vec{t}|^2). \tag{5.45}$$

This is a real, rotation invariant wavelet, originally introduced by Marr [256]. There exists also an anisotropic version, obtained by replacing in Equation (5.45) $\vec{t}$ by $A\vec{t}$, where $A = \mathrm{diag}[\epsilon^{-1/2}, 1]$, $\epsilon \geq 1$, is a 2×2 anisotropy matrix. However, this wavelet is of little use in practice, because it still acts as a second order operator and detects singularities in all directions. Indeed it is *not* a directional wavelet, in the technical sense defined below. Hence the mexican hat will be efficient for a fine pointwise analysis, but not for detecting directions.

The 2-D Morlet Wavelet

This is the prototype of an oriented wavelet,

$$\psi_M(\vec{t}) = \exp(j\vec{k}_o \cdot \vec{t})\ \exp(-\frac{1}{2}|A\vec{t}|^2) + \mathrm{corr.} \tag{5.46}$$

The parameter $\vec{k}_o$ is the wave vector, and A the anisotropy matrix as above. The correction term enforces the admissibility condition $\Psi_M(\vec{0}) = 0$, but it is numerically negligible for $|\vec{k}_o| \geq 5.6$ and will usually be dropped. The modulus of the (truncated) wavelet ψ_M is a Gaussian, whereas its phase is constant along the direction orthogonal to $\vec{k}_o$. Thus the wavelet ψ_M smoothes the signal in all directions, but detects the sharp transitions in the direction perpendicular to $\vec{k}_o$. The angular selectivity increases with $|\vec{k}_o|$, and even more so if, in addition, one introduces some anisotropy by taking $\epsilon > 1$. Then the modulus becomes a Gaussian elongated in the t_1 direction. Clearly this wavelet will detect preferentially singularities (edges) in the t_1 direction, and its efficiency increases with ϵ. The best selectivity will be obtained by combining the two effects, i.e. by taking $\vec{k}_o = (0, k_o)$. The resulting complex wavelet is shown in Figure 8, for $k_o = 5.6$ and $\epsilon = 1$.

Directional Wavelets

If one wants to detect oriented features such as segments, edges, and vector fields, one needs a wavelet that is directionally selective. To be precise, we will say that a given wavelet ψ is ***directional*** if the effective support of its Fourier transform Ψ is contained in a convex cone in spatial frequency space $\{\vec{k}\}$, with apex at the origin, or a finite union of disjoint cones. In that case, one will usually call ψ ***multidirectional***.

According to this definition, the anisotropic mexican hat is not directional, since the support of Ψ_H is centered at the origin. This is true regardless of how big its anisotropy is, as illustrated in Figure 9(a). Indeed, detailed tests confirm its poor performances in selecting directions [28].

Two good examples of directional wavelets are the Morlet and conical wavelets.

(a) The Morlet wavelet.
From Equation (5.46), we get

$$\Psi_M(\vec{k}) = \sqrt{\epsilon}\, \exp\left(-\frac{1}{2}[\epsilon k_{t_1}^2 + (k_{t_2} - k_o)^2]\right). \qquad (5.47)$$

The function Ψ_M, with $\epsilon = 5$ and rotated by $\theta = 45^o$, is shown (in level curves) in Figure 9b. Its effective support is centered at $\vec{k}_o$ and is contained in a convex cone, that becomes narrower as ϵ increases. This is indeed the archetype of a directional wavelet.

(b) Conical wavelet.
In order to achieve a directionally oriented wavelet, it suffices to consider a smooth function $\Psi_S(\vec{k})$ with support in a strictly convex cone S in spatial frequency space and behaving inside S as $P(k_{t_1}, k_{t_2})e^{-(\vec{\alpha}\cdot\vec{k})}$, with $\vec{\alpha} \in S$, or $P(k_{t_1}, k_{t_2})e^{-|k|^2}$, where $P(\cdot)$ denotes a polynomial in two variables.

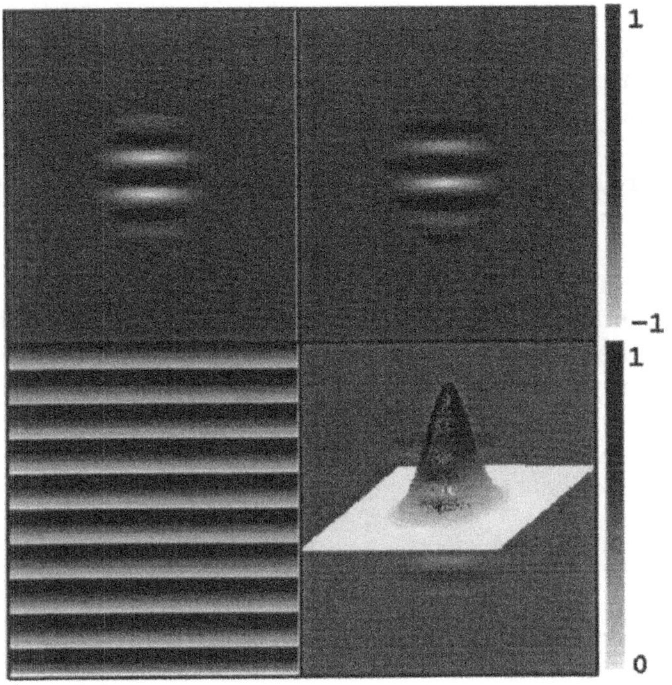

Figure 8 The Morlet wavelet ψ_M with $\vec{k}_o = (0, 5.6)$, $\epsilon = 1$ $(a = 1, \theta = 0)$: (top) real and imaginary part; (bottom) phase and modulus.

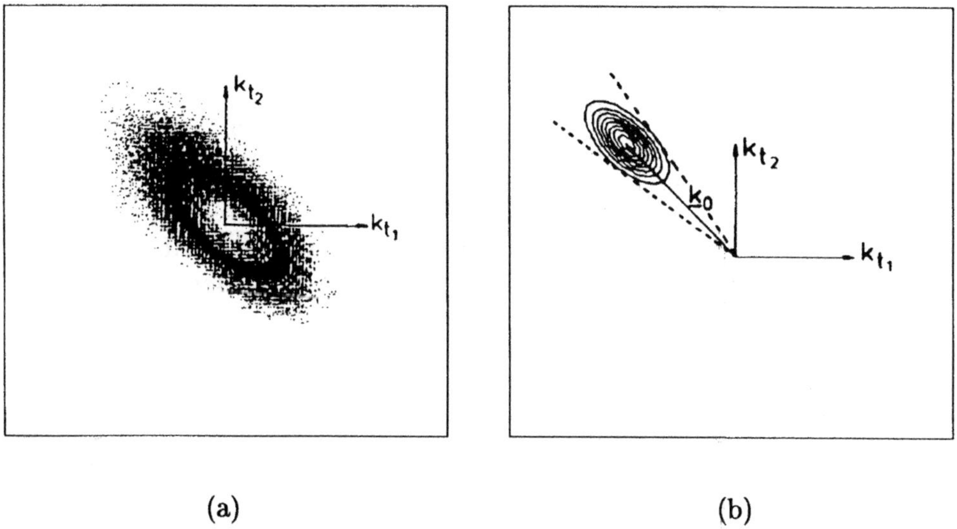

Figure 9 Frequency space realization of two basic wavelets ($\epsilon = 5, \theta = 45^o$): (a) The anisotropic mexican hat Ψ_H; (b) The Morlet wavelet Ψ_M.

Note that other wavelets, although not directional in the sense of the above definition, may have some capabilities of directional filtering. Such are, for instance, the ***gradient*** wavelets $\partial_{t_1} \exp(-|\vec{t}|^2)$ or $\partial_{t_1} \partial_{t_2} \exp(-|\vec{t}|^2)$.

6.4 Performance Evaluation of the 2-D CWT

Given a wavelet, what is the resolving power of the associated wavelet microscope? In particular what is its angular and scale selectivity? What is the minimal sampling grid for the reconstruction formula (5.42) that guarantees that no information is lost? The answer to these questions resides in a *quantitative* knowledge of the properties of the wavelet, that is, the tool must be ***calibrated.***

To that effect, one takes the WT of particular standard signals. Three such tests have proven useful, and in each case the outcome may be viewed either at fixed (a, θ) (position representation) or at fixed $\vec{b}$ (scale-angle representation).

- *Point signal:* for a snapshot at the wavelet itself, one takes as the signal a delta function, i.e. one evaluates the impulse response of the filter

$$\langle \psi_{a,\theta,\vec{b}} | \delta \rangle = a^{-1}\ \bar{\psi}(a^{-1} r_{-\theta}(-\vec{b})). \tag{5.48}$$

- *Benchmark signals:* for testing particular properties of the wavelet, such as its ability to detect a discontinuity or its angular selectivity in detecting a particular direction, one may use appropriate 'benchmark' signals.

- *Reproducing kernel:* taking as the signal the wavelet ψ itself, one obtains the reproducing kernel K, which measures the *correlation length* in each variable a, θ, $\vec{b}$,

$$c_\psi\, K(a, \theta, \vec{b} | 1, 0, \vec{0}) = \langle \psi_{a,\theta,\vec{b}} | \psi \rangle = a^{-1} \int \bar{\psi}(a^{-1} r_{-\theta}(\vec{t} - \vec{b}))\ \psi(\vec{t})\ d^2\vec{t}. \tag{5.49}$$

  A detailed analysis of K yields a definition of the resolving power of the wavelet ψ in each variable (see end of Section 6.4), and the result fits perfectly with empirical definitions [28].

Calibration of a Wavelet with Benchmark Signals

As we said above, particular properties of the wavelet may be tested on appropriate benchmark signals. For instance, its capacity at detecting a discontinuity may be measured on the signal consisting of an infinite rod (see [28] for the full discussion). The result is that both the mexican hat and the Morlet wavelet are efficient in this respect.

For testing the angular selectivity of a wavelet, one computes the WT of a segment, as a function of the difference in orientation, $\Delta\theta$, between the wavelet and the segment. The conclusion is that the Morlet wavelet is highly sensitive to orientation, but the mexican hat is not. For an eccentricity $\epsilon = 5$, ψ_M detects the orientation of a segment with a precision of the order of 5°. That is, the WT of the segment is a wall of constant height as long as $\Delta\theta$ is smaller than 5°. For $\Delta\theta$ greater, the height of the wall decreases rapidly and becomes negligible for $\Delta\theta > 15^o$. However the tips remain always visible (these are point singularities), so that, in the end, the image of the segment reduces to two peaks corresponding to the two endpoints. The same test performed with an anisotropic mexican hat gives a result almost independent of $\Delta\theta$.

Let us illustrate these features by two academic examples, both of which have direct physical applications.

(a) Detection of contours with the mexican hat.
Exactly as in the 1-D case, the WT is especially useful for detecting ***discontinuities*** in images, for instance the ***contour*** or the ***edges*** of an object [281, 28]. For that purpose, an isotropic wavelet may be chosen, such as the radial mexican hat ψ_H given in Equation (5.45). In that case the effect of the WT consists of smoothing the signal with a Gaussian and taking the Laplacian of the result. Thus large values of the amplitude will appear at the location of the discontinuities, in particular the contour of objects (which is a discontinuity in luminosity). In order to test this property, we compute the WT of a simple object, namely a set with the shape of the letter 'L', represented by its characteristic function, for various values of the scale parameter a. Then, for large values of a, the WT sees only the object as a whole, thus allowing the determination of its position in the plane. When a decreases, increasingly finer details appear. In this simple case, the WT vanishes both inside and outside the contour, since the signal is constant there, and thus only the contour remains, and it is perfectly seen at $a = 0.0625$ (Figure 10). Of course, if one takes values of a that are too small, numerical artifacts (aliasing) appear and spoil the result.

Note that the corners of the figure are highlighted in the WT by sharp peaks. The amplitude is larger at these points, since the signal is singular there in *two* directions, as opposed to the edges. In addition the WT detects the ***convexity*** of each corner. The five convex corners give rise to positive peaks, whereas the concave one yields a negative peak. Here we see the advantage of using a real wavelet and plotting the WT itself, ***not*** its modulus, as is commonly done. The same effect would appear in 1-D if one would consider, for instance, the full WT of a square pulse.

(b) Directional filtering with the Morlet wavelet.
As a consequence of its good directional selectivity, the Morlet wavelet is quite efficient for directional filtering. In order to illustrate the point, we analyze in Figure 11 a pattern made of rods in many different directions shown on the left. Applying the CWT with a fixed horizontal direction, we see it selects all those rods with roughly the same direction as shown in (b), whereas the other ones, which are misaligned, yield only a faint signal corresponding to their tips. Since this is noise, one performs a thresholding to remove it, thus getting an improved picture (c). The same two operations are then repeated with various successive orientations of the wavelet. In this way, one can count the number of objects that lie in any particular direction.

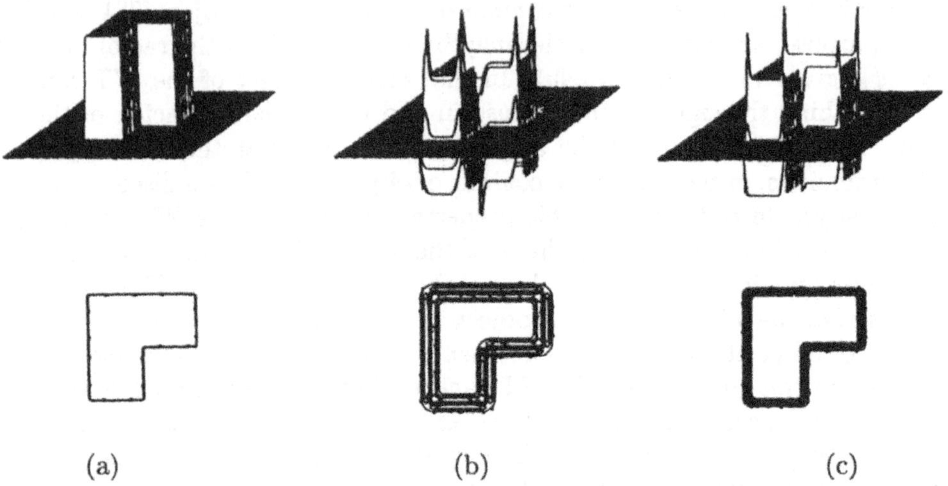

Figure 10 Detecting a contour with the mexican hat: (a) the object, in 3-D perspective (top) and in level curves (bottom); (b) the CWT at $a = 0.125$; (c) the same at $a = 0.0625$.

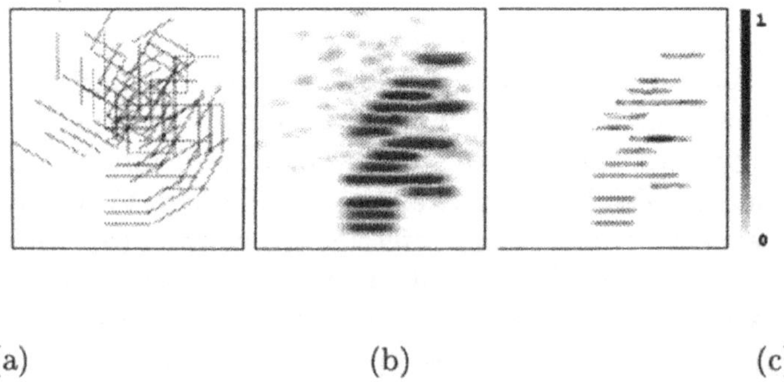

Figure 11 Directional filtering with the CWT: (a) the pattern; (b) CWT with a Morlet wavelet ($k_o = (0, 5.6), \epsilon = 5$) oriented at 0°; (c) the same after thresholding at 25%.

The Reproducing Kernel and the Resolving Power of the Wavelet

The best way of testing the correlation length of the wavelet is to analyze systematically its reproducing kernel K. First one can plot K itself in the scale-angle representation, that is, in polar coordinates (a, θ). On such a plot, one can read off directly the angular width $\Delta\theta$ of K and its scale range $\Delta a = a_{max}/a_{min}$, where $[a_{min}, a_{max}]$ is the effective support of K in the variable a. On the other hand, let the wavelet ψ have its effective support in spatial frequency in a cone of aperture $\Delta\varphi = \varphi_2 - \varphi_1$, between scales ρ_1 and ρ_2. Then an easy calculation shows that the effective support of the reproducing kernel K is given by $a_{min} = \rho_1/\rho_2$ and $a_{max} = \rho_2/\rho_1$ for the scale variable, and $-\Delta\varphi \leq \theta \leq \Delta\varphi$ for the angular variable. Thus we may define the wavelet parameters (or resolving power) in terms of those of K as: 1) scale width or scale resolving power, $\Delta\rho = \rho_2/\rho_1 = \sqrt{\Delta a}$; and 2) angular width or angular resolving power, $\Delta\varphi = \frac{1}{2}\Delta\theta$. This result may be exploited for determining the minimal sampling grid needed for the numerical evaluation of the reconstruction integral (5.42). Suppose we have a signal x that ranges in spatial frequency between ρ_m and ρ_M and we want to analyze it with the wavelet ψ just discussed. Then the extreme scales that are needed are, respectively, $a_m = \rho_m/\rho_1$ and $a_M = \rho_M/\rho_2$, where ρ_1, ρ_2 are the wavelet parameters. Thus the scale range will be

$$\frac{a_M}{a_m} = \frac{\rho_M/\rho_m}{\rho_2/\rho_1} = \frac{\Delta\rho_{\text{signal}}}{\Delta\rho_{\text{wavelet}}}. \tag{5.50}$$

Therefore, if the sampling is performed on a dyadic scale, the signal x will be completely analyzed by the family of wavelets $\{\Psi_{a_j,\theta_\ell}(\vec{k})\}$, where

$$\begin{aligned} a_j &= \frac{\rho_m}{\rho_1} \cdot 2^j, \quad j = 0, 1, \ldots, M-1, \\ \theta_\ell &= \Delta\varphi \cdot \ell, \quad \ell = 0, 1, \ldots, L-1, \end{aligned}$$

with

$$M = \text{integer part of } \log_2 \left(\frac{\Delta\rho_{\text{signal}}}{\Delta\rho_{\text{wavelet}}} \right), \quad L = \text{integer part of } \frac{2\pi}{\Delta\varphi}. \tag{5.51}$$

The resulting scheme is illustrated in Figure 12. It is gratifying that these analytical estimations fit very well with the empirical parameters determined previously [28]. We emphasize that such an analysis is only possible within the scale-angle representation. Thus it requires the use of the CWT, and it is completely outside of the scope of the discrete WT, which is essentially limited to a Cartesian geometry, since orthogonal wavelet bases are not available in polar coordinates.

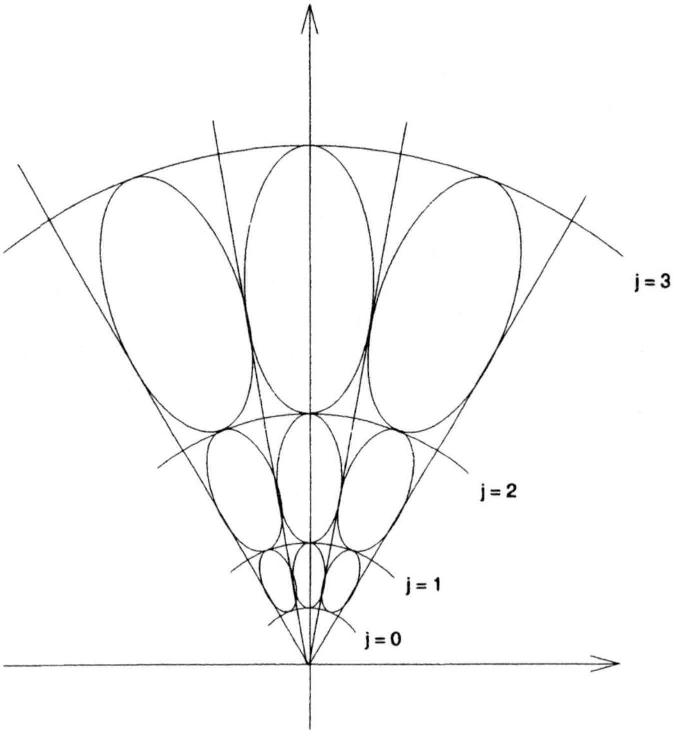

Figure 12 Tiling the spatial frequency plane with the wavelet filter bank.

6.5 Physical Applications of the 2-D CWT

The 2-D CWT has been used by a number of authors in a wide variety of physical problems [98, 264, 266]. In all cases, its main use is for the *analysis* of images. It can be used for the detection of specific features, such as a hierarchical structure, edges, filaments, contours, boundaries between areas of different luminosity, etc. Of course, the type of wavelet chosen depends on the precise aim. In particular the detection of directions requires the use of an oriented wavelet (Morlet), whereas an isotropic one (mexican hat) suffices for pointwise analysis.

It is instructive to highlight a few applications for discussion. First, in the area of analysis of astronomical images, the CWT has been used for noise filtering (background sky), and unraveling of the hierarchical structure of a galactic nebula.

In the area of analysis of 2-D fractals, the 2-D CWT has been used for the measurement of the fractal dimensions and the unraveling of universal laws (mean angle between branches, azimuthal Cantor structures, etc.). Artificial fractals such as numerical snowflakes and diffusion limited aggregates as well as natural fractals such as electrodeposition clusters and various arborescent phenomena have been thoroughly investigated with the 2-D CWT [32, 31].

In medical physics and psychophysics the CWT has been used for modeling human vision, e.g. definition of local contrast in images [132], 2-D NMR imaging and tomography, and analysis of moving patterns.

Finally, in the area of fluid dynamics the CWT has been used for 2-D analysis of turbulence in fluids, especially localization of small scales in the distribution of energy or enstrophy [142]. Other applications in fluid dynamics include the visualization and measurement of a velocity field with the help of an oriented wavelet.

Among the applications of 2-D wavelets in fluid dynamics, we would like to mention two, which both rely on the possibility of directional filtering with directional wavelets. The first one is a straightforward application of the method described in Section 6.4 [445, 444]. The aim is to measure the velocity field of a 2-D turbulent flow around an obstacle. Velocity vectors are materialized by small segments. Then the WT with a Morlet wavelet is computed twice. First the WT selects those vectors that are closely aligned with the wavelet. Then the analysis is repeated with a wavelet oriented in the orthogonal direction, thus completely misoriented with respect to the selected vectors ($\Delta\theta \simeq 90^o$). Now the WT sees only the tips of the vectors and their length may be easily measured. Using appropriate thresholdings, the complete velocity field may thus be obtained, in a totally automated fashion.

A second example concerns the disentangling of a wave train, represented by a linear superposition of damped plane waves. The problem originates from underwater acoustics. When a point source emits a sound wave above the surface of water, the wave hitting the surface splits into several components of very different characteristics (called respectively 'direct', 'lateral' and 'transient'). The goal is to measure the parameters of all components. This phenomenon has been analyzed successfully with the WT both in 1-D [354] and in 2-D [27]. In the latter case, the signal representing the underwater wave train is taken as a linear superposition of damped plane waves,

$$f(\vec{t}) = \sum_{n=1}^{N} c_n \, e^{j\vec{k}_n \cdot \vec{t}} \, e^{-\vec{l}_n \cdot \vec{t}}, \tag{5.52}$$

where, for each component, $\vec{k}_n$ is the wave vector, $\vec{l}_n$ is the damping vector, and c_n a complex amplitude. Then, using successively the scale-angle and the position representations described in Section 6.2, one is able to measure all the $6N$ parameters of this signal with surprising ease and precision. And the 3-D version is no more difficult, except for the numerical cost.

7 GOING TO HIGHER DIMENSIONS

7.1 The 3-D Case

Of course, some physical phenomena are intrinsically multiscale and three-dimensional. Typical examples may be found in fluid dynamics, for instance the appearance of coherent structures in turbulent flows, or the disentangling of a wave train in (mostly underwater) acoustics, as discussed above. In such cases, a 3-D wavelet analysis is clearly more adequate and likely to yield a deeper understanding [35]. Hence we will also describe briefly the 3-D CWT.

We consider 3-D signals of finite energy, represented by complex-valued, square integrable functions on $\mathbb{R}^3$, $x \in L^2(\mathbb{R}^3, d^3\vec{t})$. We may consider also signals with infinite energy, represented by a generalized function (such as a δ function or a plane wave), or by a fractal measure.

Given a finite energy signal $x \in L^2(\mathbb{R}^3)$, one may act on it by translation, dilation and rotation,

$$x_{a,r,\vec{b}}(\vec{t}) = [U(a, r(\gamma), \vec{b})x](\vec{t}) = a^{-\frac{3}{2}} x(a^{-1} r(\gamma)^{-1}(\vec{t} - \vec{b})), \qquad (5.53)$$

where $a > 0$, $\gamma \in SO(3)$, $\vec{b} \in \mathbb{R}^3$ and $r(\gamma) \in SO(3)$ is a 3×3 rotation matrix. The element $\gamma \in SO(3)$ may be parametrized, for instance, in terms of three Euler angles. These three operations generate the 3-D Euclidean group with dilations, that is, the similitude group of $\mathbb{R}^3$, $SIM(3) = (\mathbb{R}_*^+ \times SO(3)) \wedge \mathbb{R}^3$. Then (5.53) is a unitary representation of $SIM(3)$ in $L^2(\mathbb{R}^3, d^3\vec{t})$, which is irreducible and square integrable. Hence it generates a CWT exactly as before.

Wavelets are taken in $L^2(\mathbb{R}^3)$ and the admissibility condition is now

$$(2\pi)^{-4} c'_\psi \equiv \int |\Psi(\vec{k})|^2 \frac{d^3\vec{k}}{|\vec{k}|^3} < \infty. \qquad (5.54)$$

Also the two familiar wavelets have a 3-D realization.

- *The 3-D mexican hat* is given by

$$\psi(\vec{t}) = (3 - |A\vec{t}|^2) e^{-\frac{1}{2}|A\vec{t}|^2}, \qquad (5.55)$$

where $A = \mathrm{diag}[\epsilon_1{}^{-1/2}, \epsilon_2{}^{-1/2}, 1]$, $\epsilon_1 \geq 1, \epsilon_2 \geq 1$, is a 3×3 anisotropy matrix. We identify three cases:
1) If $\epsilon_1 \neq \epsilon_2 \neq 1$, one has the fully anisotropic 3-D mexican hat;
2) If $\epsilon_1 = \epsilon_2 = 1$, one has the isotropic, $SO(3)$-invariant, 3-D mexican hat;
3) If $\epsilon_1 = \epsilon_2 \equiv \epsilon \neq 1$, the wavelet is *axisymmetric*, i.e., $SO(2)$-invariant, but not isotropic.

- *The 3-D Morlet wavelet* is given by

$$\psi(\vec{t}) = e^{j\vec{k}_o\vec{t}} e^{-\frac{1}{2}|A\vec{t}|^2} + corr, \qquad (5.56)$$

where A is the same 3×3 anisotropy matrix as in the first example. Here again, for $\epsilon_1 = \epsilon_2 \equiv \epsilon \neq 1$, the wavelet ψ is invariant under $SO(2)$.

Then given a signal $x \in L^2(\mathbb{R}^3)$, its CWT with respect to the admissible wavelet ψ (with $c'_\psi = 1$ as before) is given as

$$W_x(a, \gamma, \vec{b}) = a^{-3/2} \int \bar{\psi}(a^{-1} r(\gamma)^{-1}(\vec{t} - \vec{b}))\, x(\vec{t})\, d^3\vec{t}. \qquad (5.57)$$

As compared with Equation (5.37), the only differences are in the normalization factors and the rotation matrices. Since the structure of the formulas is the same as before, so is the interpretation and the consequences (local filtering, reproducing kernel, reconstruction formula, etc.). Thus the CWT (5.57) may be interpreted as a mathematical *camera* with *magnification* $1/a$, *position* $\vec{b}$ and *directional selectivity* given, in the axisymmetric case, by the rotation parameters $\varpi \equiv (\theta, \varphi)$. As for the visualization, the full CWT $W_x(a, \gamma, \vec{b})$ is a function of seven variables. However, if the wavelet ψ is chosen to be *axisymmetric*, i.e. $SO(2)$-invariant, W_x depends on six variables only, $a > 0$, $\varpi \in S^2 \simeq SO(3)/SO(2)$, the unit sphere in $\mathbb{R}^3$, and $\vec{b} \in \mathbb{R}^3$. In this case again, (a^{-1}, ϖ) may be interpreted as polar coordinates in spatial frequency space. This is in fact true in any number of dimensions. It follows that, here too, there are two natural representations for the visualization of the WT, the position representation (a, ϖ fixed) and the scale-orientation (or spatial frequency) representation ($\vec{b}$ fixed).

In conclusion, let us discuss briefly two simple examples, the analysis of a cubic body and the detection of 3-D objects in a cluttered medium. These are the 3-D analogs of those described in [28], and the results are similar.

We consider first a cubic body centered at the origin, and we analyze it with the isotropic mexican hat. The wavelet transform $W_x(a, \vec{b})$ is shown in Figure 13, for $a = 1, 0.5, 0.25$ and 0.125. For large a, the WT sees only the cubic body as a whole, thus allowing us to determine its position in space. When a decreases, finer details appear. For very small a, only the contour remains.

Let us now consider a scene with 3-D objects (targets) immersed in a cluttered medium, modeled by the signal

$$x(\vec{t}) = \sum_{m=1}^{N} s_m(\vec{t}) + n(\vec{t}), \tag{5.58}$$

where $s_m(\vec{t})$ denotes the density of the target m, and $n(\vec{t})$ the density of the medium.

Since the density of the targets is very different from that of the medium, there will be a high density gradient at the boundary between the objects and the medium. In this situation, the wavelet transform $W_x(a, \theta, \varphi, \vec{b})$ may be used to extract the 3-D objects and determine their characteristics, position (range and orientation), and spatial frequency.

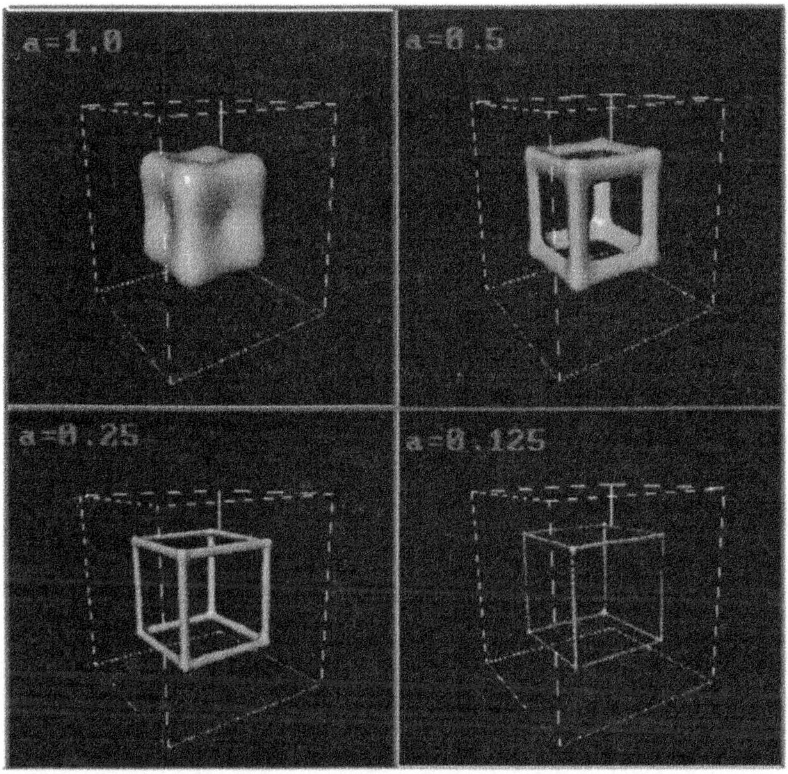

Figure 13 Wavelet transform of a cubic body, for $a = 1, 0.5, 0.25, 0.125$.

7.2 Spatio-Temporal Wavelet Transform: A Tool for Motion Tracking

An important aspect of image processing is the analysis of time-dependent or moving images, e.g. in television, and the CWT may be extended to this case too [133]. Let $x(t_1, t_2)$ be a signal in one space and one time dimension, with t_1 denoting the space variable and t_2 the time variable. On such a signal, the natural operations are space-time translations (b, τ), global dilation (a) and speed tuning (c):

$$x_{a,c,\tau,b}(t_1, t_2) = \frac{1}{a}\, x\left(\frac{1}{ac^{1/2}}(t_1 - b), \frac{c^{1/2}}{a}(t_2 - \tau)\right). \qquad (5.59)$$

The extension to 2 or 3 space dimensions is straightforward. This formulation, which is based on the physiological characteristics of motion perception, allows a group-theoretical analysis following exactly the same steps as before. It yields also a clear-cut physical interpretation. The new parameter c adjusts the speed of the wavelet to that of the target. Thus the spatio-temporal CWT is a tool for motion tracking. Clearly there are plenty of applications in which such a technique might be used.

8 OUTCOME

As we have seen in Section 4, the discretization of the CWT leads to the theory of frames. For many practical purposes in signal processing, a tight frame is almost as good as an orthonormal basis. And if one stays with the standard wavelets, as we have done here, there is no alternative, since these wavelets do not generate any orthonormal basis (like the usual coherent states [232]). However it turns out that there exist wavelets that *do* generate orthonormal bases, and this leads to the *discrete WT* (DWT). The key step was the discovery that almost all examples of orthonormal bases of wavelets can be associated with a multiresolution analysis, and furthermore that the whole construction may be transcripted into the language of QMF filters, i.e. subband filter banks. This development then led to many generalizations, such as biorthogonal wavelet bases, wavelet packets, etc. (see [265] for a nice survey). The same is true in 2-D as well as in 1-D. However, we will refrain from any further discussion of the DWT, since the more general case of filter banks is treated extensively in the first four chapters.

Yet we want to emphasize that the CWT and the DWT have almost opposite properties, hence their ranges of application differ widely too. The CWT is very efficient at detecting specific features in signals or images, such as in pattern recognition or directional filtering. On the other hand, the DWT and its generalizations are extremely fast and economical. Both are powerful tools, and very flexible ones, thanks to their adaptive character. And both have become a significant element in the standard toolbox of signal processing, which finds its way into an increasing number of reference books and software packages.

Clearly wavelets are here to stay, and they have a bright future. Of course they don't solve every difficulty, and they must be continually developed and enriched, as has been the case over the last few years. In particular, one should expect a proliferation of specialized wavelets, each dedicated to a particular type of problem, and an increasingly diverse spectrum of physical applications. This trend is only natural. It follows from the very structure of the wavelet transform. By comparison, the conventional STFT is much more rigid, since the shape of the window is the only feature that can be varied.

6

TIME-FREQUENCY LOCALIZATION WITH NON-STATIONARY WAVELET PACKETS

Albert Cohen and Eric Séré

CEREMADE-Université Paris IX

1 INTRODUCTION

Wavelet packets constitute a useful tool for the decomposition of complicated functions into a small number of elementary waveforms that are localized both in time and frequency. They have been introduced by R.Coifman, Y.Meyer and V.Wickerhauser. We recall here the main ideas of their construction.

One starts with the data of a multiresolution analysis, i.e. a nested sequence of approximation subspaces

$$\{0\} \to \dots V_{-2} \subset V_{-1} \subset V_0 \subset V_1 \subset V_2 \dots \to L^2(\mathbb{R}) \tag{6.1}$$

generated by a scaling function $\varphi \in V_0$ in the sense that for each $j \in \mathbb{Z}$ the family $\{\varphi_{j,k}\}_{k\in\mathbb{Z}} = \{2^{j/2}\varphi(2^j x - k)\}_{k\in\mathbb{Z}}$ is an orthonormal basis of V_j.

Such a function can either be expressed as the solution of a dilation equation

$$\varphi(x) = 2\sum_{n\in\mathbb{Z}} h_n \varphi(2x - n), \tag{6.2}$$

or as the limit of a subdivision scheme, i.e. iterative refinements of an initial Dirac sequence by convolutions with the coefficients h_n, since we have

$$\hat{\varphi}(\omega) = \prod_{k=1}^{+\infty} m_0(2^{-k}\omega), \tag{6.3}$$

with $m_0(\omega) = \sum_n h_n e^{-in\omega}$ (for this reason the function φ is said to be "refinable" and (6.2) is also called refinement equation).

The sequence h_n and its Fourier series $m_0(\omega)$ need to satisfy specific constraints that are related to the properties of the function φ:

- If $\varphi \in L^1$, then $\sum_n h_n = m_0(0) = 1$.
- φ is compactly supported in $[p, q]$ if and only if $h_n = 0$ for $n < p$ or $n > q$.
- When the two previous constraints are satisfied, the orthonormality of $\{\varphi(x-k)\}_{k\in\mathbb{Z}}$ is ensured if and only if

$$|m_0(\omega)|^2 + |m_0(\omega+\pi)|^2 = 1 \tag{6.4}$$

  and there exists a compact set K, congruent to $[-\pi, \pi]$ modulo 2π, such that $m_0(2^{-j}\omega) \neq 0$ for all $\omega \in K$ and $j > 0$ (see [93]).

- Additional constraints can be put on $m_0(\omega)$ so that $\varphi(x)$ has some regularity. In particular, a family of trigonometric polynomials given by

$$m_{0,L}(\omega) = \Big(\frac{1+e^{i\omega}}{2}\Big)^L p_L(\omega), \quad L \in \mathbb{N}^*, \tag{6.5}$$

  with

$$|p_L(\omega)|^2 = \sum_{j=0}^{L-1} \binom{L-1+j}{j} \sin^{2j}(\omega/2), \tag{6.6}$$

  for all $L \leq 1$, has been constructed by Daubechies [113] in such a way that the smoothness of the associated scaling functions φ_L grows linearly with L, in the asymptotic sense. Numerous contributions have been made on the more general problem of evaluating the smoothness of the limit function generated by a subdivision scheme: Dyn and Levin [136], Micchelli and Prautzsch [267], Daubechies and Lagarias [117, 118], Rioul [340] and Villemoes [434] among others.

From this framework, one derives an orthonormal wavelet basis by defining

$$\psi(x) = 2\sum_n g_n \varphi(2x-n), \tag{6.7}$$

with $g_n = (-1)^n h_{1-n}$. The family $\{\psi_{j,k}\}_{k\in\mathbb{Z}}$ is then an orthonormal basis of the orthogonal complement W_j of V_j into V_{j+1}. It characterizes the missing details between two successive levels of approximation. It follows that both $\{\varphi_{J,k}\}_{k\in\mathbb{Z}} \cup \{\psi_{j,k}\}_{j\geq J, k\in\mathbb{Z}}$ are orthonormal bases of $L^2(\mathbb{R})$.

The decomposition of V_1 into V_0 and W_0, as expressed by Equations (6.2) and (6.7), reveals a more general splitting trick. If $\{e_n\}_{n\in\mathbb{Z}}$ is an orthonormal basis of a Hilbert space H, and if h_n and g_n are coefficients that satisfy the previous constraints, then the sequences $u_n = \sqrt{2}\sum_k h_{k-2n}e_k$ and $v_n = \sqrt{2}\sum_k g_{k-2n}e_k$ are orthonormal bases of two orthogonal closed subspaces U and V such that $H = U \oplus V$.

Wavelet packets are obtained by using this trick to split the W_j spaces. More precisely, one can define a family $\{w_n\}_{n\in\mathbb{N}}$ by taking $w_0 = \varphi$, $w_1 = \psi$ and applying the following recursion

$$w_{2n} = 2\sum_k h_k w_n(2x-k) \text{ and } w_{2n+1} = 2\sum_k g_k w_n(2x-k). \qquad (6.8)$$

The families $\{w_n(x-m)\}_{m\in\mathbb{Z}} = \{w_{m,n}\}$ for $2^j \leq n < 2^{j+1}$ are the results of j splittings of the space W_j. Consequently, $\{w_{m,n}\}_{n\in\mathbb{N},m\in\mathbb{Z}}$ is an orthonormal basis of $L^2(\mathbb{R})$.

This is only a particular example of a wavelet packet basis. Many other bases can be obtained by splitting W_j more or less than j times. However, all the elements of these bases have the general form

$$w_{j,m,n}(x) = 2^{j/2}w_n(2^j x - m). \qquad (6.9)$$

This construction yields fast algorithms for the computation of the wavelet packet coefficients of any sampled function or discrete signal. One assumes that the initial data $\{s_k\}_{k\in\mathbb{Z}}$ represents a function in V_j, i.e., $f = \sum_k s_k w_{j,k,0}$. The splitting trick indicates that the coordinates of f in the basis $\{w_{m,n}\}_{m\in\mathbb{Z},0\leq n<2^j}$ can be obtained by j iterative convolutions of s_k with the discrete filters h_n and g_n, followed by a decimation of one sample out of two. Again, one has the possibility of splitting more or less than j times and using entropy tests to select the sparsest decomposition (see [97]). The computational cost is of order $CN\log N$, where N is the size of the initial data and C is the length of the filters h_n and g_n.

The wavelet packets have a simple expression in the frequency domain. If we define $m_1(\omega) = \sum_n g_n e^{-in\omega} = e^{-i\omega}\overline{m_0(\omega+\pi)}$, we can derive from (6.3) and (6.8), for all $n < 2^j$,

$$\hat{w}_n(\omega) = [\prod_{k=1}^{n} m_{\epsilon_k}(2^{-k}\omega)]\hat{w}_0(2^{-j}\omega) \qquad (6.10)$$

with $n = \sum_{k=1}^{j} \varepsilon_k 2^{k-1}$. Two observations can be made from this formula.

- If $m_0(\omega)$ is a trigonometric polynomial of degree L, then all the functions w_n have a compact support of length L.
- In the case where $m_0(\omega) = \chi_{[-\pi/2,\pi/2]}(\omega)$, one has $\hat{w}_n(\omega) = \chi_{-J_n \cup J_n}(\omega)$ where the interval $J_n = [\pi g(n), \pi(g(n)+1)]$ is determined by the Gray code $g(n) = \sum_{k=1}^{j} \gamma_k 2^{k-1}$, $\gamma_j = \varepsilon_j$ and $\gamma_k = |\gamma_{k+1} - \varepsilon_k|$ for $1 \leq k < j$.

The last observation suggests that if $m_0(\omega)$ is a trigonometric polynomial close to the ideal cut-off filter $\chi_{[-\pi/2,\pi/2]}(\omega)$, the function $w_{m,n}$ is localized both in space (around m) and in frequency (around $\pm\pi g(n)$), with a variance that does not depend on m and n in both domains. Unfortunately, this is not true. Several recent works (e.g. [96, 364]) have stated negative results concerning the frequency localization. For example, it was proved by Coifman, Meyer and Wickerhauser [96] that the variances

$$\sigma_n = \min_{\omega_0 \geq 0} \int_0^{+\infty} |\hat{w}_n(\omega)|^2 |\omega - \omega_0|^2 d\omega \tag{6.11}$$

are not bounded on average, in the sense that $2^{-j} \sum_{n=0}^{2^j-1} \sigma_n$ goes to $+\infty$ as j increases. This type of result reveals the limitations of wavelet packet analysis in terms of time-frequency localization.

On the other hand, we remark that it is not necessary to use constantly the same pair of filters in the different splitting stages, since each of them is performed independently. The goal of this chapter is to show how this freedom can be used to recover some positive results on time-frequency localization.

We are encouraged in this direction by the properties of non-stationary subdivision schemes that were at first studied by Dyn, Derfel and Levin [135]. In such schemes, one uses a different filter at each scale, so that the limit function has the more general expression

$$\hat{\varphi}(\omega) = \prod_{k=1}^{+\infty} m_0^k(2^{-k}\omega). \tag{6.12}$$

In the particular case where the trigonometric polynomials $m_0^k(\omega)$ are those of Daubechies construction (given by (6.5) and (6.6)), φ is a compactly supported

C^∞ function that was studied in detail in [94]. In that sense, the linear growth of the filter length $2k$ ensures a better frequency localization ($\hat{\varphi}$ has rapid decay at infinity) without ruining the space localization since this growth is beaten by the exponential decay of the corresponding scale 2^{-k}. This idea is the starting point of our work.

It is interesting to point out that the concept of using different filters at different levels in a subband tree was a common practice in the signal processing community for many years. Most of this work was motivated by subband coding applications. Subband trees were often designed with constituent filters that attempted to optimize performance [41, 439, 395]. In this chapter, we approach the problem from a wavelet perspective.

This chapter is organized as follows. Section 2 establishes the general setting of non-stationary wavelet packets and their basic properties. Two different types of results on frequency localization are then presented. Sections 3 and 4 discuss the L^1 norms and the variances of wavelet packets in the frequency domain, respectively. In both cases, we show that non-stationary wavelet packets outperform standard wavelet packets.

2 NON - STATIONARY WAVELET PACKETS

Let $\{m_0^k(\omega)\}_{k>0}$ be a family of trigonometric polynomials of degree $d(k)$ that satisfy the following properties:

$$|m_0^k(\omega)|^2 + |m_0^k(\omega+\pi)|^2 = 1, \tag{6.13}$$

$$m_0^k(0) = 1, \tag{6.14}$$

$$|\omega| \leq \pi/2 \Rightarrow m_0^k(\omega) \neq 0, \tag{6.15}$$

and

$$\sum_{k>0} 2^{-k} d(k) < +\infty. \tag{6.16}$$

(Condition (6.15) could be replaced by a weaker one: the existence of a compact set K congruous to $[-\pi,\pi]$ modulo 2π such that $m_0^k(2^{-l}\omega) \neq 0$ for all $k, l \geq 1$ and $\omega \in K$. See [91] for more details.) With these hypotheses the following results hold:

Theorem 2.1 For all $j \leq 0$, the tempered distributions φ^j defined by

$$\hat{\varphi}^j(\omega) = \prod_{k=1}^{+\infty} m_0^{k+j}(2^{-k}\omega) \tag{6.17}$$

are compactly supported L^2 functions. The families $\{\varphi^j(x-k)\}_{k\in\mathbb{Z}}$ are orthonormal. Moreover, the spaces V_j generated by $\{2^{j/2}\varphi^j(2^j x - k)\}_{k\in\mathbb{Z}}$ constitute a half multiresolution analysis in the sense that $V_j \subset V_{j+1}$ and the projection $P_j(f)$ of any L^2 function f onto V_j goes to f in L^2 as j goes to $+\infty$.

Theorem 2.2 If $\{m_0^k(\omega)\} = \{m_{0,L(k)}(\omega)\}$ is a subset of the family given by (6.5) and (6.6), and if $\lim_{k\to+\infty} L(k) = +\infty$, then the functions φ^j are smooth, i.e. contained in C^∞.

The proofs of these results are straightforward consequences of the results in [94], which are not the central issue in this chapter. We shall thus only sketch the main ideas here.

The proof of Theorem 2.1 goes in four steps:

- Using the Bernstein inequality, one remarks that for a fixed ω, the sequence

  $$s_k = |m_0^{k+j}(2^{-k}\omega) - 1| \leq Cd(k)2^{-k}|\omega| \tag{6.18}$$

  is absolutely summable and thus the infinite product (6.17) converges pointwise and in the sense of tempered distributions. By (6.16), φ^j is a tempered distribution with compact support.

- As in the standard construction of wavelets (see [93]), one defines the approximants φ_n^j of φ^j by

  $$\hat{\varphi}_n^j(\omega) = \prod_{k=1}^{n} m_0^{k+j}(2^{-k}\omega)\chi_{[-2^n\pi,2^n\pi]}(\omega) \tag{6.19}$$

  and checks by recursion, using (6.13), that $\{\varphi_n^j(x-k)\}_{k\in\mathbb{Z}}$ is an orthonormal system. By Fatou's lemma, it follows that φ^j is in $L^2(\mathbb{R})$.

- By (6.15), we have

  $$|\omega| \leq \pi \Rightarrow |\hat{\varphi}^j(\omega)| > C > 0 \tag{6.20}$$

  and $|\hat{\varphi}_n^j(\omega)| \leq C^{-1}|\hat{\varphi}^j(\omega)|$. By dominated convergence, it follows that $\{\varphi^j(x-k)\}_{k\in\mathbb{Z}}$ is an orthonormal system.

- By the definition of φ^j, it is clear that $V_j \subset V_{j+1}$. To check that $P_j(f)$ goes to f in L^2 for any f, it suffices to prove that $\|P_j(\chi_I)\|$ goes to $|I|$ for any finite interval I. This follows from $\int \varphi^j = 1$ for all j and from the decay of the support length of $\varphi^j(2^j x)$ to zero as j increases.

The proof of Theorem 2.2 is based on an important estimate on the trigonometric polynomials given by (6.5) and (6.6) that we shall also use in the next sections:

Lemma 2.3 The trigonometric polynomials constructed by Daubechies satisfy

$$|m_{0,L}(\omega)| \leq [m(\omega)]^{L-1}, \tag{6.21}$$

with $m(\omega) = 1$ if $|\omega| \leq \pi/2$ and $m(\omega) = |\sin(\omega)|$ if $\pi/2 \leq |\omega| \leq \pi$.

Proof This is obvious for $|\omega| \leq \pi/2$. For $\pi/2 \leq |\omega| \leq \pi$, we have $2\sin^2(\omega/2) \geq 1$ and consequently

$$\begin{aligned}
|p_L(\omega)|^2 &= \textstyle\sum_{j=0}^{L-1} \binom{L-1+j}{j} \sin^{2j}(\omega/2) \\
&= \textstyle\sum_{j=0}^{L-1} \binom{L-1+j}{j} [2\sin^2(\omega/2)]^j 2^{-j} \\
&\leq [2\sin^2(\omega/2)]^{L-1} \textstyle\sum_{j=0}^{L-1} \binom{L-1+j}{j} 2^{-j} \\
&= [2\sin^2(\omega/2)]^{L-1} |p_L(\pi/2)|^2 = [4\sin(\omega/2)]^{L-1}.
\end{aligned}$$

We have $|p_L(\pi/2)|^2 = 2^{L-1}$ by (6.13). It follows that

$$\begin{aligned}
|m_{0,L}(\omega)| &= \left(\cos^{2L}(\omega/2)|p_L(\omega)|^2\right)^{1/2} \\
&\leq \left([4\cos^2(\omega/2)\sin^2(\omega/2)]^{L-1}\right)^{1/2} = |\sin(\omega)|^{L-1}.
\end{aligned}$$

□

The proof of Theorem 2.2 goes then in two steps:

- By using a technique introduced in [113], one checks that the function $P(\omega) = \prod_{k=1}^{+\infty} m(2^{-k}\omega)$ satisfies

$$|P(\omega)| \leq C(1 + |\omega|)^{-s}, \tag{6.22}$$

with $s > 0$. It can be shown that the largest value for s is $1 - \frac{\log 3}{2\log 2}$.

- For a fixed j and for any $A > 0$, it is possible to find $j_A > j$ such that, for all $k > j_A$, $s(L(k)-1) \geq A$. It follows that φ^{j_A} satisfies

$$\begin{aligned}|\hat{\varphi}^{j_A}(\omega)| &= \prod_{k=1}^{+\infty} |m_{0,L(k+j_A)}(2^{-k}\omega)| \\ &\leq \prod_{k=1}^{+\infty} |m(2^{-k}\omega)|^{L(k+j_A)-1} \\ &\leq \prod_{k=1}^{+\infty} |m(2^{-k}\omega)|^{A/s} \\ &\leq |P(\omega)|^{A/s} \\ &\leq C_A(1+|\omega|)^{-A}.\end{aligned}$$

Since $\hat{\varphi}^j(\omega) = \hat{\varphi}^{j_A}(2^{j-j_A}\omega)\prod_{k=1}^{j_A-j} m_{0,L(k+j)}(2^{-k}\omega)$, it follows that we have

$$|\hat{\varphi}^j(\omega)| \leq C'_A(1+|\omega|)^{-A}, \tag{6.23}$$

and since this holds for any $A > 0$, it follows that φ^j is a C^∞ function.

These results allow us to construct our non-stationary wavelet packets with the following definition:

Definition 2.4 Let $\{m_0^k\}_{k>0}$ be a family of positive trigonometric polynomials that satisfy the set of conditions (6.13)-(6.16) and define $m_1^k(\omega) = e^{-i\omega}\overline{m_0^k(\omega+\pi)}$. The non-stationary wavelet packets associated to $\{m_0^k\}_{k>0}$ are a sequence $\{w_n\}_{n\geq 0}$ of tempered distributions given by

$$\hat{w}_n(\omega) = \prod_{k=1}^{+\infty} m_{\varepsilon_k}^k(2^{-k}\omega), \tag{6.24}$$

with $n = \sum_{k>0} 2^{k-1}\varepsilon_k$, $\varepsilon_k \in \{0,1\}$.

From Theorems 2.1 and 2.2, we can easily derive the following proposition.

Proposition 2.5 The distributions $\{w_n\}_{n\geq 0}$ are compactly supported L^2 functions and the length of their support is independent of n.

The family $w_{n,m}(x) = w_n(x-m)$, $n \geq 0$, $m \in \mathbb{Z}$, constitutes an orthonormal basis of $L^2(\mathbb{R})$.

If $\{m_0^k(\omega)\} = \{m_{0,L(k)}(\omega)\}$ is a subset of the family given by (6.5) and (6.6) and if $\lim_{k\to+\infty} L(k) = +\infty$, then the functions w_n are infinitely smooth, i.e. contained in C^∞.

Proof From the definition of w_n, we see that for $0 \leq n < 2^j$, the families $\{w_n(x-m)\}_{m\in\mathbb{Z}}$ are the result of j splittings of the space V_j.

It follows that these functions are finite combinations of the functions $\varphi^j(2^j x - k)$. As a consequence, they are in L^2. They are also in C^∞ as soon as the hypotheses of Theorem 2.2 are satisfied.

From the denseness of the spaces V_j, it follows that the family $w_{n,m}(x) = w_n(x-m)$, $n \geq 0$, $m \in \mathbb{Z}$, constitutes an orthonormal basis of $L^2(\mathbb{R})$. Finally, since m_1^k has the same degree as m_0^k, the support of w_n cannot be larger than $L = \sum_{k>0} 2^{-k} d(k) < +\infty$. □

Remarks

Since the generating functions of the V_j spaces are different at every scale, one cannot use a recursion formula such as (6.8) to derive the non-stationary wavelet packets. However, fast algorithms can still be used to compute the non-stationary wavelet packet coefficients. The principle is exactly the same as in the standard case except that one uses different filters at every stage of the decomposition. Note that in the case where the degree $d(k)$ of m_0^k is increasing as k goes to $+\infty$, the decomposition starts at the finest scale with a long filter and ends at a coarser scale with a smaller filter. An opposite approach was used by Hess-Nielsen [185] to study the frequency spreading of wavelet packets. The choice of using longer filters for finer scales offers three important advantages:

- Since the functions φ^j are in C^∞, the reconstruction from the low frequency components will have a smooth aspect.
- In the algorithm, one needs to deal properly with the boundaries of the signal. Specific methods have been developed by Cohen, Daubechies and Vial [93] to adapt the filtering process near the edges. These methods require in particular that the size of the filter is smaller than the size of the signal. This can only be achieved by using smaller filters in the coarse scales.
- Finally, these techniques can be used to combine time and frequency localization properties, as we shall now investigate.

3 FREQUENCY LOCALIZATION AND L^1 FOURIER ESTIMATES

Estimates on the L^1 norms of the Fourier transforms of the wavelet packets were used by Coifman, Meyer and Wickerhauser in [96] to show the following result:

Theorem 3.1 The variances

$$\sigma_n = \min_{\omega_0} \int_0^{+\infty} |\hat{w}_n(\omega)|^2 |\omega - \omega_0|^2 d\omega \tag{6.25}$$

of the standard wavelet packets are not bounded in average, in the sense that $2^{-j} \sum_{n=0}^{2^j-1} \sigma_n$ goes to $+\infty$ as j increases.

Proof First one remarks that for any $n \geq 0$ and $\omega_0 \in \mathbb{R}$, one has

$$\begin{aligned} \|\hat{w}_n\|_1 &= 2\int_0^{+\infty} |\hat{w}_n(\omega)| d\omega \\ &\leq 2\left[\int_0^{+\infty} \frac{d\omega}{1+|\omega-\omega_0|^2} \int_0^{+\infty} (1+|\omega-\omega_0|^2)|\hat{w}_n(\omega)|^2 d\omega\right]^{1/2} \\ &\leq C(1+\int_0^{+\infty} |\hat{w}_n(\omega)|^2|\omega-\omega_0|^2)^{1/2} d\omega, \end{aligned}$$

where we assume that the functions w_n are real valued so that their Fourier transforms $\hat{w}_n$ are even functions. This leads to the inequality

$$\|\hat{w}_n\|_1 \leq C(1+\sigma_n)^{1/2}, \tag{6.26}$$

where C is independent of n. It is thus sufficient to show that as j increases, $2^{-j} \sum_{n=0}^{2^j-1} \|\hat{w}_n\|_1$ goes to $+\infty$.

For the sake of simplicity, we assume here that $m_0(\omega)$ does not vanish on $[-\pi/2, \pi/2]$, so that $\hat{\varphi}$ does not vanish on $[-\pi, \pi]$. It follows that we have, if $0 \leq n \leq 2^j - 1$,

$$D \int_{-2^j\pi}^{2^j\pi} |\prod_{k=1}^{j} m_{\varepsilon_k}(2^{-k}\omega)| d\omega \leq \|\hat{w}_n\|_1, \tag{6.27}$$

with $D = \inf_{[-\pi,\pi]} |\hat{\varphi}(\omega)|$ and $\sum 2^{k-1}\varepsilon_k = n$. Summing on all $n \leq 2^j - 1$, we obtain,

$$2^{-j} \sum_{n=0}^{2^j-1} \|\hat{w}_n\|_1 \geq D 2^{-j} \int_{-2^j\pi}^{2^j\pi} \prod_{k=1}^{j} p(2^{-k}\omega) d\omega, \tag{6.28}$$

with $p(\omega) = |m_0(\omega)| + |m_1(\omega)|$. Using the concavity of the logarithm, we get

$$\begin{aligned}\log(2^{-j}\sum_{n=0}^{2^j-1}\|\hat{w}_n\|_1) &\geq \log\big(D\int_{-\pi}^{\pi}\prod_{k=0}^{j-1}p(2^k\omega)d\omega\big)\\ &\geq \int_{-\pi}^{\pi}\log\big(\prod_{k=0}^{j-1}p(2^k\omega)\big)d\omega + \log D\\ &= \sum_{k=0}^{j-1}\int_{-\pi}^{\pi}\log(p(2^k\omega))d\omega + \log D\\ &= j\int_{-\pi}^{\pi}\log(p(\omega))d\omega + \log D.\end{aligned}$$

But (6.4) implies $p(\omega) \geq 1$ and since m_0 is a smooth function, $p(\omega) > 1$ on a set of nonzero measure. It follows that $\int_{-\pi}^{\pi}\log(p(\omega))d\omega > 0$ and thus the L^1 norms of the wavelet packets in the Fourier domain diverge on average, as well as their variances, according to (6.26). □

Let us now examine the behavior of the L^1 norms in the case of non-stationary wavelet packets. We assume here that the hypotheses (6.13)-(6.16) are satisfied and that $\{m_0^k(\omega)\} = \{m_{0,L(k)}(\omega)\}$ is a subset of the family given by (6.5) and (6.6), and $\lim_{k\to+\infty} L(k) = +\infty$ so that proposition 2.5 can be applied. We shall now prove that the L^1 norms of the non-stationary wavelet packets can be bounded on average if we impose the condition that $L(k)$ grows sufficiently fast.

Theorem 3.2 Suppose that $L(k) \geq Ck^{3+r}$ for some $r > 0$. Then, there exists a constant B such that, for all $j > 0$,

$$2^{-j}\sum_{n=0}^{2^j-1}\|\hat{w}_n\|_1 \leq B. \qquad (6.29)$$

Proof First we remark that for $0 \leq n \leq 2^j - 1$, we have

$$\|\hat{w}_n\|_1 \leq C_j\int_{-2^j\pi}^{2^j\pi}|\prod_{k=1}^{j}m_{\varepsilon_k}^k(2^{-k}\omega)|d\omega, \qquad (6.30)$$

with $C_j = \sup_\omega \sum_{l\in\mathbb{Z}}|\hat{\varphi}^j(\omega+2l\pi)|$. Since the functions $\hat{\varphi}^j$ have rapid decay at infinity, all the C_j are finite. Furthermore, they are uniformly bounded: as in the proof of Theorem 2.2, we can find j_A such that, for all $j \geq j_A$, we have

$$\begin{aligned}|\hat{\varphi}^j(\omega)| &\leq |P(\omega)|^A\\ &\leq C(1+|\omega|)^{-\varepsilon A}.\end{aligned}$$

By choosing A such that $\varepsilon A > 1/2$, we see that C_j is a bounded sequence.

We can thus write

$$\begin{aligned} 2^{-j}\sum_{n=0}^{2^j-1}\|\hat{w}_n\|_1 &\leq C2^{-j}\sum_{n=0}^{2^j-1}\int_{-2^j\pi}^{2^j\pi}|\prod_{k=1}^{j} m_{\varepsilon_k}^k(2^{-k}\omega)|d\omega \\ &= C2^{-j}\int_{-2^j\pi}^{2^j\pi}\prod_{k=1}^{j} p^k(2^{-k}\omega)d\omega \\ &= C\int_{-\pi}^{\pi}\prod_{k=0}^{j-1} p^{j-k}(2^k\omega)d\omega \\ &= C\int_{-\pi}^{\pi} S_j(\omega)d\omega, \end{aligned}$$

with $S_j(\omega) = \prod_{k=0}^{j-1} p^{j-k}(2^k\omega)$ and $p^k(\omega) = |m_0^k(\omega)| + |m_1^k(\omega)|$. We shall now prove by induction that $S_j(\omega)$ is uniformly bounded in L^1. For this, we divide the integral of S_j into

$$\begin{aligned} I_j &= \int_{-\pi}^{\pi}\chi_{|\omega-\pm\pi/2|<\varepsilon_j}(\omega)S_j(\omega)d\omega + \int_{-\pi}^{\pi}\chi_{|\omega-\pm\pi/2|>\varepsilon_j}(\omega)S_j(\omega)d\omega \\ &= I_{j,1} + I_{j,2}, \end{aligned}$$

where ε_j is a sequence that we shall choose in an appropriate way. To estimate $I_{j,1}$, we remark that we have

$$\begin{aligned} I_{j,1} &\leq \max_\omega p^j(\omega)\int_{|\omega-\pm\pi/2|<\varepsilon_j}\prod_{k=1}^{j-1} p^{j-k}(2^k\omega)d\omega \\ &= \sqrt{2}\int_{|\omega-\pm\pi/2|<\varepsilon_j}\prod_{k=1}^{j-1} p^{j-k}(2^k\omega)d\omega \\ &= \sqrt{2}\int_{|\omega-\pm\pi/2|<\varepsilon_j} S_{j-1}(2\omega)d\omega. \end{aligned}$$

Let us assume that $\varepsilon_j = 2^{-n(j)}\pi$ with $1 < n(j) < j$. We shall prove that

$$\int_{|\omega-\pm\pi/2|<\varepsilon_j} S_{j-1}(2\omega)d\omega \leq \frac{2\varepsilon_j}{\pi}\int_{-\pi}^{\pi} S_{j-1}(2\omega)d\omega = \frac{2\varepsilon_j}{\pi}I_{j-1}, \quad (6.31)$$

i.e. that $S_{j-1}(2\omega)$ is smaller on average on $\{|\omega - \pm\pi/2| < \varepsilon_j\}$ than on the whole interval $[-\pi,\pi]$. For this, we note that for all $k > 0$, the functions $p^k(\omega) = |m_{L(k),0}(\omega)| + |m_{1,L(k)}(\omega)|$ are even, π-periodic and are increasing on $[0,\pi/2]$ and decreasing on $[\pi/2,\pi]$. This last statement can be easily derived from the fact that $|m_{0,L}(\omega)|^2$ is a decreasing function (it can also be expressed as $C_L\int_\omega^\pi \sin^{2L-1}(\omega)d\omega$, see [113]) and that the function $x+\sqrt{1-x^2}$ is increasing on $[0,\sqrt{2}]$ and decreasing on $[\sqrt{2},1]$. We now write

$$\begin{aligned} S_{j-1}(2\omega) &= \prod_{k=1}^{n(j)-1} p^{j-k}(2^k\omega)\prod_{k=n(j)}^{j-1} p^{j-k}(2^k\omega) \\ &= R_{j-1}(\omega)T_{j-1}(\omega). \end{aligned}$$

Let us subdivide the interval $J = [-\pi - \varepsilon_j, \pi - \varepsilon_j]$ into

$$J = \bigcup_{m=-2^{n(j)-1}}^{2^{n(j)-1}-1} [(2m-1)\varepsilon_j, (2m+1)\varepsilon_j] = \bigcup_{m=-2^{n(j)-1}}^{2^{n(j)-1}-1} J_m. \qquad (6.32)$$

We remark that the function $T_{j-1}(\omega)$ has period ε_j and that, by the properties of $p^k(\omega)$, for all $\omega \in [\pi/2 - \varepsilon_j, \pi/2 + \varepsilon_j]$ and for all $m \in \mathbb{Z}$, we have $R_{j-1}(\omega) \leq R_{j-1}(\omega + 2m\varepsilon_j)$. It follows that, if we define $m_0 = 2^{n(j)-2}$, we have for all $m \in \mathbb{Z}$,

$$\int_{J_{m_0}} S_{j-1}(2\omega) = \int_{J_{-m_0}} S_{j-1}(2\omega) \leq \int_{J_m} S_{j-1}(\omega). \qquad (6.33)$$

But since $J_{-m_0} \cup J_{m_0} = \{|\omega - \pm\pi| < \varepsilon_j\}$, we thus have

$$\int_J S_{j-1}(\omega)d\omega \geq 2^{n(j)-1} \int_{|\omega - \pm\pi| < \varepsilon_j} S_{j-1}(\omega)d\omega, \qquad (6.34)$$

which is exactly equivalent to (6.31). We thus have

$$I_{j,1} \leq \frac{2\sqrt{2}}{\pi}\varepsilon_j I_{j-1}. \qquad (6.35)$$

The estimate for $I_{j,2}$ is easier to derive, using lemma 2.3:

$$\begin{aligned} I_{j,2} &\leq \max_{|\omega - \pm\pi/2| > \varepsilon_j} p^j(\omega) \textstyle\int_{-\pi}^{\pi} S_{j-1}(2\omega)d\omega \\ &= p^j(\pi/2 + \varepsilon_j)I_{j-1} \\ &\leq (1 + |\sin(\pi/2 + \varepsilon_j)|^{L(j)})I_{j-1} \end{aligned}$$

Combining the two estimates, we find,

$$I_j \leq (1 + \frac{2\sqrt{2}}{\pi}\varepsilon_j + |\sin(\pi/2 + \varepsilon_j)|^{L(j)})I_{j-1}. \qquad (6.36)$$

Consequently, I_j will be a bounded sequence if we choose $\varepsilon_j \in \ell^1$ and $L(j)$ such that $|\sin(\pi/2 + \varepsilon_j)|^{L(j)}$ is also absolutely summable. As j goes to $+\infty$, we have

$$\begin{aligned} \log(|\sin(\pi/2 + \varepsilon_j)|^{L(j)}) &\approx L(j)\log(1 - \varepsilon_j^2/2) \\ &\approx -L(j)\varepsilon_j^2/2. \end{aligned}$$

Absolute summability of $|\sin(\pi/2 + \varepsilon_j)|^{L(j)}$ will be ensured in particular if $L(j)\varepsilon_j^2/2$ grows faster than tj for some $t > 0$ as j goes to $+\infty$. Since we can take $\varepsilon_j \geq Cj^{-1-r}$ for $j > j_0$ with arbitrarily large C and small $r > 0$, a sufficient condition for convergence is thus the linear growth of $L(j)j^{-2-2r}$. This concludes the proof of the theorem.

□

Unfortunately, there is no uniform bound on $\|\hat{w}_n\|_1$:

Theorem 3.3 Suppose that $m_\varepsilon^k = m_{\varepsilon,L(k)}$ satisfies (6.13)-(6.16), and that $L(k) \to \infty$ as $k \to \infty$. Then the sequence $\|\hat{w}_n\|_1$ is not bounded.

Proof

Since $L(k) \to \infty$ as $k \to \infty$, we have

$$\inf_{[-\frac{\pi}{2},\frac{\pi}{2}]} \prod_{k\geq 1} |m_0^{j+k}(2^{-k}\xi)| \to 1 \quad \text{as } j \to \infty. \tag{6.37}$$

By Bernstein's inequality, we have

$$|(m_0^j)'(\xi)| \leq d(j)|m_0^j(\xi)|. \tag{6.38}$$

So, since $p^j(\pi/2) = 2|m_0^j(\pi/2)| = \sqrt{2}$, there is $C > 0$ such that

$$p^j(\xi) \geq \frac{1+\sqrt{2}}{2} \quad \text{for any } \xi \in [\frac{\pi}{2}, \frac{\pi}{2} + \frac{1}{C\,d(j)}]. \tag{6.39}$$

Let $a(j)$ be the integer-valued function defined by $2^{a-1} < C\,d(j) \leq 2^a$. We denote $I_j = 2^j[\frac{\pi}{2} - 2^{-a}\pi, \frac{\pi}{2} + 2^{-a}\pi]$, $I_j' = 2^j[\frac{\pi}{2}, \frac{\pi}{2} + 2^{-a-1}\pi]$.

For $n = \sum_{k\geq 0} 2^k \varepsilon_k$ and $j \geq 0$, we define

$$l_n^j = \int_{-2^j\pi}^{2^j\pi} |\prod_{k=1}^{j} m_{\varepsilon_k}^k(2^{-k}\omega)|d\omega \tag{6.40}$$

For n fixed, the sequence l_n^j is non-decreasing. For a proof in the stationary case (see [364], Lemma 1), the extension to non-stationary filters is immediate.

From (6.14) and (6.15), there is $\lambda > 0$ such that

$$\prod_{j\geq 1} |m_0^j(2^{-j}\xi)| \geq \lambda \quad \text{for any } \xi \in [-\pi, \pi]. \tag{6.41}$$

It follows that $|\prod_{k=1}^{j} m_{\varepsilon_k}^k(2^{-k}\omega)|$ is bounded on $[-2^j\pi, 2^j\pi]$ by $|\hat{w}_n(\omega)|/\lambda$ for all $j > \log_2(n)$. So, from Lebesgue's convergence theorem, one has

$$\|\hat{w}_n\|_1 = \lim_{n\to\infty} l_n^j = \sup_{n\geq 0} l_n^j. \tag{6.42}$$

This allows us to work with l_n^j instead of $\|\hat{w}_n\|_1$.

If $(\varepsilon_{j-a(j)},...,\varepsilon_j) = (0,0,...,0,1,0)$, then

$$\begin{aligned} l_n^j &\geq 2\int_{I_j} |m_0^j(2^{-j}\omega)|\,|m_1^{j-1}(2^{1-j}\omega)| \\ &\qquad |\textstyle\prod_{k=j-a+1}^{j-2} m_0^k(2^{-k}\omega)|\,|m_0^{j-a}(2^{a-j}\omega)|\,F(\omega)\,d\omega\,. \end{aligned}$$

Here, $F(\omega) = |\prod_{k=1}^{j-a(j)-1} m_{\varepsilon_k}^k(2^{-k}\omega)|$ is non-negative, even, 2^{j-a} π-periodic, and $G(\omega) = |m_0^{j-a}(2^{a-j}\omega)|\,F(\omega)$ satisfies

$$2\int_0^{2^{j-a}\pi} G(\omega)\,d\omega = l_n^{j-a(j)}. \tag{6.43}$$

From (6.38), as $j \to \infty$, we have

$$c_j = \inf_{I_j}\big(|m_1^{j-1}(2^{1-j}\omega)| \prod_{k=j-a+1}^{j-2} |m_0^k(2^{-k}\omega)|\big) \to 1. \tag{6.44}$$

From (6.37),

$$|p^j(2^{-j}\omega)| \geq \frac{1+\sqrt{2}}{2}, \quad \forall\omega \in I_j. \tag{6.45}$$

Clearly, F is symmetric with respect to $\alpha_j = 2^{j-1}\pi + 2^{j-a-1}\pi$ and $|m_0^{j-a}(2^{a-j}\omega)|$ is decreasing on $2^j[\frac{\pi}{2}, \frac{\pi}{2} + 2^{-a}\pi]$. As a consequence,

$$G(\omega) \geq G(2\alpha_j - \omega) \quad \text{for any } \omega \in I_j'. \tag{6.46}$$

Now, by definition of c_j and from (6.45), we get

$$\begin{aligned} l_n^j &\geq 2c_j \int_{I_j} |m_0^j(2^{-j}\omega)| G(\omega)\,d\omega \\ &= 2c_j \int_{I_j'} \big(p^j(2^{-j}\omega)G(\omega) + p^j(2^{-j}(2\alpha_j - \omega))G(2\alpha_j - \omega)\big)\,d\omega \\ &\geq 2c_j \int_{I_j'} \big(\tfrac{1+\sqrt{2}}{2}G(\omega) + G(2\alpha_j - \omega)\big)\,d\omega. \end{aligned}$$

If $x \geq 1$, and $y \geq z \geq 0$, then $xy + z \geq \frac{1+x}{2}(y+z)$. Applying this inequality to $x = \frac{1+\sqrt{2}}{2}$, $y = G(\omega)$, $z = G(2\alpha_j - \omega)$, we find

$$\begin{aligned} l_n^j \geq\ & \tfrac{3+\sqrt{2}}{2} c_j \int_{I_j'} \big(G(\omega) + G(2\alpha_j - \omega)\big)\,d\omega \\ =\ & \tfrac{3+\sqrt{2}}{2} c_j \int_0^{2^{j-a}\pi} G(\omega)\,d\omega = \tfrac{3+\sqrt{2}}{4} c_j\, l_n^{j-a(j)}. \end{aligned}$$

So from (6.44), there is $\rho > 1$ such that, for j large enough we have

$$l_n^j \geq \rho\, l_n^{j-a(j)}. \tag{6.47}$$

We now construct, by induction, a sequence $\{j_q\}_{q\geq 0}$ such that $j_0 = 2$ and $j_{q-1} = j_q - a(j_q)$. This is possible because, from (6.16), $\frac{L(k)}{2^k} \to 0$ as $k \to \infty$.

We define a sequence $(\varepsilon_k, k \geq 1)$ by $\varepsilon_k = 1$ if there is $q \geq 0$ such that $k = j_q - 1$, and $\varepsilon_k = 0$ otherwise. Let $n_q = \sum_{k=1}^{j_q} 2^k \varepsilon_k$. From (6.46), we see that for some $c > 0$,

$$l_{n_q}^{j_q} \geq c\,\rho^q \tag{6.48}$$

and finally, $\|\hat{w}_{n_q}\|_1 \geq c\,\rho^q \to \infty$ as $q \to \infty$.

□

4 ESTIMATES ON THE VARIANCE OF $\hat{\omega}_N$

In the introduction, we have called Gray code the mapping which associates to each integer $n = \sum_{k\geq 1} \varepsilon_k 2^{k-1}$ an integer $g(n) = \sum_{k\geq 1} \gamma_k 2^{k-1}$, with $\gamma_k = |\gamma_{k+1} - \varepsilon_k|$ for $k \geq 1$, and we have defined the variance of $\hat{w}_n$ as

$$\sigma_n = \min_{\omega_0 \geq 0} \int_0^{+\infty} |\hat{w}_n(\omega)|^2 |\omega - \omega_0|^2 d\omega\ .$$

In general, we do not know which value of ω_0 achieves this minimum, but in the case of ideal cut-off filters, one has $\hat{w}_n(\omega) = \chi_{-J_n \cup J_n}(\omega)$ with $J_n = \pi[g(n), g(n)+1]$. This suggests taking $\xi_n = \pi g(n)$ as an approximation of the minimizer, and we are led to define, for general filters,

$$\tilde{\sigma}_n = \int_0^{+\infty} |\hat{w}_n(\omega)|^2 |\omega - \xi_n|^2 d\omega. \tag{6.49}$$

Obviously, one has $\sigma_n \leq \tilde{\sigma}_n$. In this part, we give lower and upper estimates on $\tilde{\sigma}_n$, which are easier to manipulate than σ_n.

If we consider stationary wavelet packets generated by smooth filters, we have already seen that the variances are not bounded in average (see Section 3, Theorem 3.1). In fact, we have the following (negative) result, which is stronger than Theorem 3.1.

Theorem 4.1 Let w_n be the standard wavelet packets associated to smooth filters m_0 and m_1. Then for any $M > 0$, we have

$$\lim_{n\to\infty} \frac{1}{n} \operatorname{card}\{k \le n \,/\, \tilde{\sigma}_k > M\} = 1. \tag{6.50}$$

Proof

Let $n = \sum_{k\ge 1} \varepsilon_k 2^{k-1}$ and $\hat{w}_n^j(\xi) = \Pi_{k=1}^j m_{\varepsilon_k, L(k)}(2^{-k}\xi)\chi_{[-2^j\pi, 2^j\pi]}(\xi)$.

We define

$$\tilde{\sigma}_n^j = \int_0^{2^j\pi} |\hat{w}_n^j(\omega)|^2 |\omega - \xi_n^j|^2 d\omega \tag{6.51}$$

where $\xi_n^j = \pi g(n_j)$, $n_j = \sum_{k=1}^j \varepsilon_k 2^{k-1}$.

For n fixed, the sequence $(\tilde{\sigma}_n^j, j \ge 0)$ is non-decreasing. Indeed, one either has $\varepsilon_j = 0$ and $\xi_n^j = \xi_n^{j-1}$, or $\varepsilon_j = 1$ and $\xi_n^j = 2^j\pi - \xi_n^{j-1}$, hence

$$\begin{aligned}
\tilde{\sigma}_n^j &= \int_0^{2^{j-1}\pi} \left(|\hat{w}_n^j(\omega)|^2 |\omega - \xi_n^j|^2 + |\hat{w}_n^j(2^j\pi - \omega)|^2 |2^j\pi - \omega - \xi_n^j|^2\right) d\omega \\
&= \int_0^{2^{j-1}\pi} |\hat{w}_n^{j-1}(\omega)|^2 \left(|\omega - \xi_n^{j-1}|^2 |m_0(2^{-j}\omega)|^2 \right.\\
&\qquad \left. + |2^j\pi - \omega - \xi_n^{j-1}|^2 |m_0(2^{-j}\omega + \pi)|^2\right) d\omega \\
&\ge \int_0^{2^{j-1}\pi} |\hat{w}_n^{j-1}(\omega)|^2 |\omega - \xi_n^{j-1}|^2 \left(|m_0(2^{-j}\omega)|^2 + |m_0(2^{-j}\omega + \pi)|^2\right) d\omega \\
&= \tilde{\sigma}_n^{j-1}.
\end{aligned}$$

So, arguing as in the proof of Theorem 3.3, from formula (6.41), we find the following analog of formula (6.42):

$$\tilde{\sigma}^j = \lim_{n\to\infty} \tilde{\sigma}_n^j = \sup_{n\ge 0} \tilde{\sigma}_n^j \tag{6.52}$$

and we are going to study $\tilde{\sigma}_n^j$ instead of $\tilde{\sigma}_n$.

We now take $\alpha > 0$. For $j \ge 0$, we define

$$\mathcal{I}_j^\alpha = 2^{j-1}\pi[\alpha, 1-\alpha]. \tag{6.53}$$

We are going to prove that there is $j(M,\alpha)$ such that, for any n, j satisfying $\xi_n^{j-1} \in \mathcal{I}_j^{2\alpha}$ and $j \ge j(M,\alpha)$, the estimate $\tilde{\sigma}_n^j > M$ holds.

If $\tilde{\sigma}_n^{j-1} > M$, then we also have $\tilde{\sigma}_n^j > M$, by monotonicity of $(\tilde{\sigma}_n^j, j \geq 0)$. So let us assume that $\xi_n^{j-1} \in \mathcal{I}_j^{2\alpha}$ and $\tilde{\sigma}_n^{j-1} \leq M$. We denote

$$I_{n,j} = [\xi_n^{j-1} - 2^{j-2}\pi\alpha, \xi_n^{j-1} + 2^{j-2}\pi\alpha]. \tag{6.54}$$

Then, by the classical Chebyshev inequality,

$$\|\hat{w}_n^{j-1}\|_{L^2([0,2^j\pi]\setminus I_{n,j})} \leq \frac{M}{2^{j-2}\pi\alpha}. \tag{6.55}$$

As a consequence, there is $i(M,\alpha)$ such that $\|\hat{w}_n^{j-1}\|_{L^2(I_{n,j})} \geq \frac{1}{2}$ for $j \geq i(M,\alpha)$. Now, we define

$$J_{n,j} = [2^j\pi - \xi_n^j - 2^{j-2}\pi\alpha, 2^j\pi - \xi_n^j + 2^{j-2}\pi\alpha]. \tag{6.56}$$

If $\varepsilon_j = 1$, then $J_{n,j} = I_{n,j}$, otherwise $J_{n,j} = 2^j\pi - I_{n,j}$. Moreover, the assumption $\xi_n^{j-1} \in \mathcal{I}_j^{2\alpha}$ implies that $I_{n,j} \subset \mathcal{I}_j^{\alpha}$. So we have

$$\begin{aligned} \tilde{\sigma}_n^j &\geq \textstyle\int_{J_{n,j}} |\hat{w}_n^j(\omega)|^2 |\omega - \xi_n^j|^2 d\omega \\ &\geq 2^{2j}\alpha^2\pi^2 \inf_{\omega\in\mathcal{I}_j^{\alpha}} |m_1(2^{-j}\omega)|^2 \textstyle\int_{I_{n,j}} |\hat{w}_n^{j-1}(\omega)|^2 d\omega \\ &\geq 2^{2j-2}\alpha^2\pi^2 \inf_{\xi\in\mathcal{I}_0^{\alpha}} |m_1(\xi)|^2 = 2^{2j}\beta, \quad \beta > 0. \end{aligned}$$

So, taking $j(M,\alpha) \geq i(M,\alpha)$ large enough, we get the estimate $\tilde{\sigma}_n^j \geq M$ for $j \geq j(M,\alpha)$.

To end the proof of the theorem, consider a positive integer n. Let j be the unique integer such that $2^j \leq n < 2^{j+1}$. We have $\frac{1}{n}\text{card}\{k \leq n \;/\; \xi_n^{j-1} \in \mathcal{I}_j^{2\alpha}\} \geq 1 - 4\alpha$. Thus, if $j \geq j(M,\alpha)$, i.e. if $n \geq 2^{j(M,\alpha)}$, then

$$\frac{1}{n}\,\text{card}\{k \leq n \;/\; \tilde{\sigma}_k \geq M\} \geq 1 - 4\alpha. \tag{6.57}$$

□

If we work with non-stationary wavelet packets, the situation is much better. We assume once again that the hypotheses (6.13)-(6.16) are satisfied and that $\{m_0^k(\omega)\} = \{m_{0,L(k)}(\omega)\}$ is a subset of the family given by (6.5) and (6.6). We have:

Theorem 4.2 Assume that $L(k) \geq Ck^{3+r}$ for some $r > 0$. Then, for any $\varepsilon > 0$ there are $M_\varepsilon > 0$ and $n_\varepsilon \in \mathbf{N}$ such that, if $n \geq n_\varepsilon$, then

$$\frac{1}{n}\text{card}\,\{k \leq n \,/\, \tilde{\sigma}_k > M_\varepsilon\,\} \leq \varepsilon. \tag{6.58}$$

Proof

Choosing $C = 1 + \frac{r}{4}$, one has $\frac{L(j)}{j^{2C+1}} \to \infty$. So, from Lemma 4.3 below, for any $j_0 > 0$, there is a $K(j_0)$ such that, if $n \sum_{k\geq 1} \varepsilon_k 2^{k-1}$, and $(\varepsilon_{j_0}, ..., \varepsilon_k, ...)$ contains no subsequence of the form $(\varepsilon_{k-[C\,\log_2 k]}, ..., \varepsilon_k) = (0, ..., 0, 1)$, then $\tilde{\sigma}_n \leq K(j_0)$.

Given $j_0 > 0$, the probability $P^C_{j_0}$ for a sequence $(\varepsilon_{j_0}, ..., \varepsilon_k, ...)$ to contain a subsequence $(\varepsilon_{k-[C\,\log_2 k]}, ..., \varepsilon_k) = (0, ..., 0, 1)$ is smaller than $\sum_{k=j_0}^{\infty} \frac{1}{k^C}$.

Given $\varepsilon > 0$, we choose j_ε such that $P^C_{j_\varepsilon} \leq \varepsilon$, and $M_\varepsilon = K(j_\varepsilon)$. The inequality (6.58) is satisfied. □

Lemma 4.3 For any $C > 0$, there exists $l(C) > 0$ such that, if $L(j) \geq l(C)\,j^{2C+1}$, the following holds:

Let $n = \sum_{k\geq 1} \varepsilon_k 2^{k-1}$ and $j_0 \geq 1$ be given. There exists $K(j_0) > 0$ such that, if the sequence $(\varepsilon_{j_0}, ..., \varepsilon_k, ...)$ contains no subsequence of the form $(\varepsilon_{k-[C\,\log_2 k]}, ..., \varepsilon_k) = (0, ..., 0, 1)$, then $\tilde{\sigma}_n \leq K(j_0)$.
Here, we have denoted $[x]$ as the integer part of x.

Proof

By (6.52), we just have to find estimates on $\tilde{\sigma}^j_n$ independent of j. We proceed by induction on j.

We assume that $(\varepsilon_{j-[C\,\log_2 j]}, ..., \varepsilon_j) \neq (0, ..., 0, 1)$ and we are going to estimate $\tilde{\sigma}^{j+1}_n$ by a function of j and $\tilde{\sigma}^j_n$.

Since $|m_0|$ is even and 2π-periodic, and $|m_1(\xi)| = |m_0(\xi + \pi)|$, we see that the value of $\tilde{\sigma}^{j+1}_n$ is independent of ε_{j+1}. So we may assume that

$\varepsilon_{j+1} = 0$. Then $\xi_n^{j+1} = \xi_n^j$, and we write, for $\delta > 0$,

$$\begin{aligned}\tilde{\sigma}_n^{j+1} &= \int_0^{2^j(\pi-2\delta)} |\hat{w}_n^{j+1}(\omega)|^2 |\omega - \xi_n^j|^2 d\omega \\ &+ \int_{2^j(\pi-2\delta)}^{2^j(\pi+2\delta)} |\hat{w}_n^{j+1}(\omega)|^2 |\omega - \xi_n^j|^2 d\omega \\ &+ \int_{2^j(\pi+2\delta)}^{2^{j+1}\pi} |\hat{w}_n^{j+1}(\omega)|^2 |\omega - \xi_n^j|^2 d\omega \\ &= I_1 + I_2 + I_3 .\end{aligned}$$

Clearly, $I_1 \le \tilde{\sigma}_n^j$. Now, if δ is small enough, one has $\sin(\frac{\pi}{2}+\delta) \le 1-\frac{\delta^2}{3}$. So, from lemma 2.3, $|m_{0,L}(\omega)| \le (1-\frac{\delta^2}{3})^{L-1}$ for any $\omega \in [\frac{\pi}{2}+\delta, \pi]$, hence

$$\begin{aligned}I_3 &\le \int_{2^j(\pi+2\delta)}^{2^{j+1}\pi} |\hat{w}_n^j(2^j - \omega)|^2 |m_{0,L(j+1)}(2^{-j-1}\omega)|^2 |\omega - \xi_n^j|^2 d\omega \\ &\le \|\hat{w}_n^j\|_{L^2}^2 (1-\tfrac{\delta^2}{3})^{2L(j+1)-2} 2^{2j+2}\pi^2 \\ &\le c_1 2^{2j} (1-\tfrac{\delta^2}{3})^{2L(j+1)} .\end{aligned}$$

The remaining term is I_2.
By the assumption on $(\varepsilon_{j-[C \log_2 j]}, \ldots, \varepsilon_j)$, we have

$$|\xi_n^j - 2^j\pi| \ge \frac{2^j\pi}{j^C} \tag{6.59}$$

and there is an integer k between $j-[C \log_2 j]$ and j, such that

$$m_{\varepsilon_k}^{L(k)}(2^{j-k}\pi) = 0. \tag{6.60}$$

From Lemma 3.2, for any $\rho > 0$, we have

$$\sup\{|m_{1,L}(\xi)| \,;\, 0 \le \xi \le \rho\} \le \rho^{L-1}. \tag{6.61}$$

So, if we assume that δ satisfies

$$2^{j+1}\delta \le 2^k \rho \tag{6.62}$$

then $|m_{\varepsilon_k, L(k)}(2^k\xi)| \le c_2 \rho^{L(k)}, \quad \forall \xi \in [2^j(\pi - 2\delta_j), 2^j(\pi + 2\delta_j)]$.
Since $k \ge j - [C \log_2 j]$, condition (6.62) is satisfied by taking

$$\delta = \frac{\rho}{2j^C} \tag{6.63}$$

and it gives the estimate

$$I_2 \le c_3 \rho^{2L(k)} 2^{2j} \le c_3 \rho^{2L(j-C \log j)} 2^{2j}. \tag{6.64}$$

To obtain (6.64), we have assumed for simplicity that $L(j)$ is an increasing function of j.

Using the estimates on I_1, I_2, I_3, we get

$$\tilde{\sigma}_n^{j+1} \leq \tilde{\sigma}_n^j + c_4 2^{2j} \left(\rho^{2L(j-C \log j)} + (1 - \frac{\rho^2}{12 j^{2C}})^{2L(j+1)}\right). \qquad (6.65)$$

Clearly, the series $\sum 2^{2j} \rho^{2L(j-C \log j)}$ is convergent if $\frac{L(j)}{j} \to \infty$ as $j \to \infty$ and $0 < \rho < 1$.

Let us study the convergence of $\sum 2^{2j}(1 - \frac{\rho^2}{12j^{2C}})^{2L(j+1)}$. It is ensured if $2^{2j}(1 - \frac{\rho^2}{12j^{2C}})^{2L(j+1)} \leq a^{2j}$ for some $a \in (0,1)$. This can be written in the form

$$j \log 2 + L(j+1) \log (1 - \frac{\rho^2}{12 j^{2C}}) \leq -j \log (\frac{1}{a}). \qquad (6.66)$$

This inequality is true if $j(\log 2 + \log(\frac{1}{a})) \leq \frac{\rho^2}{12j^{2C}} L(j)$, i.e.

$$L(j) \geq c_5 j^{1+2C}. \qquad (6.67)$$

So $\tilde{\sigma}_n^j$ is bounded by $K(j_0)$ for $j \geq j_0$, if (4.19) is satisfied. This proves the lemma.

□

It would be nice to obtain $\tilde{\sigma}_n$ bounded independently of n, or at least bounded on average, for a suitably chosen polynomial growth of L. Unfortunately, this is impossible:

Theorem 4.4 Assume that the hypotheses (6.13)-(6.16) are satisfied and that $d(k) \leq Ck^r$ for some $r > 0$. Then the sequence $s_n = \frac{1}{n}\sum_{k=0}^{n-1} \tilde{\sigma}_k$ is not bounded.

Proof

We recall that there is $\lambda > 0$ such that

$$\prod_{j \geq 1} |m_0^j(2^{-j}\xi)| \geq \lambda \quad \text{for any } \xi \in [-\pi, \pi]. \qquad (6.68)$$

By (6.38), there is a polynomial $p(j)$ of degree less than or equal to $[r]+1$ such that

$$|m_0^j(\xi)| \geq \frac{1}{2} \quad \text{for any } \xi \in [\frac{\pi}{2}, \frac{\pi}{2} + \frac{\pi}{p(j)}]. \tag{6.69}$$

Let $a(j)$ be the integer-valued function defined by $2^{a-1} < p \leq 2^a$. We denote $I_j = 2^j[\frac{\pi}{2}, \frac{\pi}{2} + 2^{-a}\pi]$.

If $(\varepsilon_{j-a(j)}, ..., \varepsilon_j) = (1, 0, ..., 0, 1, 0)$, then

$$\begin{aligned} \tilde{\sigma}_n^j &\geq \int_{I_j} |\hat{w}_n^j(\omega)|^2 |\omega - \xi_n^j|^2 \\ &= \int_{I_j} |m_0^j(2^{-j}\omega)|^2 |m_1^{j-1}(2^{1-j}\omega)|^2 \\ &\qquad |\textstyle\prod_{k=j-a+1}^{j-2} m_0^k(2^{-k}\omega)|^2 |f(\omega)|^2 |\omega - \xi_n^j|^2 \, d\omega, \end{aligned}$$

where f is $2^{j-a}\pi$-periodic and satisfies

$$\int_0^{2^{j-a}\pi} |f(\omega)|^2 = 1. \tag{6.70}$$

From (6.68), one has

$$|m_1^{j-1}(2^{1-j}\omega)| \prod_{k=j-a+1}^{j-2} |m_0^k(2^{-k}\omega)| \geq \lambda, \quad \forall \omega \in I_j \tag{6.71}$$

and from (6.69),

$$|m_0^j(2^{-j}\omega)| \geq \frac{1}{2}, \quad \forall \omega \in I_j. \tag{6.72}$$

Moreover,

$$\xi_n^j \leq 2^j(\frac{\pi}{2} - 2^{-a-1}\pi). \tag{6.73}$$

Combining these estimates, we get

$$\tilde{\sigma}_n^j \geq \frac{1}{4}\lambda^2 2^{2(j-a-1)}\pi^2 \geq c\frac{2^{2j}}{p(j)^2} \tag{6.74}$$

for some $c > 0$.

Now, the probability that $(\varepsilon_{j-a(j)}, ..., \varepsilon_j) = (1, 0, ..., 0, 1, 0)$ is

$$\mathcal{P}_j = \frac{1}{2^{a+1}} \geq \frac{1}{4p(j)}. \tag{6.75}$$

From (6.52), (6.74) and (6.75), we see that

$$\frac{1}{2^j} \sum_{0 \leq k < 2^j} \tilde{\sigma}_k \geq c \frac{2^{2j}}{4p(j)^3} \to \infty \text{ as } j \to \infty. \tag{6.76}$$

□

5 CONCLUSION

We have shown in this chapter how the use of different filters at each scale can help in recovering some properties of time-frequency localization for the wavelet packets. Note that the results can be quite different when one uses the L^1 norm (Theorem 3.2) or the variance (Theorem 4.4) as a localization criterion.

An interesting perspective coming out from this work is the application of the same techniques to build stable biorthogonal wavelet packets. Indeed it was shown by Cohen and Daubechies [92] that the generalization of wavelet packets to the biorthogonal case leads to an unstable decomposition and that this instability is essentially related to their spreading in the frequency domain.

7

SUBBAND IMAGE CODING

Bernd Girod, Frank Hartung, and Uwe Horn

Lehrstuhl für Nachrichtentechnik,
Universität Erlangen-Nürnberg,
Cauerstraße 7, 91058 Erlangen, Germany

Visual information plays an important role in almost all areas of our life. Due to the vast amount of data associated with images, compression is a key technology for their digital transmission and storage. As an example, consider a single still color image of 512×512 pixels with 3 color components and 8 bits per sample per component resolution, which represents a data volume of 768 kBytes. For video, the problem is even more severe; a one-hour movie in a resolution according to the CCIR 601 studio standard [75] would need as much as 74.7 GBytes, when stored in uncompressed PCM format. These examples clearly demonstrate the need for efficient compression algorithms for images.

All methods for compression of still images or video are based on two fundamental principles. One principle is to exploit the properties of the signal source and to remove redundancy from the signal. The other principle is to exploit the properties of the signal receiver (usually the human visual system) and to omit parts or details of the signal that will not be noticed by the receiver. These principles are often referred to as redundancy reduction and irrelevancy reduction, respectively. The theory of subband decomposition provides an efficient framework for the implementation of schemes for redundancy and irrelevancy reduction.

It has been demonstrated repeatedly that subband and wavelet based schemes outperform other waveform based coding schemes [7, 447, 56]. Nevertheless, most of today's image and video coding standards use the discrete cosine transform (DCT) [2] for signal decomposition [207, 208, 76]. It can, however, be shown that the DCT, like all orthogonal transforms, is just a special case of subband decomposition [418]. Though the image coding standards are restricted to that special case, there is room for subband coding schemes in proprietary applications, which will become more important with the feasibility of software-only image codecs.

An important motivation for the use of subband decomposition schemes (rather than e.g. DCT-based schemes) is the demand for "scalable" image representations [193, 166, 393]. In the context of image coding, scalability means that the transmitted bit-stream can be decoded hierarchically, i.e., a low-resolution version of the transmitted image can be decoded with few operations, and the full resolution image will only be decoded if necessary or desired. This is, for example, useful for database browsing applications. For broadcasting applications, the low resolution signal can be transmitted with better error protection. This preserves at least a low resolution version of the image or video, if the channel characteristics vary, and the threshold behavior of digital transmission (where the receiver can decode the signal either in full digital quality or, if the bit error rate exceeds the threshold for the given error protection, nothing at all) is avoided [432]. Subband coding schemes are especially suitable for applications where scalability and "graceful degradation" are important issues.

This chapter treats the compression of still images by means of subband filter banks, whereas the next chapter is dedicated to subband compression of video. Section 1 reviews the theoretical foundations of subband coding. Rate-distortion theory is introduced, and rate-distortion functions for different signal models are presented. From rate-distortion theory, we can conclude that for stationary image models it is appropriate to decompose a signal into frequency bands and to encode those bands separately.

Section 2 discusses various approaches to subband decomposition of images. Equal band splitting and non-uniform band splitting are explained. The discrete cosine transform (DCT) is shown to be a special case of subband decomposition. Section 3 treats the compression of the subband signals after decomposition. Compression can be achieved by applying scalar quantization or vector quantization (VQ) to the subband coefficients, in combination with fixed or variable wordlength encoding. The concepts of scalar quantization, fixed-wordlength VQ, entropy-constrained VQ and lattice VQ are explained and compared experimentally. Optimal bit-rate allocation among the subbands is explained in Section 3.4.

1 THEORETICAL FOUNDATIONS OF SUBBAND IMAGE CODING

Before we concern ourselves with specific subband image coding algorithms we would like to know how far we can expect to lower the transmission bit-rate for a given required picture quality. Are there fundamental limits that cannot be exceeded by any coding scheme? Such limits would be extremely helpful to judge the relative performance of a practical image coding scheme such as subband coding.

1.1 The Rate Distortion Function

Rate distortion theory is a branch of information theory that allows us to calculate performance bounds without consideration of a specific coding method. In particular, rate-distortion theory yields the minimum transmission bit-rate R, if a distortion D between the original image x at the transmitter and the reconstructed image $\hat{x}$ at the receiver shall not exceed a maximum acceptable distortion D^*. Unfortunately, the theory does not provide us with a method for constructing a practical optimum coder and decoder. We will see that rate-distortion theory can nevertheless provide us with very important hints about the properties of an optimum codec.

Rate distortion theory is based on two central concepts: "mutual information" and "distortion." Mutual information is a symmetric measure of the information that symbols x and $\hat{x}$ convey about each other. While the framework is entirely general, let's think of x representing an original image at the transmitter and $\hat{x}$ a reconstructed image at the receiver. We are mainly interested in average quantities in information theory. The average mutual information between ensembles X and $\hat{X}$, representing amplitude-continuous vector-valued random variables x and $\hat{x}$ with joint probability density function $p_{X\hat{X}}(x, \hat{x})$ and marginal probability density functions $p_X(x)$ and $p_{\hat{X}}(\hat{x})$ is defined as

$$I(X;\hat{X}) = \int_{\hat{x}} \int_{x} p_{X\hat{X}}(x,\hat{x}) \log \frac{p_{X\hat{X}}(x,\hat{x})}{p_X(x) p_{\hat{X}}(\hat{x})} dx\, d\hat{x}. \tag{7.1}$$

Mutual information is related to the differential entropy of an ensemble,

$$h(X) = \int_x p_X(x) \log \frac{1}{p_X(x)} \, dx, \tag{7.2}$$

and the conditional differential entropy of X given $\hat{X}$,

$$h(X|\hat{X}) = \int_{\hat{x}} \int_x p_{X\hat{X}}(x, \hat{x}) \log \frac{1}{p_{X|\hat{X}}(x|\hat{x})} \, dx \, d\hat{x} \tag{7.3}$$

by

$$I(X;\hat{X}) = h(\hat{X}) - h(\hat{X}|X) = h(X) - h(X|\hat{X}). \tag{7.4}$$

In Equation (7.3), $p_{X|\hat{X}}(x|\hat{x})$ is the probability of an original image x, if a decoded image $\hat{x}$ is observed. Clearly, average mutual information is symmetric and non-negative

$$0 \leq I(X;\hat{X}) = I(\hat{X};X). \tag{7.5}$$

The average mutual information $I(X;\hat{X})$ between the original image at the coder and the reconstructed image at the decoder is related to the channel capacity C available between X and $\hat{X}$. The channel capacity is the maximum number of bits per symbol for a given symbol-rate that a transmission channel can accommodate without bit-errors. The channel capacity can be shown to be the maximum of the average mutual information between transmitter and receiver [154], i.e.

$$I(X;\hat{X}) \leq C. \tag{7.6}$$

The distortion measure $d(x, \hat{x})$ is a scalar quantity that should reflect the fidelity of the reproduction of an original image x by an image $\hat{x}$. To qualify as a proper distortion measure, we require

$$d(x, \hat{x}) \geq 0 \tag{7.7}$$

with equality if $x = \hat{x}$. The average distortion then is

$$D = \int_{\hat{x}} \int_{x} p_{X\hat{X}}(x, \hat{x}) d(x, \hat{x}) \, dx \, d\hat{x}. \tag{7.8}$$

The 'rate-distortion function' is defined as

$$R(D^*) = \inf_{p_{\hat{X}|X}} \{I(X; \hat{X}) : D \leq D^*\}, \tag{7.9}$$

i.e., $R(D^*)$ is the greatest lower bound of the average mutual information, subject to the constraint that the average distortion D may not exceed D^*. When there is no risk of confusion, we will drop the asterisk in the following and write $R(D)$ for the rate-distortion function. It can be shown that $R(D)$ is a monotonically decreasing convex function [154, 51]. At $R = 0$, it assumes a maximum distortion $D_{\max}$. For an amplitude-continuous source, the rate required for an exact reproduction $\hat{x} = x$ is infinite, nevertheless the rate $R(0)$ can be finite, if the distortion measure contains a threshold below which $d(x, \hat{x}) = 0$, even if $x \neq \hat{x}$. Since $R(D)$ is monotonically decreasing, we can also use the distortion rate function $D(R)$ when it is more convenient.

The rate-distortion function is a performance bound that no source coder can beat. Conversely, rate-distortion theory shows that a source coder with a performance arbitrarily close to the rate-distortion function exists. The typical theoretically optimum source coder encodes a very large number of symbols jointly, thus requiring a very large memory and introducing a very large delay. While this might be impractical, it suggests that a good coder jointly encodes many symbols. We will make use of this idea in the following.

1.2 Shannon Lower Bound

The Shannon lower bound is a useful lower bound of the rate-distortion function. With a simple coordinate transformation in Equation (7.3), it can be shown that

$$h(X - \hat{X}|\hat{X}) = h(X|\hat{X}), \tag{7.10}$$

where $X - \hat{X}$ is the reconstruction error ensemble representing the amplitude-continuous vector-valued difference $x - \hat{x}$ [51]. With this, we can rewrite the rate-distortion function as

$$\begin{aligned} R(D^*) &= \inf_{p_{\hat{X}|X}} \{h(X) - h(X|\hat{X}) : D \leq D^*\} \\ &= h(X) - \sup_{p_{\hat{X}|X}} \{h(X|\hat{X}) : D \leq D^*\} \end{aligned}$$

$$= h(X) - \sup_{p_{\hat{X}|X}} \{h(X - \hat{X}|\hat{X}) : D \leq D^*\}. \tag{7.11}$$

Observing that

$$h(X - \hat{X}|\hat{X}) \leq h(X - \hat{X}) \tag{7.12}$$

we arrive at the Shannon lower bound

$$R(D^*) \geq h(X) - \sup_{p_{\hat{X}|X}} \{h(X - \hat{X}) : D \leq D^*\}. \tag{7.13}$$

Equality in Equations (7.12) and (7.13) holds for statistical independence between $X - \hat{X}$ and $\hat{X}$. Thus, ideally, the source coding scheme would introduce a reproduction error $x - \hat{x}$ that is statistically independent of the reconstructed signal $\hat{x}$. Note that this is not always possible, particularly not at low rates. Nevertheless, it provides another guideline for the design of an efficient coding scheme.

It is possible to draw additional conclusions from the Shannon lower bound, if the distortion measure is given. Let us consider a single letter distortion measure

$$d = (x - \hat{x})^2, \tag{7.14}$$

i.e., the squared difference between the original and reconstructed images is calculated on a sample-by-sample basis. With the mean squared error $D \leq D^*$, the differential entropy of the reconstruction error is bounded according to [55]

$$h(X - \hat{X}) \leq \frac{1}{2} \log(2\pi e D^*). \tag{7.15}$$

Equality in Equation (7.15) holds for a Gaussian pdf of $x - \hat{x}$. Successive values of $x - \hat{x}$ should all be independent, identically distributed random variables. Thus, an optimum source coder for a mean squared error distortion measure should produce white Gaussian noise independent from the reconstructed signal.

1.3 Memoryless Gaussian Source

In general, it is difficult to calculate the rate-distortion function. However, there are some important cases where the result can be stated analytically. Consider a memoryless Gaussian source with variance σ^2. For this case, the rate-distortion function is

$$R(D) = \frac{1}{2} \max\{\log \frac{\sigma^2}{D}, 0\}. \tag{7.16}$$

If log $=$ $\log_2$ in Equation (7.16), the unit for the rate is "bits". The rate-distortion curve is plotted as curve "ML" in Figure 1, with distortion shown on a logarithmic axis. With signal-to-noise ratio defined as

$$\mathrm{SNR} = 10\log_{10}\frac{\sigma^2}{D}\ \mathrm{dB} \tag{7.17}$$

we encounter the information theoretic rule-of-thumb that 6 dB corresponds to 1 bit. Remarkably, the rate-distortion curves for non-Gaussian sources with the same variance σ^2 are always below the Gaussian $R(D)$ curve. Thus, Gaussian sources are the most demanding sources for coding.

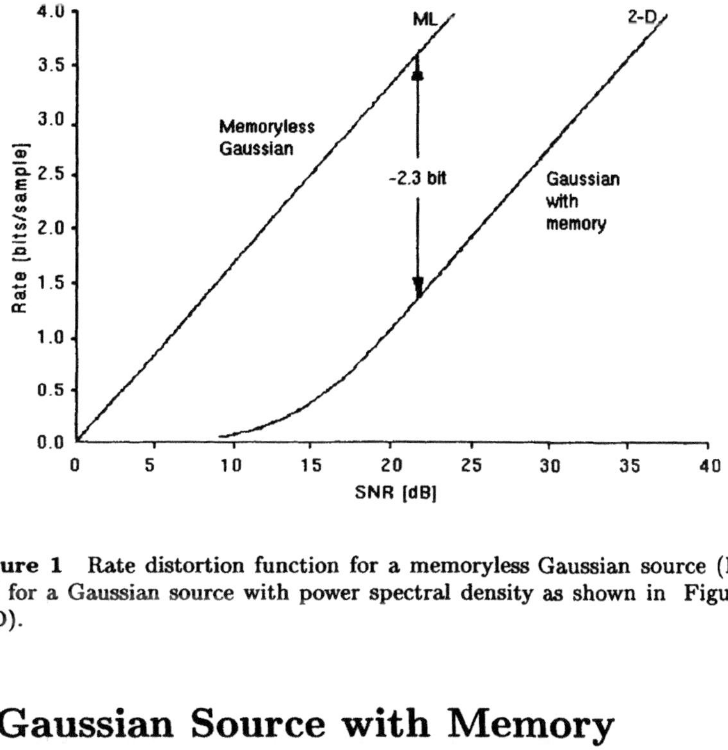

Figure 1 Rate distortion function for a memoryless Gaussian source (ML) and for a Gaussian source with power spectral density as shown in Figure 3 (2-D).

1.4 Gaussian Source with Memory

For sources with memory, the correlation between neighboring samples can be exploited and a lower rate can be achieved. Let us consider an ergodic two-dimensional space and amplitude continuous Gaussian source with power spectral density $\Phi_{xx}(\Omega_1, \Omega_2)$. Again, we use the squared error distortion measure (7.14). While the rate-distortion function cannot be given in closed form for this source and distortion measure, it can be stated in parametric form as

$$D(\theta) = \frac{1}{4\pi^2}\int_{\Omega_2}\int_{\Omega_1} \min(\theta, \Phi_{xx}(\Omega_1, \Omega_2))\,d\Omega_1\,d\Omega_2 \tag{7.18a}$$

$$R(\theta) \quad = \quad \frac{1}{8\pi^2} \int_{\Omega_2} \int_{\Omega_1} \max(0, \log \frac{\Phi_{xx}(\Omega_1, \Omega_2)}{\theta}) d\Omega_1 \, d\Omega_2. \qquad (7.18b)$$

Each value of θ in Equation (7.18a) (within an appropriate range) produces a point $R(D)$ of the rate-distortion curve. Again, for non-Gaussian sources with the same power spectral density, the rate-distortion curve is always below the Gaussian case.

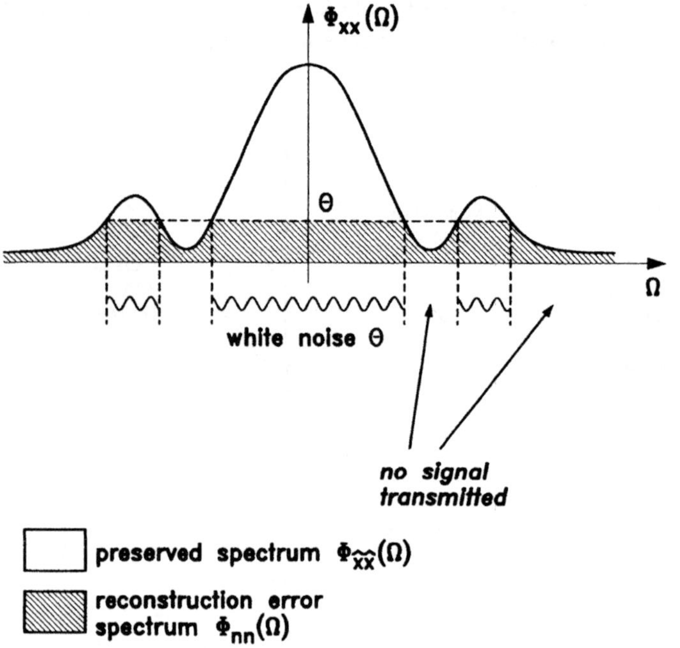

Figure 2 Interpretation of the rate-distortion function for a one-dimensional Gaussian source with memory.

The parametric solution (7.18a) can be interpreted as illustrated in Figure 2 for a one-dimensional case. Let $\Phi_{nn}(\Omega_1, \Omega_2)$ denote the power spectral density of the reconstruction error $x - \hat{x}$ (compare Figure 23). In the frequency range where $\Phi_{xx}(\Omega_1, \Omega_2) > \theta$, white noise with power spectral density $\Phi_{nn}(\Omega_1, \Omega_2) = \theta$ is introduced. With the same arguments as used in the context of the Shannon lower bound, it can be shown that this noise is uncorrelated with the reconstructed signal $\hat{x}$. Thus

$$\Phi_{\hat{x}\hat{x}}(\Omega_1, \Omega_2) = \Phi_{xx}(\Omega_1, \Omega_2) - \theta, \forall \Omega_1, \Omega_2 : \Phi_{xx}(\Omega_1, \Omega_2) > \theta. \qquad (7.19)$$

In the frequency range where $\Phi_{xx}(\Omega_1, \Omega_2) < \theta$ the signal power itself is smaller than the coding noise introduced. Thus, it leads to the minimum mean squared error to transmit no signal at all in these parts of the spectrum, and

$$\left.\begin{array}{rcl} \Phi_{\hat{x}\hat{x}}(\Omega_1, \Omega_2) & = & 0 \\ \Phi_{nn}(\Omega_1, \Omega_2) & = & \Phi_{xx}(\Omega_1, \Omega_2) \end{array}\right\} \forall \Omega_1, \Omega_2 : \Phi_{xx}(\Omega_1, \Omega_2) < \theta. \tag{7.20}$$

Since the overall rate in Equation (7.18ab) is the integral over the rate contributions dR of individual frequency components, an optimum coder can be built by splitting the original signal into spectral components of infinitesimal bandwidth $d\Omega_1\, d\Omega_2$ and encoding these spectral components independently [283]. This suggests an approach where the subband components with energy above a threshold are encoded with a number of bits that is proportional to the logarithm of their energy, while the remaining subbands are suppressed.

1.5 R(D) for a Gaussian Image Model with Exponentially Decaying Autocorrelation Function

We can compute the rate-distortion function (7.18a), if we know the power spectrum $\Phi_{xx}(\Omega_1, \Omega_2)$. A commonly used model for the autocorrelation function of an image is the isotropic, exponentially decaying autocorrelation function

$$R_{xx}(\Delta_1, \Delta_2) = e^{-\Omega_0 \sqrt{{\Delta_1}^2 + {\Delta_2}^2}}, \tag{7.21}$$

where Δ_1 and Δ_2 are the horizontal and vertical distances between samples of x, and Ω_0 is a constant. The corresponding power spectral density is computed as the 2-D Fourier transform of (7.21),

$$\Phi_{xx}(\Omega_1, \Omega_2) = \frac{2\pi}{\Omega_0^2}\left(1 + \frac{\Omega_1^2 + \Omega_2^2}{\Omega_0^2}\right)^{-\frac{3}{2}}. \tag{7.22}$$

The model power spectrum (7.22) is shown in Figure 3 with parameters adapted to a video signal sampled in the "Common Intermediate Format" (CIF, 288 lines × 352 pixels per line for a video frame with 4/3 aspect ratio). The rate-distortion curve is computed by numerically integrating Equation (7.18a). The result is shown in Figure 1 as curve "2D". The Gaussian source with a power spectrum (7.22) requires about 2.3 bits/sample less than the memoryless Gaussian source (7.16) at high rates. The slope is again 6 dB/bit at high rates where the rate-distortion curve coincides with its Shannon lower bound.

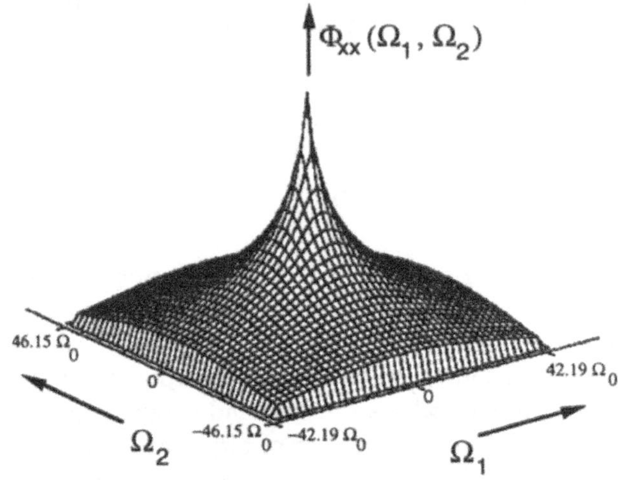

Figure 3 Power spectral density model of an image (vertical axis in logarithmic scale).

2 IMAGE SUBBAND DECOMPOSITION

Rate-distortion theory suggests that an efficient source coder splits the image into frequency bands and independently encodes the individual subband signals. While images are two-dimensional signals, we discuss one-dimensional subband coding first. Most two-dimensional subband decompositions are built by cascading one-dimensional subband filter banks.

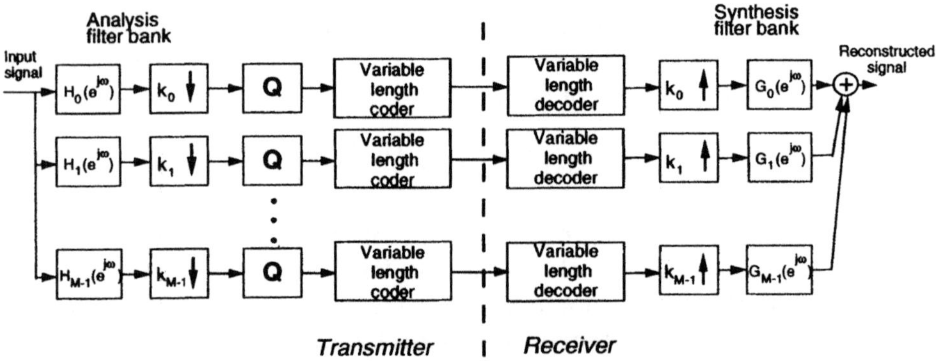

Figure 4 Diagram of a one-dimensional M-channel subband coding scheme.

A general one-dimensional subband coding scheme is shown in Figure 4. The sampled input signal is convolved in a bank of M bandpass analysis filters with frequency responses $H_i(e^{j\omega})$ and downsampled by factors k_i that correspond to the bandwidths of $H_i(e^{j\omega})$. The frequency responses and downsampling factors are usually chosen such that the number of samples before and after subband decomposition is the same, i.e.,

$$\frac{1}{k_0} + \frac{1}{k_1} + \ldots + \frac{1}{k_{M-1}} = 1. \tag{7.23}$$

Equation (7.23) ensures that there is no additional redundancy introduced by the subband decomposition. The subband signals are quantized and transmitted using an appropriate fixed or variable length code. At the decoder, the subband signals are decoded, upsampled by factors k_i and passed through a bank of synthesis filters with frequency responses $G_i(e^{j\omega})$. The output signals of the synthesis filters are finally summed up to yield the reconstructed signal.

2.1 Subband Filter Banks

As discussed in Chapters 1–4, there are many filter banks available for general use. However, when choosing filters for subband image decomposition, there are additional requirements that are specific to image coding [379]. Analysis filters should have a short impulse response to preserve the localization of image features. Synthesis filters should also have a short impulse response in order to prevent spreading of artifacts resulting from quantization errors at edges and other local features. Long synthesis filters often have very good mean squared error performance but lead to annoying ringing effects around edges. In addition, linear phase filters are desirable for subband image coding. Filters with nonlinear phase introduce subjectively unpleasant waveform distortions, when the lowpass channel is viewed by itself.

Many filters have been suggested for subband image coding [216, 425, 162, 237, 153, 1, 198]. We have used the 9-tap QMFs

$$\begin{aligned} H_0(e^{j\omega}) = G_0(e^{j\omega}) &= 0.5645751 \\ &+ 0.2927051(e^{j\omega} + e^{-j\omega}) \\ &- 0.05224239(e^{j2\omega} + e^{-j2\omega}) \\ &- 0.04270508(e^{j3\omega} + e^{-j3\omega}) \\ &+ 0.01995484(e^{j4\omega} + e^{-j4\omega}) \\ H_1(e^{j\omega}) &= H_0(e^{j(\omega+\pi)}) \\ G_1(e^{j\omega}) &= -G_0(e^{j(\omega+\pi)}). \end{aligned} \tag{7.24}$$

The reconstruction error introduced by the QMF bank is at -50 dB and thus smaller than the typical quantization error. The stopband attenuation of the analysis and synthesis filters is about 35 dB.

It should be noted that these filters require that the subsampling points be staggered between the two subbands, i.e., the low band is sampled at positions $0, 2, 4, \ldots$, and the high band is sampled at the positions $1, 3, 5, \ldots$. The alternating sampling preserves the information more uniformly and gives better reconstruction than even length QMFs with coinciding sampling in both bands.

2.2 Equal Band Splitting

Subband decomposition of images into more than two bands can conveniently and efficiently be carried out by cascading horizontal and vertical two-band filter banks. As an example, Figures 5 and 6b illustrate a decomposition into 4×4 subbands. Any two-band filter bank, such as the short-kernel filters from [153] or the QMFs (7.24) can be employed in this structure. Note that the filter bank in Figure 5 is separable in horizontal and vertical directions, i.e. the resulting frequency responses of the individual subbands can be written as a product of 1-D functions of ω_1 and ω_2.

Rather than cascading two-band filter banks, we can also use an M-channel filter bank for image subband decomposition. An example of a separable filter bank with $M \times M$ channels is the Lapped Orthogonal Transform [254]. The majority of filter banks investigated for image coding are separable into horizontal and vertical filter banks. Such filter banks perform better on strictly horizontal and vertical edge features than on diagonal ones. While horizontal and vertical edges fall into distinct subbands in decompositions such as the one shown in Figures 5 and 6b, diagonal edges with 45° and 135° orientations fall into the same subbands. Non-separable subband filter banks with directional characteristics have been investigated by [369, 37]. However, the computational effort required for non-separable filter banks [369] is often prohibitively large.

2.3 Discrete Cosine Transform as a Subband Coding Technique

The discrete cosine transform (DCT), introduced by Ahmed, Natarajan, and Rao in 1974 [2, 333] has been adopted in several image coding standards in

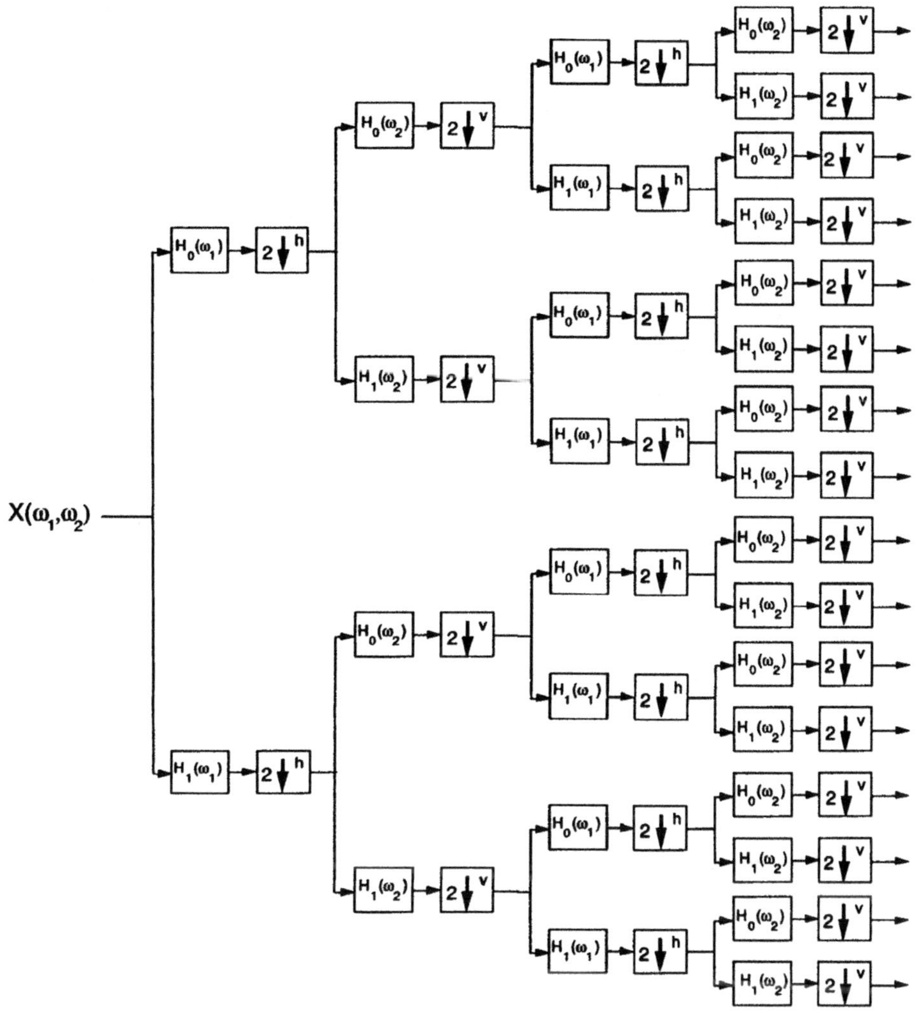

Figure 5 Illustration of (4×4)-band analysis filter bank built out of two-band filter banks. The symbol $2{\downarrow}^h$ denotes horizontal subsampling by a factor of 2, and $2{\downarrow}^v$ denotes vertical subsampling by a factor of 2.

the last years. We will show in this section that the DCT can be interpreted as a special subband coding technique. In transform image coding, the image is subdivided into square blocks, and each block undergoes an orthonormal transformation $\mathbf{A}$ (Figure 7). We can write the transformation as a matrix multiply

$$\mathbf{c} = \mathbf{A}\mathbf{x} \tag{7.25}$$

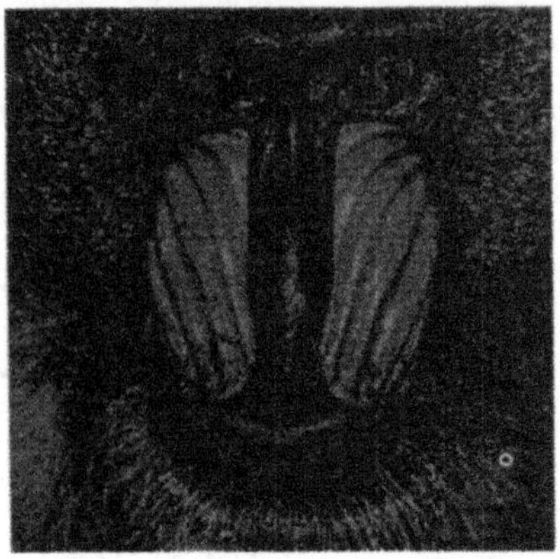

Figure 6a Original image Mandrill.

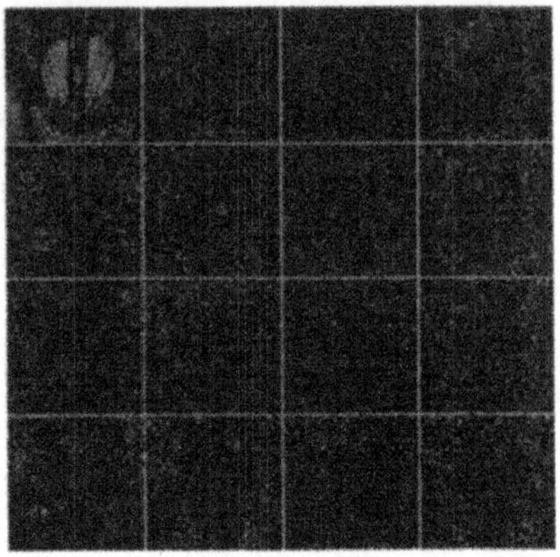

Figure 6b Mandrill decomposed into 4 × 4 bands (using short-kernel filters from [153]).

where $\mathbf{x}$ is a column vector formed by the $M \times M$ samples of an image block, $\mathbf{A}$ is an $M^2 \times M^2$ matrix, and $\mathbf{c}$ is the vector of M^2 transform coefficients. For an orthonormal transform,

$$\mathbf{x} = \mathbf{A^{-1}c} = \mathbf{A^{T}c}, \tag{7.26}$$

i.e. the inverse transform matrix is the transpose of the forward transform matrix. The equation $\mathbf{x} = \mathbf{A^{-1}c}$ can be interpreted as a representation of an image block by a superposition of "basis functions" (columns of $\mathbf{A^{-1}}$) which are weighted by coefficients c_k (elements of vector $\mathbf{c}$).

A two-dimensional $M \times M$ DCT is separable into a horizontal and a vertical one-dimensional M-point DCT. The elements a_{kn} of the transform matrix $\mathbf{A}$ for the DCT ("DCT-II" after the classification in [333]) of length M are

$$a_{kn} = \alpha_k \cos\frac{\pi(2n+1)k}{2M} \quad \text{with} \quad \alpha_0 = \sqrt{\frac{1}{M}},\ \alpha_k = \sqrt{\frac{2}{M}}\ \forall k \neq 0. \tag{7.27}$$

Figure 8 shows the 64 basis functions for an 8×8 DCT.

The DCT is closely related to the discrete Fourier transform (DFT). It can be shown that the DFT is nearly optimum for images with stationary signal statistics when the block size is very large [313]. The argument is similar to the rate-distortion analysis presented in Section 1.4 that provided the insight that ergodic sources should be decomposed and coded in independent subbands. For

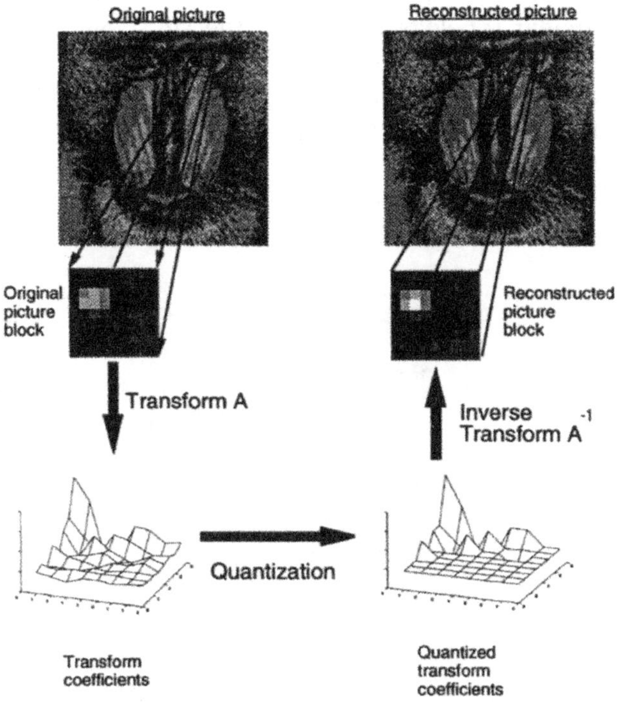

Figure 7 Principle of transform coding.

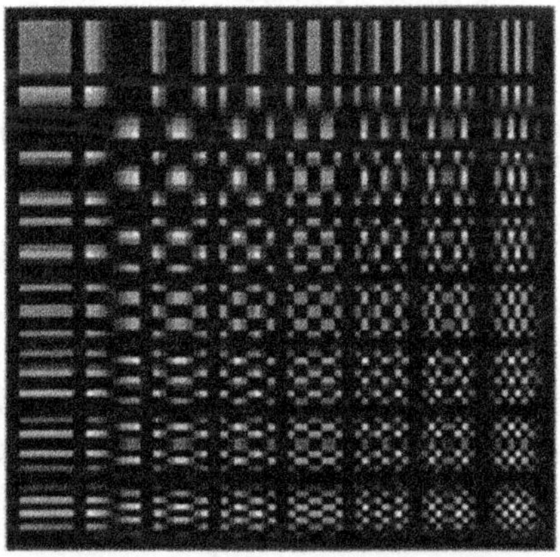

Figure 8 Basis functions of an 8×8 DCT.

finite block sizes, block boundary effects with DFT coding are very annoying owing to the circular topology underlying the DFT. Since, with a DFT, the right edge of a block is "connected" to the left edge and the top edge to the bottom edge, the signal discontinuity at the block edge leads to severe Gibbs ringing when higher DFT coefficients are suppressed in a coder. This problem is greatly reduced by mirroring an $M \times M$ block horizontally and vertically such that a $2M \times 2M$ block results (Figure 9).

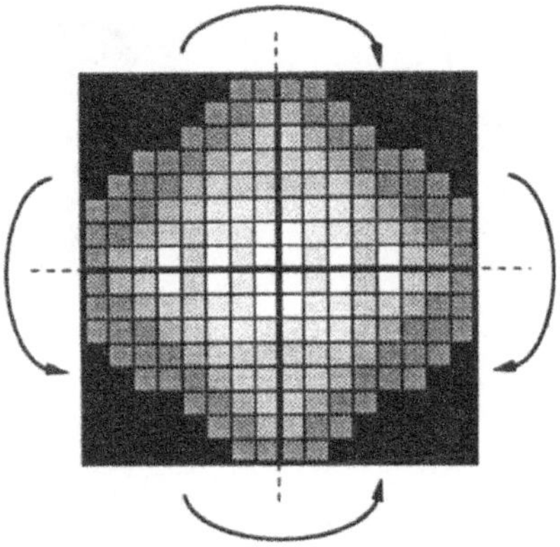

Figure 9 Horizontal and vertical block mirroring leads to an interpretation of the DCT in terms of the DFT.

The $2M \times 2M$ block is then transformed using the DFT. Due to the symmetry of the mirrored block, the transform has only $M \times M$ degrees of freedom and only $M \times M$ non-zero coefficients remain. They are the coefficients of a new transform, the discrete cosine transform, with the transform matrix **A** (7.27).

An $M \times M$ DCT can be viewed as a subband decomposition into $M \times M$ subbands. The impulse responses of the analysis and synthesis filters corresponding to an M-point DCT are exactly M samples long. In the analysis filter bank, the image is convolved separately with each row of the transform matrix **A** (7.27), and then each subband signal is downsampled by a factor $k_i = M$. Values that are omitted in the downsampling stage need not be calculated in the analysis filter convolutions. Since the impulse responses are not longer than the downsampling factors, the input signal can be subdivided into non-overlapping portions and processed in independent blocks. A similar argument applies to the inverse DCT as a synthesis filter bank. Note that the DCT is in fact a perfect reconstruction filter bank with identical analysis and synthesis filters, except for a horizontal and vertical reversal of the filter impulse responses. Figure 10 shows the frequency response of a forward (and, at the same time, an inverse) 8-point DCT when viewed as a filter bank. The frequency selectivity of the DCT is remarkably poor. On the other hand, the compact support of the filter impulse responses is a very desirable property. Exact image subband decomposition and reconstruction using the DCT are described in [126]. Approaches combining the DCT and subband filter banks are described in [46, 111].

2.4 Nonuniform Band Splitting

When a signal is spectrally flat within a band, memoryless coding is optimal. Figure 3 shows the model of a typical image power spectral density (7.22). Apparently, the spectrum varies more rapidly at low frequencies than at high frequencies. When decomposing the spectrum (7.22) into equal bands, the high bands will be spectrally flatter than the low bands. This suggests a non-uniform band splitting that decomposes the lower frequency part into narrower bands than the higher frequency range.

A signal decomposition into infinitesimally narrow subbands is suggested by the rate-distortion theoretical argument in Section 1.4. This argument applies to ergodic Gaussian sources. Images, however, contain spatially localized features, such as edges, and very narrow subbands do not yield good coding results. The impulse responses of filters producing narrow subbands are necessarily long, and the localization of image features is poor after filtering. In principle, the same features can occur in an image on all scales, due to the perspective projection of objects at all distances from the camera onto the image plane. This suggests using short filter impulse responses for high frequency channels and long impulse responses for low frequency channels.

A powerful non-uniform subband decomposition is the subband pyramid illustrated in Figure 11. The image spectrum is split into four bands containing horizontal lows/vertical lows (LL), horizontal lows/vertical highs (LH), horizontal highs/vertical lows (HL), and horizontal highs/vertical highs (HH). This four-band decomposition can be accomplished by cascading a horizontal and a vertical two-band filter bank. In the next stage, the same four band decomposition is applied to the LL component only. The LH, HL, and HH components are left alone. The procedure is repeatedly applied to each resulting low frequency band, resulting in a decomposition into octave bands.

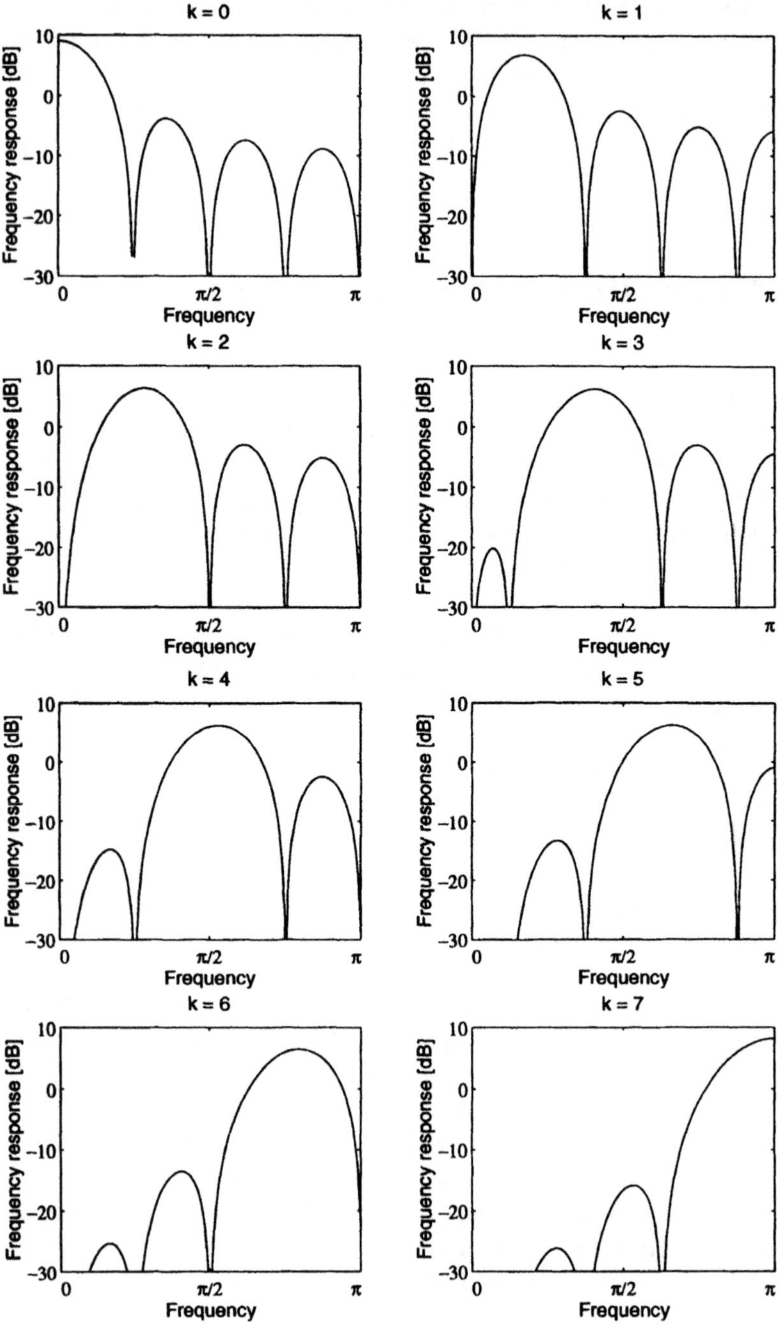

Figure 10 Subband frequency response of 8-point DCT.

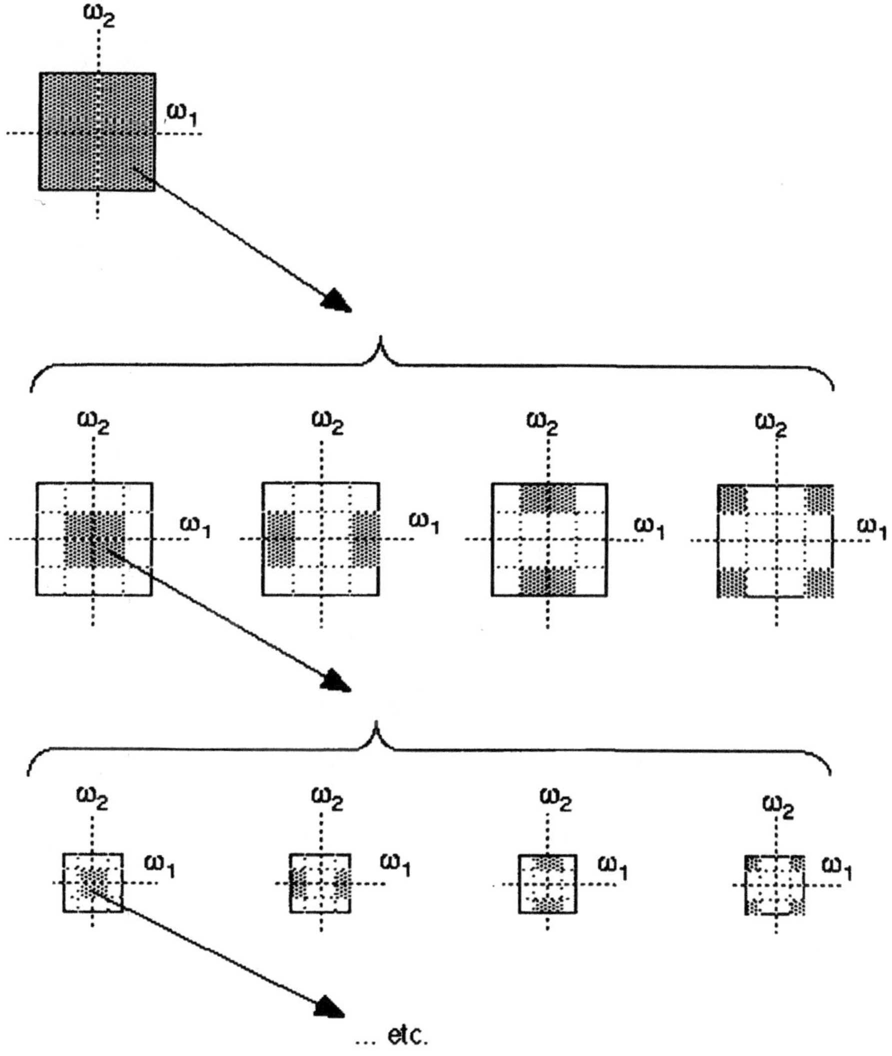

Figure 11 Subband pyramid decomposition in the frequency domain.

Figure 12 Image Mandrill (Figure 6a) decomposed by a four level subband pyramid.

Figure 12 shows an image decomposed into a subband pyramid with four levels. The QMFs (7.24) were used for the two-band decomposition at each level of the subband pyramid.

Besides critically sampled subband pyramids (Equation (7.23)), oversampled pyramid decompositions can be employed for image coding. They were first introduced as bandpass pyramids by Burt and Adelson in 1983 [67] and successively refined, e.g. by [193, 432]. An input picture is first lowpass filtered and downsampled. By interpolating the low resolution image back to its original resolution, a prediction of the original image is obtained. The resulting prediction error image is a bandpass signal containing the missing detail of the prediction from the lower resolution. This decomposition into a lower resolution image and a prediction error image can be repeated for the lower resolution image to obtain more pyramid layers. Note that the number of samples increases by up to $1/3$ for two-to-one two-dimensional subsampling. On the other hand, one gains complete freedom in choosing appropriate filters. Lower resolution images within predictive resolution pyramids often have better subjective image quality than those obtained from critically sampled subband pyramids.

3 COMPRESSION OF IMAGE SUBBAND SIGNALS

After the decomposition stage, the image is split into subimages that contain spectral components of the original image. If the subimages are critically sampled, the total number of samples is the same as in the original image. The decomposition is still fully reversible, and no compression has been accomplished. Quantization can be performed on single subband coefficients [201, 422, 240] ("scalar quantization"), or on several coefficients together ("vector quantization (VQ)"). Vector quantization techniques have been proved to be very powerful for quantization of subbands [46, 438, 231, 363, 152, 134, 276, 29]. In Section 3.1, scalar quantization is explained. In Section 3.2, different approaches to vector quantization are presented. In Section 3.3, the performance of different quantization schemes is illustrated by coding experiments on images. Finally, Section 3.4 discusses bit allocation among subbands.

3.1 Scalar Quantization

A scalar quantizer rounds each sample of the input x independently to the nearest representative output level. Figure 13 shows an example of the staircase input-output characteristic of a scalar quantizer. The input signal amplitude within a certain range is mapped onto one common representative output level. The levels that separate the input signal ranges are called decision thresholds. Simultaneously finding the N representative levels and $N-1$ decision thresholds of a quantizer yielding minimum mean squared quantization error for a given probability density function (pdf) $p_X(x)$ of the input signal x is a classical problem that has been solved by Lloyd [248] and Max [258] independently. The solution, the "Lloyd–Max quantizer," has the following properties:

- The decision thresholds lie exactly half-way between the representative levels.
- The representative levels are the centroids of the pdf between successive decision thresholds.

Since the optimum decision thresholds depend on the representative levels, while the optimum representative levels depend on the decision thresholds, a closed form solution of the Lloyd–Max quantization problem can usually not be found except for some trivial cases. Iterative schemes have to be used that converge to the optimum solution.

For sufficiently fine quantization and a smooth pdf, an approximate solution of the Lloyd–Max quantization problem can be obtained according to

$$\Delta\hat{x}(x) = c \cdot (p_X(x))^{-\frac{1}{3}}, \tag{7.28}$$

where $\Delta\hat{x}(x)$ is the distance between two successive quantizer representative levels. Interestingly, the approximation (7.28) was proposed by Panter and Dite [309] before the exact solution was known. The Panter and Dite approximation also provides an estimate of the resulting quantization error variance

$$\sigma_q^2 = \frac{1}{12N^2}\Big(\int_x (p_X(x))^{\frac{1}{3}} dx\Big)^3, \tag{7.29}$$

where N is again the number of representative levels.

The Lloyd–Max quantizer and the Panter and Dite approximation minimize the mean squared quantization error for a given number of representative levels.

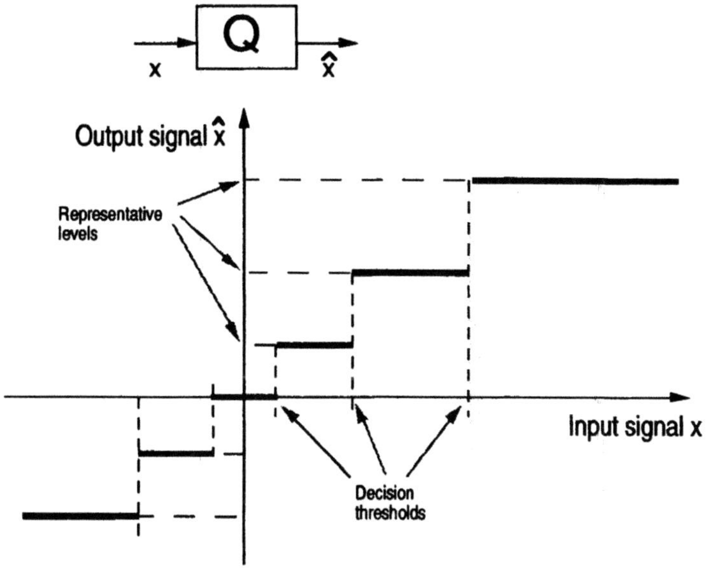

Figure 13 Example for a quantizer input-output characteristic.

The solution implies that each representative level is coded by a fixed codeword length. Often, we want to employ an entropy coder after the quantizer which generates codewords of variable length in order to approach the entropy of the quantized signal. In this case, it is not appropriate to minimize mean squared quantization error subject to a fixed number of representative levels, but rather subject to a fixed entropy of the quantized signal $\hat{x}$. It can be observed that a uniform quantizer often yields a better result than a Lloyd–Max quantizer, when followed by entropy coding. In fact, for sufficiently fine quantization, a uniform quantizer yields minimum entropy [446].

Scalar quantization followed by entropy coding is often used in the context of subband image coding [201, 422, 240]. The amplitude histogram of a typical subband signal is shown in Figure 14. It can be approximated well by a Laplacian pdf. Optimum scalar quantizers for Laplacian pdfs and Gamma pdfs have been investigated extensively by Brusewitz [66]. His work shows that uniform quantizers with a representative level at zero and a larger threshold around zero yield excellent results even at low rates (Figure 15). Uniform threshold quantizers are employed today in the DCT-based image coding standards ITU-T H.261, MPEG, and JPEG [207, 208, 76].

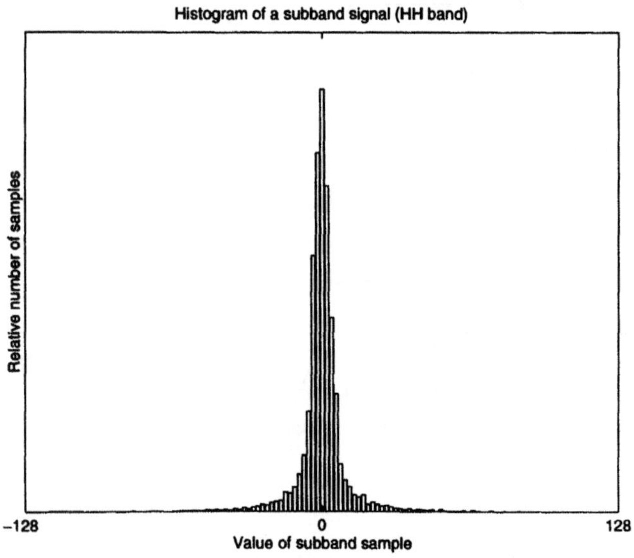

Figure 14 Histogram of a subband signal.

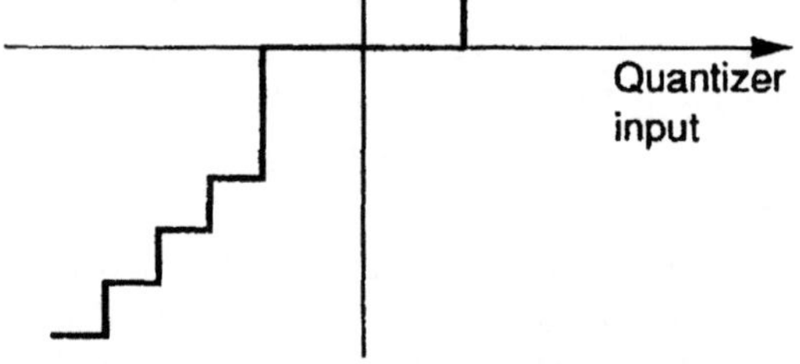

Figure 15 Uniform threshold quantizer.

3.2 Vector Quantization Techniques for Image Subband Coding

A subband decomposition reduces or even eliminates the correlation between image samples, but there are still statistical dependencies of higher order within or across subband signals unless the signal statistics are Gaussian. A vector quantizer, which jointly quantizes several samples, can also exploit these statistical dependencies of higher order. The input to a vector quantizer can be formed by combining several neighboring samples within an image subband [363], or spatially coinciding samples from different subbands [438], or a combination of both. In [134] several cases for forming vectors within a subband coding scheme for color images are investigated. In this section, we introduce some basic vector quantization (VQ) techniques as well as lattice vector quantization. A broader overview can be found in [285, 158].

Fixed-Wordlength VQ

In 1980, Linde, Buzo, and Gray generalized the Lloyd algorithm for the design of nonuniform scalar quantizers to yield a codebook for vector quantizers containing a fixed number of codevectors [247]. Their algorithm, known as the "LBG algorithm" or "generalized Lloyd algorithm," computes a codebook with minimum average distortion for a given training set and given codebook size. In the following, let us assume squared error as the distortion measure, which is, for k-dimensional vectors $\mathbf{x}$ and $\mathbf{y}$, defined as

$$d(\mathbf{x},\mathbf{y}) = \frac{1}{k}\sum_{i=1}^{k}(x_i - y_i)^2. \tag{7.30}$$

The algorithm starts with an initial codebook $C = \{\mathbf{r}_i; i = 1,\ldots,N\}$ containing N representative vectors $\mathbf{r}_i$ and a training set $T = \{\mathbf{x}_j; j = 1,\ldots,M\}$ of size $M \gg N$. In the first step for T a minimum distortion partitioning P is computed as

$$\begin{aligned} P &= \{S_i; i = 1,\ldots,N\} \\ S_i &= \{\mathbf{x} \mid d(\mathbf{x},\mathbf{r}_i) \leq d(\mathbf{x},\mathbf{r}_j),\ \ \forall j \neq i \ \wedge\ \mathbf{x} \notin S_k\ \forall k < i\}. \end{aligned} \tag{7.31}$$

Partition S_i contains all vectors $\mathbf{x}$ which can be reproduced by $\mathbf{r}_i$ with least possible distortion. Ties are broken by favoring the partition with the smallest index. P is then the best partition for codebook C, but unfortunately C is not optimal. To obtain the best codebook C^* for our partitioning P we compute for each partition S_i a centroid $\mathbf{r}_i^*$ which serves as its new representative vector:

$$C^* = \{\mathbf{r}_i^* \mid \mathbf{r}_i^* = \frac{1}{|S_i|} \sum_{j:\mathbf{x}_j \in S_i} \mathbf{x}_j\}, \tag{7.32}$$

where $|S_i|$ denotes the number of vectors belonging to partition S_i. Since we have obtained a new codebook C^* we need to recompute the partitioning of our training set. Therefore we set C to C^* and repeat the partitioning (7.31) and centroid calculation (7.32) until the average distortion

$$D_{\text{avg}} = \frac{1}{M} \sum_{j=1}^{M} \min_{\mathbf{r} \in C} d(\mathbf{x}_j, \mathbf{r}) \tag{7.33}$$

obtained with codebook C cannot be significantly decreased by further iterations.

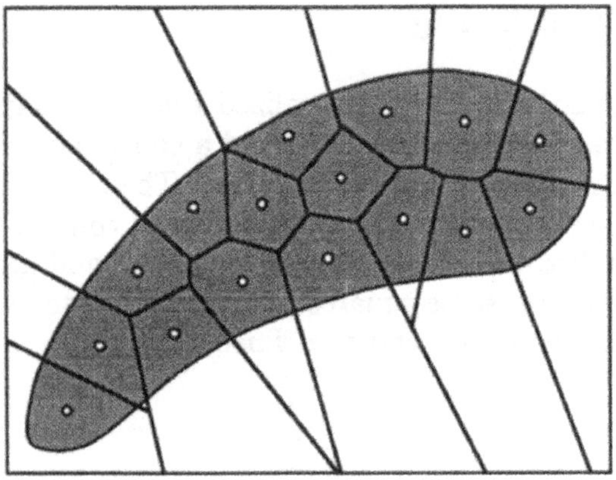

Figure 16 Two-dimensional vector space with uniformly distributed data quantized by full search LBG VQ.

Figure 16 shows how a two-dimensional vector space is partitioned by applying the LBG algorithm for $N = 16$ using Equation (7.30) as the distortion measure. The training set is represented by the shaded banana-shaped region wherein training vectors are uniformly distributed. White dots stand for the representatives of the final codebook.

A disadvantage of the codebook found by the LBG algorithm is, that it is unstructured. This implies that a codevector can only be found by a full search algorithm which requires roughly kN subtractions, multiplications, and additions. On the other hand, decoding is simply a table look-up. We will compare unstructured vector quantizers for fixed word length encoding using the LBG algorithm and the mean squared error distortion measure for subband image coding in Section 3.3.

Entropy-Constrained VQ

The LBG algorithm can be extended for codebook design under an entropy constraint [79]. Entropy-constrained vector quantization (ECVQ) minimizes distortion for a given average codeword length rather than a given codebook size. The average codeword length constraint is built into the codebook design by a Lagrange multiplier formulation. Instead of just minimizing the average distortion D, now $D + \lambda R$ is minimized, where R is the average bit-rate.

Let $l(i)$ denote the codeword length needed to encode the representative vector $\mathbf{r}_i$. Then the Lagrange formulation can be easily incorporated into the LBG algorithm described in the previous section. The LBG rule (7.31) for computing the minimum distortion partition P now contains an entropy constraint:

$$\begin{aligned} P &= \{S_i;\ i = 1, \ldots, N\} \qquad (7.34) \\ S_i &= \{\mathbf{x} \mid d(\mathbf{x}, \mathbf{r}_i) + \lambda l(i) \leq d(\mathbf{x}, \mathbf{r}_j) + \lambda l(j), \forall j \neq i \ \wedge \ \mathbf{x} \notin S_k \ \forall\ k < i\}. \end{aligned}$$

A vector $\mathbf{x}$ will be mapped to partition S_i if the distortion between $\mathbf{x}$ and $\mathbf{r}_i$ biased by the codeword length $l(i)$ is minimized. Ties are broken as in Equation (7.31). In other words, the best representative for $\mathbf{x}$ now is the one that yields the best tradeoff between distortion and codeword length. Computation of the optimal codebook C^* for P is carried out according to Equation (7.32). The original LBG algorithm for fixed codeword length is a special case with $\lambda = 0$.

In [79] it is proposed to use $l(i) = \log_2(1/p(i))$ as the "ideal" codeword length for $\mathbf{r}_i$ with $p(i) = M/|S_i|$. This assumes that noninteger codeword lengths are allowed. Another possibility incorporates the construction of a Huffman code [195] into the codebook design. In [79] it is reported that assuming noninteger codeword lengths during codebook design performs nearly identical to a system with an incorporated Huffman algorithm.

Figure 17 shows the changed subdivision of the vector space compared to Figure 16 if an entropy-constraint is built into the LBG algorithm. Since the codeword length is taken into account in the codebook design, the decision hyperplanes separating the regions in vector space with the smallest Euclidean distance around individual representatives are shifted compared to the fixed word length VQ. As an example, consider an input vector that is approximately equally close to two reproduction vectors. It is advantageous to choose

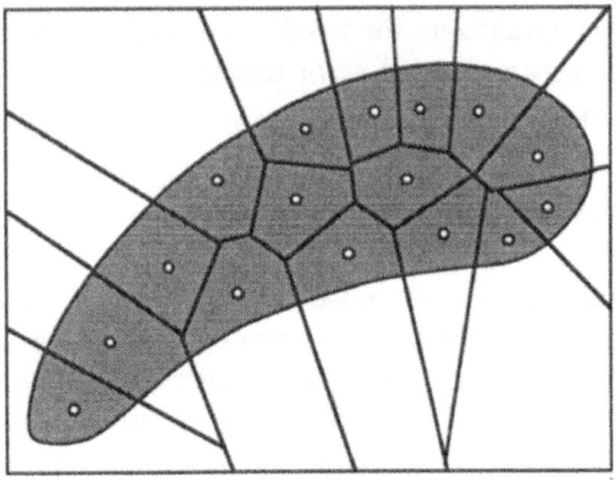

Figure 17 Two-dimensional vector space with uniformly distributed data quantized by full search LBG VQ with an entropy constraint.

the representative with the shorter codeword length, even though its associated distortion might be slightly larger. Since there are more bits remaining, the resulting overall distortion will be smaller. When designing an ECVQ codebook, the initial codebook size has to be sufficiently large. Otherwise, the resulting codebook will not achieve the theoretical minimum distortion for a given rate [79]. Redundant vectors are automatically eliminated in the iterative design process.

Subband image coding with ECVQ codebooks is compared to other quantization schemes in Section 3.3.

Lattice VQ

Another important class of vector quantizers is lattice quantizers, which are characterized by a highly regular codebook structure. We say that a regular arrangement of points in n-dimensional space is a lattice, Λ_n. A quantizer that uses the points of a lattice as codevectors is said to be a lattice vector quantizer. Compared to VQ with unstructured codebooks, lattice VQ has a much lower encoder complexity at interesting bit-rates. Fast quantization and decoding algorithms for several lattices can be found in [103] and [102]. As an example we describe later in this section a fast quantization algorithm for an eight-dimensional lattice. In Section 3.3, we shall present results for lattice vector quantizers based on various lattices.

An n-dimensional lattice, Λ_n, is defined as

$$\Lambda_n = \{\mathbf{x} \in R^m \mid \mathbf{x} = \mathbf{zG}, \mathbf{z} \in Z \times Z \times \ldots \times Z = \mathbf{Z}^n\} \tag{7.35}$$

where $\mathbf{G}$, the generator matrix of the lattice, is an $m \times n$ matrix ($m \geq n$) with n linear independent rows belonging to R^m. Some lattices can be defined in a more convenient way. For example, for $n \geq 3$ the D_n lattice can be described as

$$D_n = \{\mathbf{x} \in Z^n \mid \sum_{i=1}^{n} x_i \ \text{ even}\}. \tag{7.36}$$

It contains all points of the Z^n-lattice whose coordinate sum is even. Lattices can often be expressed as a union of cosets of simpler lattices. A coset of a lattice is a lattice resulting from adding a fixed vector to each lattice point. For instance the E_8-lattice can be defined as

$$E_8 = D_8 \bigcup \left[D_8 + (\frac{1}{2}, \frac{1}{2}, \frac{1}{2}, \frac{1}{2}, \frac{1}{2}, \frac{1}{2}, \frac{1}{2}, \frac{1}{2})\right], \tag{7.37}$$

which is the union of the D_8-lattice and a coset of the D_8-lattice obtained by adding $\mathbf{\frac{1}{2}} = (\frac{1}{2}, \ldots, \frac{1}{2})$ to each lattice point. The E_8-lattice may also be defined by the generator matrix

$$\mathbf{G}_{E_8} = \begin{bmatrix} 2 & 0 & 0 & 0 & 0 & 0 & 0 & 0 \\ 1 & 1 & 0 & 0 & 0 & 0 & 0 & 0 \\ 1 & 0 & 1 & 0 & 0 & 0 & 0 & 0 \\ 1 & 0 & 0 & 1 & 0 & 0 & 0 & 0 \\ 1 & 0 & 0 & 0 & 1 & 0 & 0 & 0 \\ 1 & 0 & 0 & 0 & 0 & 1 & 0 & 0 \\ 1 & 0 & 0 & 0 & 0 & 0 & 1 & 0 \\ \frac{1}{2} & \frac{1}{2} & \frac{1}{2} & \frac{1}{2} & \frac{1}{2} & \frac{1}{2} & \frac{1}{2} & \frac{1}{2} \end{bmatrix}. \tag{7.38}$$

Another important lattice is the A_n-lattice. It can be defined by the $((n+1) \times n)$-dimensional generator matrix

$$\mathbf{G}_{A_n} = \begin{bmatrix} -1 & 1 & 0 & 0 & \ldots & 0 & 0 \\ 0 & -1 & 1 & 0 & \ldots & 0 & 0 \\ 0 & 0 & -1 & 1 & \ldots & 0 & 0 \\ \vdots & \vdots & \vdots & \vdots & & \vdots & \vdots \\ 0 & 0 & 0 & 0 & \ldots & -1 & 1 \end{bmatrix}. \tag{7.39}$$

It can be shown, that for a smooth multidimensional probability density function and sufficiently fine quantization, a lattice vector quantizer derived from the densest multidimensional sphere packing can approach the minimum mean squared quantization error for a given entropy [452]. Therefore lattices of interest are mainly those corresponding to densest sphere packings. For instance, in two dimensions, it is known that a hexagonal partitioning of the space corresponds to the densest packing. The corresponding hexagonal lattice is called the A_2-lattice (see Figure 18). In 3-dimensional space, the A_3-lattice is known to yield the densest sphere packing. Unfortunately in higher dimensions, lattices corresponding to densest sphere packings are currently only known for 8 dimensions (E_8-lattice) and 24 dimensions (Leech lattice).

Let us now consider an example of a fast lattice quantization algorithm that finds the D_n-lattice point closest to a given input vector. For $x \in R$, let $f(x)$ be defined as the closest integer to x and $w(x) = f(x) + \text{sign}(x - f(x))$ with $\text{sign}(y) = 1$ if $y \geq 0$ and $\text{sign}(y) = -1$ otherwise. The D_n-lattice point closest to $\mathbf{x} \in R^n$ can be found in the following way: compute first $f(\mathbf{x}) = (f(x_1), \ldots, f(x_n))$, the point of Z^n which is closest to $\mathbf{x}$. If $f(\mathbf{x})$ does not have an even sum of components compute $g(\mathbf{x}) = (f(x_1), \ldots, w(x_i), \ldots, f(x_n))$, where x_i is the component furthest away from an integer. Whichever of $f(\mathbf{x})$ and $g(\mathbf{x})$ has an even sum of coordinates is the closest D_n-lattice point to $\mathbf{x}$.

If $\Phi_\Lambda(\mathbf{x})$ is the closest lattice point for a given lattice Λ then $\Phi_\Lambda(\mathbf{x} - \mathbf{t}) + \mathbf{t}$ is the closest point on the translated lattice $\Lambda + \mathbf{t}$. With this insight we now can formulate algorithms for lattices that can be defined as the union of lattice cosets, e.g. for the E_8-lattice. Knowing the algorithm for the D_8-lattice we can find a nearest E_8-lattice point by computing $\Phi_{E_8}(\mathbf{x})$ and $\Phi_{E_8}(\mathbf{x} - \frac{\mathbf{1}}{\mathbf{2}}) + \frac{\mathbf{1}}{\mathbf{2}}$ and then choosing among the two results the one closest to $\mathbf{x}$. This can be generalized for every lattice that can be described as a union of cosets.

The structure of a lattice gives an implicit partitioning of the vector space into lattice cells that are regular in contrast to the partitionings found by the LBG algorithm. We have already seen that the best codebook for a set of given partitions results from using each partition's centroid as its representative (see Equation (7.32)). This also can be applied to lattice quantizers and motivates the use of centroids as codevectors for lattice cells. The centroids are computed for a training set together with lattice cell probabilities and are stored in a special codebook. The probabilities can be used to construct for instance a Huffman code for the encoding of lattice cells. To keep memory requirements in reasonable bounds, centroids and probabilities should only be stored for the most popular lattice cells. In the other lattice cells the lattice point itself is used for reconstruction and equal probabilities are assumed.

This is illustrated in Figure 18 for an A_2 lattice vector quantizer. The use of the centroid instead of a lattice point is best seen for cells that are only partially covered by the shaded training set.

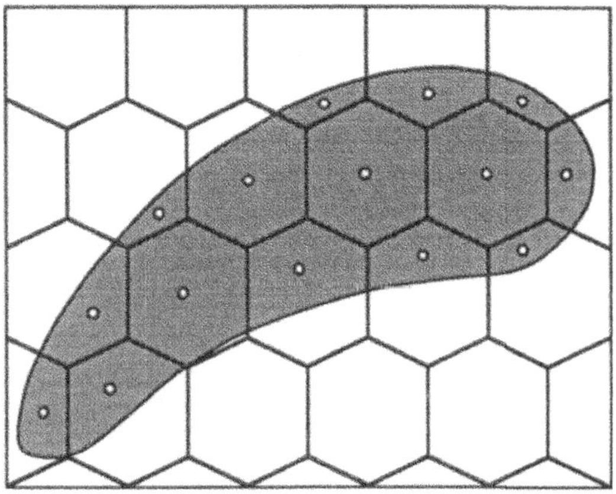

Figure 18 2-D vector space with uniformly distributed data quantized by A_2-lattice VQ.

3.3 Experimental Comparison of Vector Quantizers for Image Subbands

In this section, various vector quantization techniques as described in the previous section are compared experimentally, using both synthetic source data and natural images decomposed by a subband pyramid. In each case, we have measured the distortion over a range of bit-rates. Comparisons for other quantization techniques like pyramid vector quantization can be found in [56].

Figure 19 shows the performance of 8-dimensional VQ for a first-order Gauss-Markov source with a correlation coefficient of $r = 0.95$. This corresponds to the typical correlation of adjacent samples in a baseband image. For comparison, the Shannon lower bound of the rate-distortion function (7.13) is shown in Figure 19 as well. Unstructured VQ with or without variable length coding both perform well. It is somewhat surprising that ECVQ does not achieve any gain over fixed codeword length (CWL) VQ. The E_8 lattice vector quantizer is about 2 dB worse than unstructured VQ. Optimum scalar quantization followed by entropy coding is also shown. It does not exploit the correlation between successive samples and thus performs poorly.

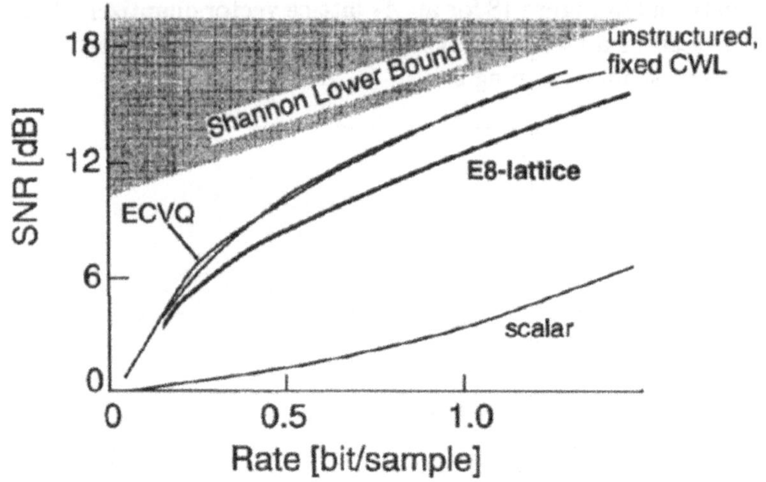

Figure 19 Distortion rate curves for 8-D vector quantization of a Gauss-Markov source with correlation coefficient $r = 0.95$.

Subband images are much less correlated than the full band image. Their amplitude distribution resembles a Laplacian distribution. We have therefore compared the various VQs for data from a memoryless Laplacian source. Results for 8-D VQ are shown in Figure 20. The curves are dramatically different from the results shown in Figure 19. Since the source is memoryless, gains beyond scalar quantization are small. Unstructured fixed word length VQ performs poorly at low bit-rates. E_8-lattice VQ performs as well as unstructured ECVQ at low bit-rates.

The following results have been obtained for the well-known image "Lenna" (of size 512×512), decomposed in a subband pyramid. The vector quantizers were obtained using a training set of four images. "Lenna" was not contained in the training set. We have measured distortion rate curves for each individual subband for each of the vector quantizers. It is beyond the scope of this chapter to present these results comprehensively. Rather, we show two typical examples in Figures 21 and 22. The presentation has been changed compared to Figures 19 and 20. It is more convenient to work with mean squared error curves rather than SNR for optimum bit allocation, as discussed in the next section.

Figure 21 shows results for 8-D vector quantization of the horizontal high band of the first pyramid level. Unstructured VQ with fixed word length performs even worse than optimum scalar quantization with (memoryless) entropy coding. As expected, unstructured VQ with entropy coding performs best. The

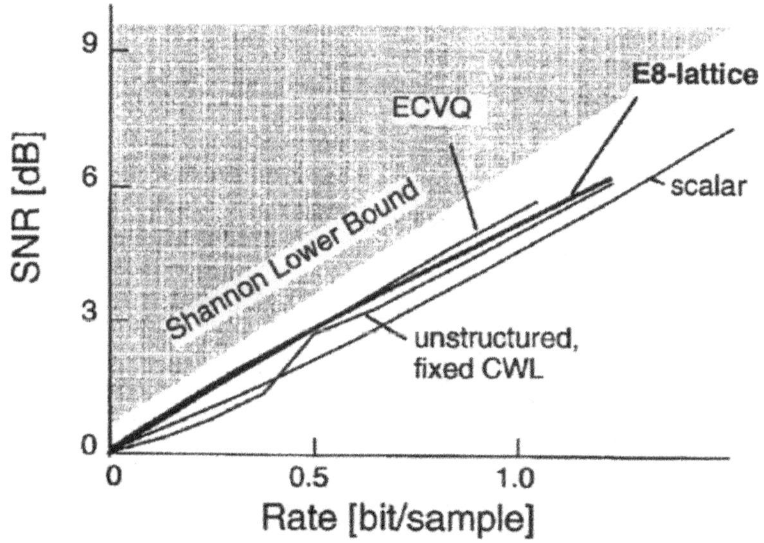

Figure 20 Distortion rate curves for 8-D vector quantization of a memoryless Laplacian source.

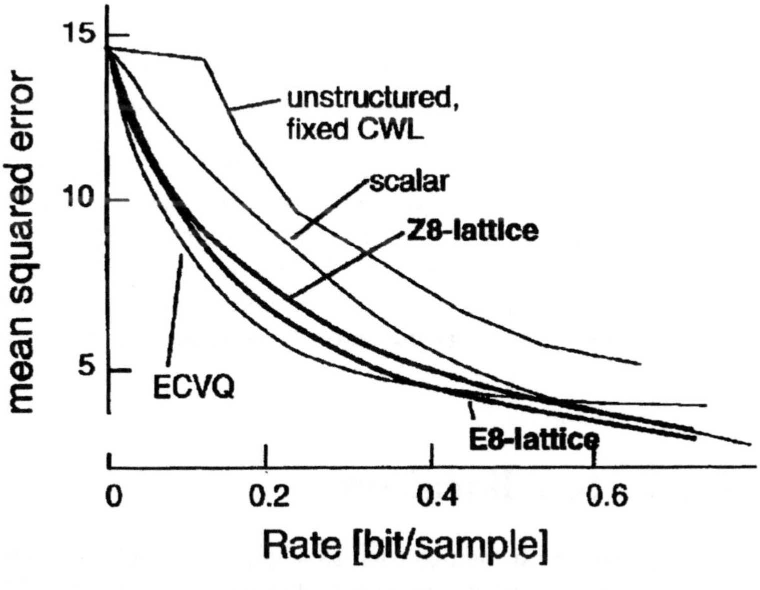

Figure 21 Distortion rate curves for 8-D vector quantization of the horizontal high frequency band of the first pyramid level.

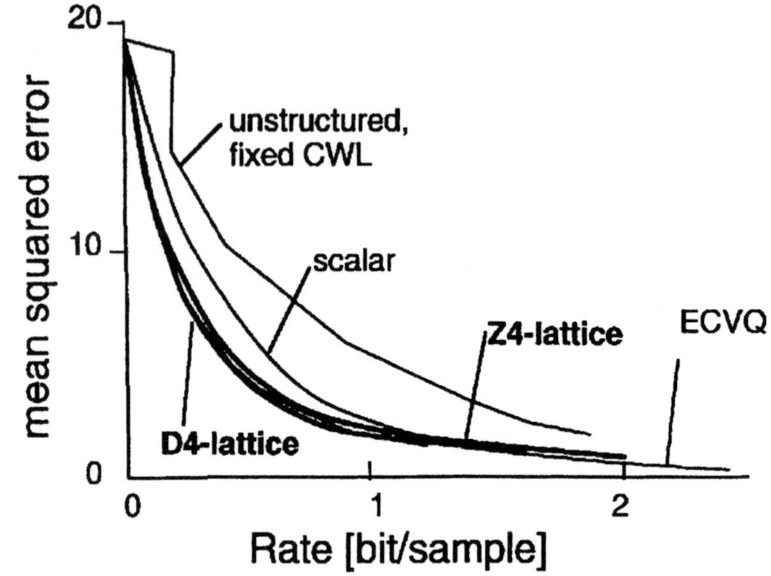

Figure 22 Distortion rate curves for 4-D vector quantization of the vertical high frequency band of the second pyramid level.

E_8-lattice VQ is close to the optimum. At higher rates the training set becomes too small to reliably train all the degrees of freedom of the unstructured ECVQ codebook, and its performance becomes slightly worse than E_8-lattice VQ. Interestingly, the simple orthogonal Z^8-lattice performs almost as well as the E_8-lattice. Since it requires even less computation than E_8, it is certainly a very interesting alternative. Figure 22 shows similar results for 4-D VQ of the vertical high band of the second pyramid level. The unstructured ECVQ performs best, but there is practically no loss when a highly structured D_4-lattice VQ or an even simpler orthogonal Z^4-lattice VQ is used. VQ with fixed word length does not perform well at all, and it is even worse than optimum scalar quantization. A more detailed discussion of these results has been presented in [363].

3.4 Bit-Rate Allocation

A subband image coder can be regarded as a special case of the system shown in Figure 23. An original image x is represented by M independent bit-streams with individual bit-rates R_i, yielding an overall rate

$$R = R_0 + R_2 + ... + R_{M-1}. \tag{7.40}$$

At the decoder, a signal $\hat{x}$ is reconstructed that deviates from x by some average distortion D.

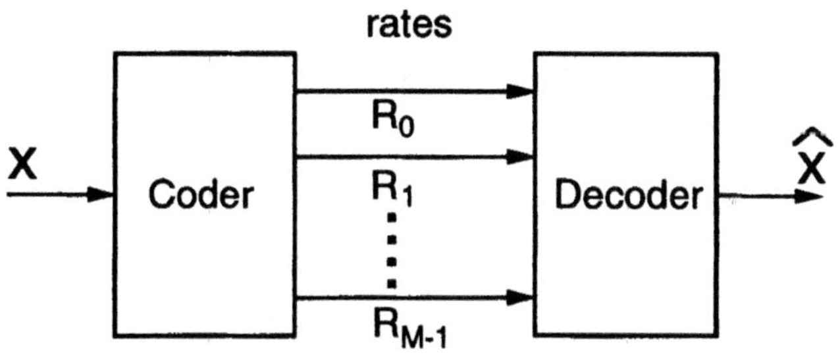

Figure 23 Codec representing signal x by M parallel independent bit-streams.

We assume that the multivariate distortion rate function $D(R_0, R_1, ..., R_{M-1})$ is strictly convex and differentiable everywhere, and that

$$\frac{\partial D}{\partial R_i} \leq 0, \forall i, \tag{7.41}$$

i.e. increasing the rate for any one of the bit-streams will decrease the distortion D. We can find the optimum bit allocation that minimizes D subject to a fixed overall rate R by setting the total differential of the distortion to zero, i.e.

$$dD = \frac{\partial D}{\partial R_0} dR_0 + \frac{\partial D}{\partial R_1} dR_1 + ... + \frac{\partial D}{\partial R_{M-1}} dR_{M-1} = 0. \tag{7.42}$$

From Equation (7.40), we obtain

$$dR_0 + dR_1 + \ldots + dR_{M-1} = 0. \tag{7.43}$$

From Equations (7.42) and (7.43) follows the optimum bit allocation condition

$$\frac{\partial D}{\partial R_0} = \frac{\partial D}{\partial R_1} = \ldots = \frac{\partial D}{\partial R_{M-1}}. \tag{7.44}$$

Equation (7.44) can be interpreted as follows. If we have one more (infinitesimal) bit to spend, we would add it to the bit-stream with the smallest $\frac{\partial D}{\partial R_i}$, since this would decrease distortion by the greatest amount. We would continue to do so, until some other bit-stream offers a greater pay-off. The optimum balance is achieved according to Equation (7.44), when it does not matter to which bit-stream we add the next bit.

If the overall distortion D is the sum of the individual subband distortions D_i, i.e.,

$$D = \sum_i D_i, \tag{7.45}$$

Equation (7.44) can be simplified to

$$\frac{\partial D_i}{\partial R_i} = \frac{\partial D_j}{\partial R_j} \qquad \text{for all } i, j. \tag{7.46}$$

In other words, we have to pick points of equal slope on the individual distortion rate curves for each subband. Equation (7.45) holds if mean-squared error is used as the distortion measure and the subbands are orthogonal. Many subband decomposition schemes are designed with an orthogonality criterion or are nearly orthogonal. Based on Equation (7.46), we have optimized a

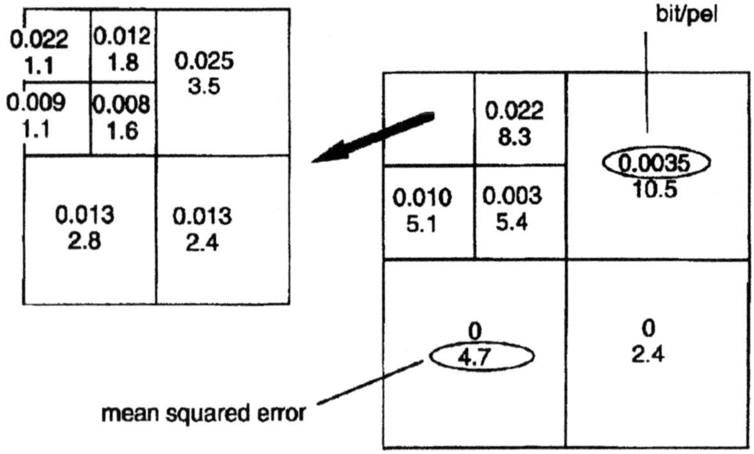

Figure 24 Bit allocation and mean squared error contributions of individual subbands.

subband pyramid coder that uses E_8-lattice VQ on the first pyramid level, D_4-lattice VQ on the second level, and the A_2-lattice on the third and fourth levels. With this coder we have compressed test image "Lenna" and have achieved a bit rate of 0.136 bpp while maintaining a peak-to-peak SNR of 30.9 dB. For a visual impression of the achieved quality, refer to [363]. The overall quality is very good for a compression ratio of approximately 60:1 (starting at 8 bpp). Figure 24 summarizes the bit allocation among subbands and the contribution of each subband to the overall distortion. Interestingly, the mean squared error (MSE) contribution of the various subbands differs considerably. It ranges from MSE $= 1.1$ for the lowest band to MSE $= 10.5$ for the horizontal high band on the first pyramid level. If we do not use rule (7.46) for bit allocation but rather allocate bits for equal mean squared error contribution of the bands (a popular text book solution motivated by the rate-distortion analysis presented in Section 1.1), overall SNR drops by 0.8 dB for the same bit-rate. We also encoded the image with orthogonal lattice quantizers, and it was confirmed that these quantizers perform only slightly worse than the denser lattices. The excellent performance of subband pyramid coding in combination with lattice VQ has been confirmed by results reported by Barlaud et al. [39, 40].

4 CONCLUSIONS

In this chapter, we have reviewed the principles of subband coding of still images and discussed a variety of algorithms employed in practical systems today. Compression can be achieved by exploiting the properties of both signal source and signal receiver. Rate-distortion theory establishes the relationship between the fidelity of the reconstructed image and the lowest transmission bit-rate for a given source model and a given distortion measure. An efficient coder jointly encodes many symbols; ideally, the reconstruction error is statistically independent from the reconstructed signal. For an ergodic Gaussian source with memory it is optimal to split the signal into frequency bands of (ideally) infinitesimal bandwidth and to encode those spectral components independently, according to the signal energy contained in the band. This is the underlying idea and the motivation for subband image coding.

Subband image coders consist of three essential stages: decomposition of the signal into frequency bands by means of subband filter banks, quantization of the subband signals, and (lossless) entropy coding.

For perfect reconstruction, subband filter banks have to obey certain design rules. Quadrature mirror filters (QMF) with linear phase can yield nearly perfect reconstruction and are widely used in image coding. The signal may be split into bands of equal or unequal bandwidths by appropriate cascading of two-band decompositions. Subband pyramids that recursively decompose the low band are now widely used. Another important decomposition is the discrete cosine transform (DCT), which is in fact a special case of a subband decomposition.

After the decomposition stage, the subband signals have to be compressed. A scalar quantizer quantizes each sample independently and is easy to implement. Better results are achieved when jointly quantizing several samples by means of vector quantization (VQ). Fixed-wordlength vector quantizers can be designed based on a training set using the LBG algorithm. The LBG algorithm can be extended to the design of vector quantizers with an entropy constraint. For a given entropy and a smooth probability density function, a lattice VQ can approach the minimum mean squared error. In Section 3.3, we have compared various vector quantizers experimentally. For the lowpass band (baseband image), VQ is far superior to scalar quantization. Unstructured VQ with variable length coding is the best choice for encoding of the subband signals. The subband signals are, however, spatially much less correlated, and VQ does not offer as much gain as for the lowpass band. For low bit-rates, unstructured VQ with an entropy constraint (ECVQ) performs best, but it is also computationally most expensive. Lattice VQ is computationally less expensive and performs almost as well.

Finally, we have reviewed conditions for optimum bit allocation among the subbands. In a coding example, we have demonstrated that a monochrome still image can be compressed by 60:1 while maintaining an acceptable level of image quality.

8

SUBBAND VIDEO CODING

Reginald L. Lagendijk, Frank Bosveld*, and Jan Biemond

Information Theory Group, Department of Electrical Engineering, Delft University of Technology, P.O. Box 5031, 2600 GA Delft, The Netherlands

**Advanced Development Centre, Philips Sound and Vision, P.O. Box 80002, 5600 JB Eindhoven, The Netherlands*

The recent advances in digital signal processing, communications, storage and integrated circuit design have caused an expansion of services and applications using digital video. We mention digital television broadcasting, video-on-demand, video-CDs, multipoint video conferencing, and multimedia communications with multiple windows [21, 30, 166, 242, 245, 410]. Among the various services and applications, a wide variety of *digital video standards* and *communication channels* with different bandwidth and reliability is considered for use.

For instance, in advanced (high-definition) television broadcasting a typical video format defines a frame rate of 25 fps (frames per second) and a spatial frame dimension of 1152 lines and 1440 pixels per line. The bandwidth or bit rate of such a video signal, when PCM encoded at 8 bit per picture element (bpp) in the luminance-chrominance (YUV) color space, is 660 Mbs (Megabits per second). With appropriate video coding techniques, the required transmission bandwidth should be reduced to approximately 50 Mbs. This calls for a compression factor of 10–20. Advanced television broadcasting would typically take place over a terrestrial, satellite or fiber-based ATM channel, each of which has specific degradation characteristics such as delay, drop-out, and bit and burst error rate.

Another example is the broadcasting or the storage of standard CCIR-601 resolution video signals (25 fps, 576 lines, 720 pixels per line, YUV representation). Straightforward PCM encoding at 8 bpp requires a bandwidth of 166 Mbs. The target bit rate for CD or magnetic tape storage, or for cable or ATM network transmission is in the range of 2 to 10 Mbs, requiring a compression factor of 15–50.

The last example is low resolution QCIF (Quarter Common Intermediate Format) video (144 lines, 180 pixels per line, 8 to 12.5 fps), which needs to be carried over mobile or small-band ISDN channels with bit rates in the range of 16 to 64 kbs (kilobits per second). Compression factors of well over 100 are not uncommon for these applications. In addition to the difference in frame size and frame rate, video frames can be represented on either a progressive or interlaced scanning lattice.

In a communications environment where such a vast variety of video standards and communication channels exists, the cry for compatibility is becoming stronger and stronger. This chapter deals with subband *video* coding, and focuses on how to take advantage of the multiresolution nature of subband coders to achieve compatibility through ***scalability*** and ***layered transmission***.

In Section 1 we define and motivate scalability and layered transmission, and briefly discuss the concept of motion-compensated video coding. Section 2 contains a concise description of the three basic forms in which subband video coding systems occur, namely the intraframe subband video coder, the motion-compensated interframe subband video coder, and the 3-D subband video coder. Scalable and layered variants of these are discussed in Sections 3–5.

Section 3 concentrates on scalable intraframe video subband coding. Although this coder does not capture all aspects of video sequences, since it does not take advantage of temporal dependencies, it is useful for introducing an essential concept in optimal scalable coding, namely multirate quantization (MRQ). Although we discuss MRQ in the context of intraframe coding, it is also applicable to the encoding approaches discussed in Sections 4 and 5.

Section 4 is devoted to scalable interframe subband video coding. Here a spatially decomposed motion-compensated prediction error is encoded. We discuss two different methods to employ motion information, namely by in-band and by out-band compensation.

Section 5 addresses scalable video coding using a 3-D subband decomposition. In addition to the spatial decomposition, the video sequence is also temporally decomposed, either with or without taking motion information into account. This yields a flexible but also complicated system concept which can support many types of scalability and layering.

1 INTRODUCTION

1.1 Scalable Coding and Layered Transmission

Compatibility is needed for the *interworking* of various video compression standards and video services [62, 138, 245, 410, 413, 432]. The idea is that no matter what the original video format or the communication channel bandwidth is, an arbitrary decoder should always be able to decode and display the incoming compressed video signal. This suggests a multiresolution system approach, in which decoding and displaying of the received compressed video are independent of

- the *spatial resolution* of the frames or fields in the sequence,
- the spatial *frequencies* contained in the frames,
- the (visual) *quality* of the decoded frames,
- the *frame rate* or the number of frames per second.

We define the four forms of flexible decoding and displaying as spatial, frequency, amplitude, and temporal *scalability*, respectively. In Figure 1 the various scalability types are illustrated, yielding various resolution and quality *scales*. The combination of amplitude and frequency scalability is commonly called "SNR scalability." In this chapter we concentrate on the two most essential forms of scalability, namely spatial and temporal scalability.

The simplest way to achieve scalability is by simulcast techniques. In simulcasting the video signal is encoded at all required resolutions and quality scales independently, and is transmitted in multiple independent subchannels. This is a highly redundant coding concept since some signal components (such as low spatial and temporal frequencies) will be present in most subchannels and are thus transmitted multiple times. A much more efficient alternative is to encode each resolution and each quality scale, except for the first scale, with respect to a previous scale, thus avoiding duplications in transmitted information.

Layered transmission is important for two reasons. In the first place, it should be possible for an arbitrary decoder to extract only those portions of an incoming bit stream, that are required for reconstructing a video signal at a scale prescribed by the decoder display size or the decoder processing capacity. A decoder with a small-sized display or with low processing capacity ought

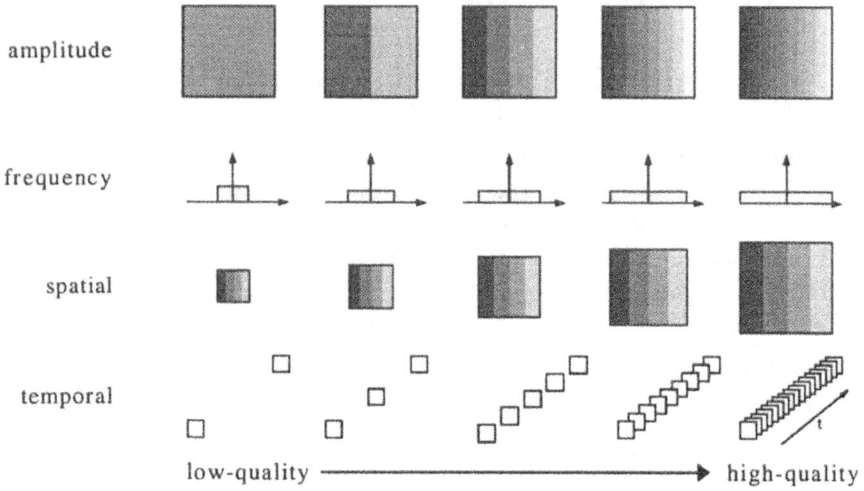

Figure 1 Four forms of scalability.

to decode only the base-layer, i.e. that bit stream portion that carries the low-resolution spatio-temporal information. On the other hand, a high-end decoder with a large display and powerful processors would utilize the entire bit stream. Therefore, a scalable video encoder should produce a bit stream that embeds appropriately defined ***substreams*** or ***layers***. Layers can be carried by either constant bit rate (CBR) or variable bit rate (CBR) subchannels. In this chapter we restrict ourselves to subchannels that have a fixed (CBR) bandwidth.

A second justification of layered transmission is that it can provide better error protection for some parts of the bit stream. Typically the base-layer carries important spatio-temporal lowpass information, but also motion vectors, the encoder parameter setting and synchronization information. This layer must be received virtually error-free by the decoder, otherwise decoding of any layer is impossible. It should therefore be transmitted in a highly reliable subchannel. Other portions of the encoded video signal, which provide additional information to proceed from one scale to the next, are inherently less important to the decoder, and can be carried by less reliable subchannels. In this way layered transmission greatly improves the system's error robustness and leads to graceful degradation under decreasing channel quality. Figure 2 illustrates the layered transmission concept.

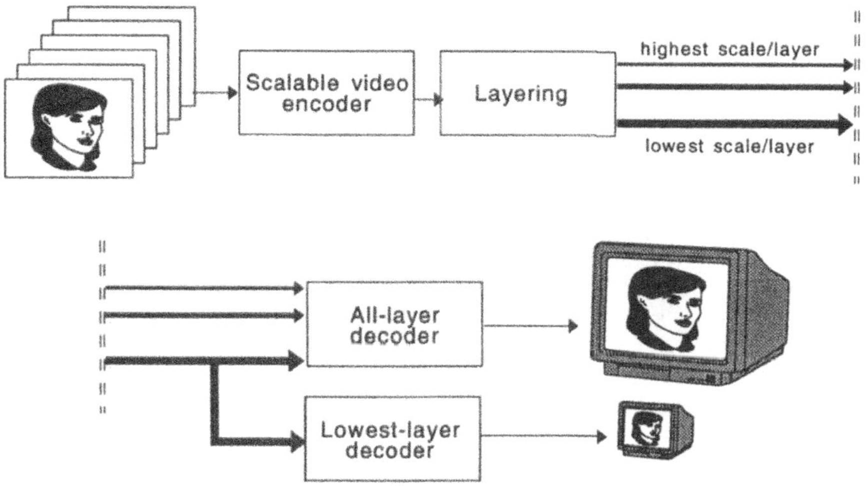

Figure 2 Layered transmission to support scalable coding and error robustness.

Many transmission and storage media allow or even require bit stream layering. We mention terrestrial transmission using a multiresolution 64-QAM modulation technique [175, 238, 357], cell-based communication in ATM (asynchronous transmission mode) networks [21, 204], and magnetic tape recording devices with trick mode support [151, 122].

1.2 Motion-Compensated Video Coding

Most of today's and near future video coding systems are based on the ITU-T H.320, ISO MPEG-1 or ISO MPEG-2 [208] standard for generic coding of moving pictures and associated audio. This class of coding standards can be seen as an extension of the JPEG still picture compression standard [207, 315] with temporal redundancy reduction. The essential ingredients are motion estimation (ME) and motion-compensated (MC) prediction, the discrete cosine transform (DCT), and uniform quantization (Q) followed by 2-D variable length encoding (VLC). The basic structure of a MC/DCT video coder and decoder is depicted in Figure 3.

The DCT, quantization and VLC encoding have been discussed in Chapter 7. Motion estimation and compensation is a powerful technique to take advantage

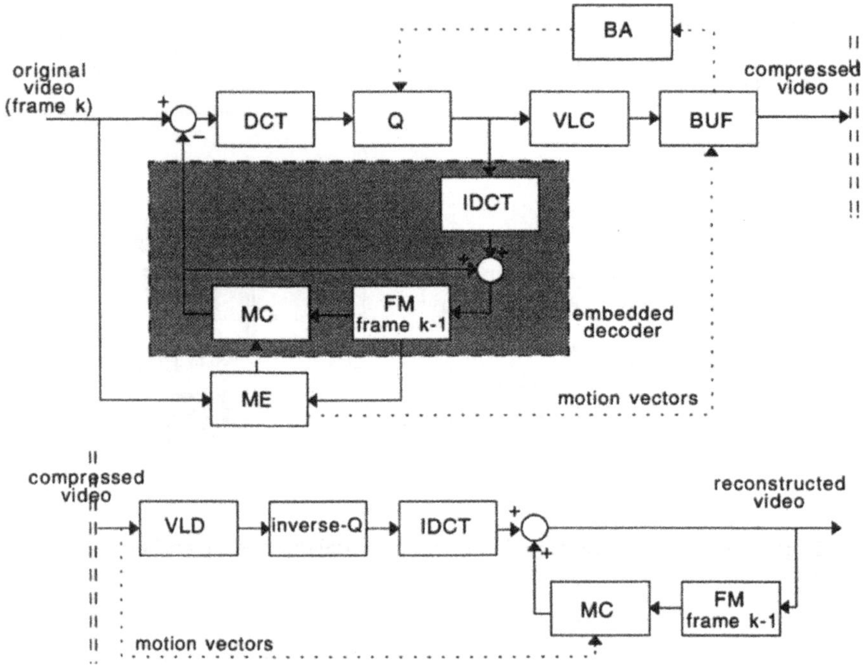

Figure 3 Motion-compensated DCT-based video coder and decoder.

of temporal dependencies between consecutive frames in a video sequence. An example of a popular motion estimator is hierarchical block matching (HBM) [52]. First, for each square subblock in the frame to be encoded, the best corresponding subblock in the previous already encoded frame in the frame memory (FM) is sought using a hierarchical search strategy. The difference in the location of the subblocks is called the ***motion vector***. Next, the motion-compensated frame difference or ***motion-compensated prediction error*** between the two subblocks is encoded. This process is repeated for all subblocks in a video frame. In general, this is a more efficient way than direct encoding of the frame information in the subblock itself, or than encoding of the temporal difference without taking motion into account.

For the encoding of the next frame, the reconstruction of the current frame needs to be available in the frame memory. For this reason the interframe encoder contains an embedded decoder, indicated by the shaded box in Figure 3.

In addition to a motion estimator and compensator, the coder also contains a bit stream ***output buffer*** (BUF) and a ***dynamic rate control*** or ***bit allocation*** (BA). The reason for both components is that the number of bits that the encoder produces per frame depends heavily on the success of the motion compensation. The resulting variable bit rate is buffered to the extent that the output bit stream of the buffer exactly satisfies the constant bit rate prescribed by the transmission channel or storage device. The rate control monitors the fullness of the buffer, and dynamically adjusts the number of bits to be allocated to subsequent frames. For instance, if the buffer becomes too full, subsequent frames will be assigned a smaller number of bits, resulting in a coarser quantization.

The performance of any motion-compensated coder depends strongly on the motion estimator used. The problem of motion estimation is not addressed in this chapter. The interested reader is referred to the abundance of literature in this field, for instance [409]. Figure 4 illustrates motion estimation and compensation.

The debate whether the above MC/DCT video coding system, on which H.261 and MPEG compression standards are based, is also sufficiently suitable for scalable coding and layered transmission, is still on-going. Alternatively, coding systems based on the Laplacian pyramid [67, 413] and on subband decompositions [57, 221, 287, 306] are proposed. The inherent multiresolution nature of subband coders sometimes leads to more attractive scalable and layered transmission structures. Further, for low-bit rate applications, subband video coders appear to have a better visual quality than MC/DCT video coders.

2 BASIC FORMS OF SUBBAND VIDEO CODING

In this section we give an overview of the three basic video coding configurations in which subband decompositions are used. These subband coders usually operate on full video ***frames*** for the case of progressively scanned sequences, or on video ***fields*** (half frames) for the case of interlaced scanning. The encoding of interlaced video frames is unattractive because of the artificial vertical high frequency components introduced by motion in a scene. These components, however, need to be encoded very accurately to avoid object blur in the decoded result. For the purpose of simplicity, we only use the term "frame" in this chapter.

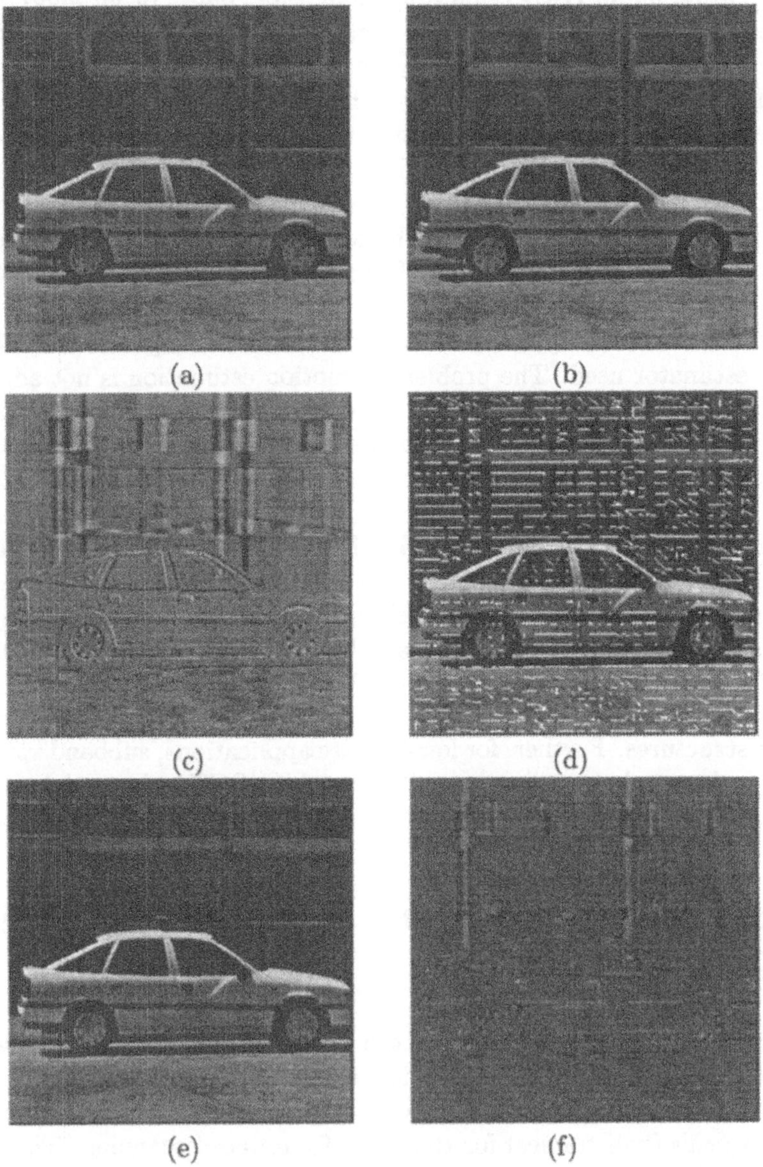

Figure 4 Motion estimation and compensation: (a) and (b) two original video frames, (c) the temporal frame difference (variance=218.0), (d) the motion vector field, (e) the motion-compensated prediction of b) using a) and the motion field, (f) the motion-compensated prediction error (variance=105.8).

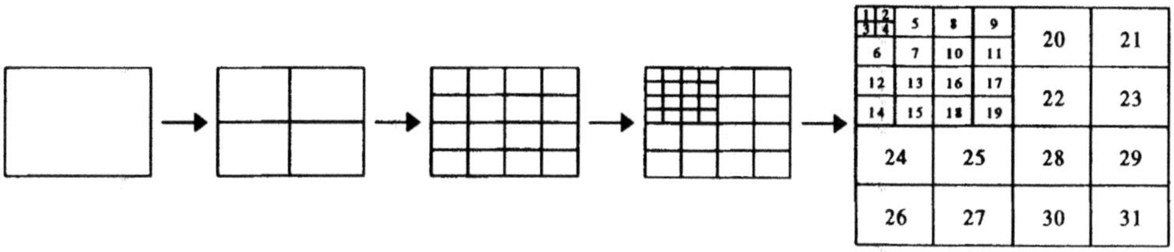

Figure 5 Spatial decomposition into 31 subbands (the result of the last decomposition stage is enlarged for the purpose of readability).

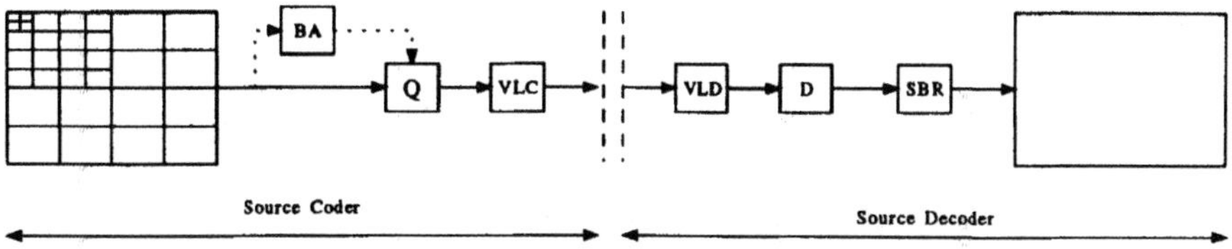

Figure 6 Basic configuration for an intraframe subband coder.

In many subband coders the color difference signals (UV) are encoded in the same way as the luminance component (Y). The available bit rate can be distributed dynamically between the luminance and chrominance components, depending on the "color richness" of the video sequence, or can be divided *a priori* using a fixed luminance-chrominance bandwidth ratio. In this chapter we only consider coding of the luminance components.

2.1 Intraframe Coder

An intraframe coder encodes all frames in a video sequence independently [21, 439, 447]. Further, it allocates a given number of bits to each frame to comply with the channel's bandwidth restrictions. Thus an intraframe subband *video* coder is basically a subband *image* coder applied to each individual frame in a sequence, without taking into account temporal redundancies. Any of the subband image coders discussed in Chapter 7 or variants of these can be used for encoding the individual frames. Figure 6 shows the basic structure of an intraframe subband video coder. The subband decomposition operates on *spatial* signals. In this chapter we use the 31 spatial subband decomposition

(SBD) shown in Figure 5. This decomposition structure can be obtained by concatenating 4-band spatial analysis filter banks.

Often the baseband (subband 1), containing lowpass spatial information, is encoded using DPCM, while the other subbands are PCM encoded. The quantizer Q is typically a uniform threshold quantizer (UTQ), followed by a lossless variable length coder (VLC) such as a Huffman or an arithmetic coder [143, 315]. The subbands are individually encoded under the supervision of a single forward bit allocation procedure (BA) [440]. On the decoder side the subbands are variable length decoded (VLD), inverse quantized (D), and combined by the spatial subband reconstruction (SBR), yielding the decoded result. For a more detailed discussion on the statistical properties of the subbands, the design of optimal scalar quantizers, bit allocation procedures, and the encoding of the subbands, the reader is referred to Chapter 7.

With the intraframe subband video coder, spatial, frequency, and amplitude scalability can be implemented, but not temporal scalability. For ***spatial scalability*** the subband reconstruction is done at a smaller frame size, thus excluding many subbands from the reconstruction process. For instance, a sequence of quarter spatial size is obtained by reconstructing only subbands 1–19 (see Figure 7c).

Frequency scalability is obtained by decoding the frames at their original size, but at the same time excluding some subbands from the reconstruction process. In other words, some subbands are effectively set to zero prior to the subband reconstruction.

Amplitude scalability is achieved by synthesizing coarser quantized subbands than produced by the encoder. Although all encoded subbands are used in the subband reconstruction, at lower scales coarser versions are used than at higher scales. Amplitude and frequency scalability are often jointly applied, which is then called ***SNR scalability***. An example of SNR scalability is shown in Figure 7d. The performance measure used in Figure 7 is the signal-to-noise (SNR) ratio:

$$\text{SNR} \quad = \quad 10\log_{10}(\frac{\sigma_x^2}{\sigma_q^2}) \quad \text{(dB)}, \tag{8.1}$$

where σ_x^2 and σ_q^2 are the variances of the original frame and the mean squared difference between the original and decoded frames, respectively.

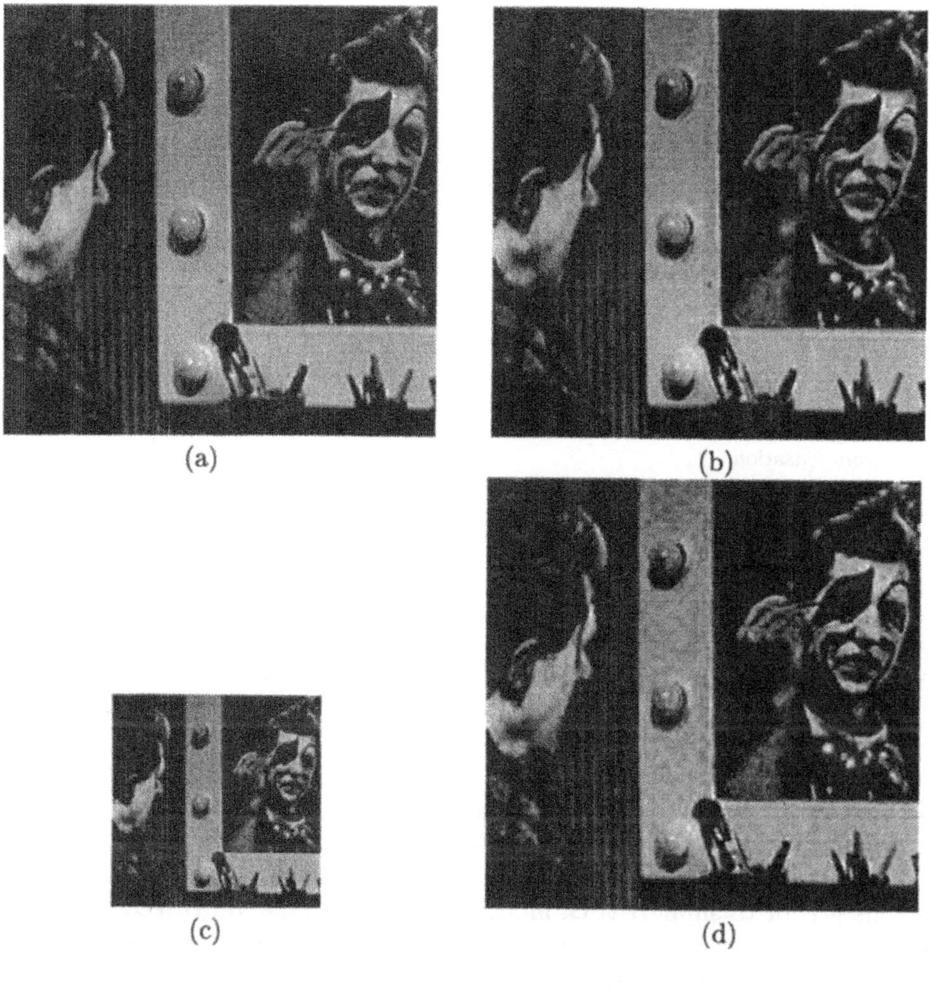

Figure 7 Scalability examples: (a) Original frame, (b) Decoded frame at highest scale (bit rate = 0.8 bpp, SNR=22.8 dB), (c) Spatial scalability (bit rate = 1.6 bpp), (d) SNR scalability (bit rate = 0.4 bpp, SNR=18.45 dB). Note that reconstructions (c) and (d) require the same total number of bits.

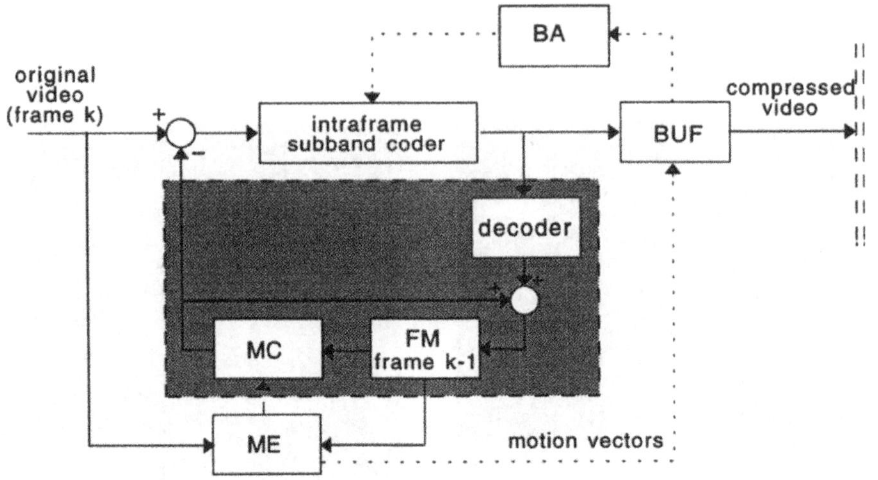

Figure 8 Basic configuration for an interframe subband coder with motion compensation.

Of the above three types of scalability, spatial scalability is the most general, but also the most complicated one. Once a method for reaching spatial scalability in a subband coder is found, frequency and amplitude scalability can be obtained by straightforward extensions. For this reason we restrict ourselves to the discussion of spatial and (for interframe coders) temporal scalability in the remainder of this chapter.

Intraframe video coders do not give the maximally achievable compression because temporal redundancies are ignored. However, in some applications intraframe coders are necessary for instance because of the severe channel degradations [47] or to support trick modes in magnetic tape video recorders [122]. By trick modes we mean the possibility to carry out slow motion, still frames, and fast forward and reverse search while still decoding and displaying a (lower quality) video signal. As a consequence of the absence of any temporal prediction or filtering in intraframe coders, errors in the bit stream have only a limited effect on the decoding of individual frames, and do not introduce temporal transients.

2.2 Interframe Coder Using Motion-Compensated Prediction

As discussed in Section 1.2, motion-compensated prediction is a powerful way to employ temporal redundancies in video sequences. The basic configuration for a motion-compensated interframe subband coder is shown in Figure 8 [160, 362, 442]. We point out the similarity to the MC/DCT video coder in Figure 3.

First, motion estimation is carried out on the current frame at time k and the previously decoded frame at time $k-1$, which is contained in the frame memory (FM). Using the estimated motion vectors, a motion-compensated frame difference is formed, which is next passed through a standard intraframe subband coder. Thus an interframe subband coder is basically built around the spatial subband decomposition of the motion-compensated frame difference. Since for the encoding of frame $k+1$ the reconstruction of frame k is required in the frame memory, the encoder contains an embedded intraframe subband decoder. There is a high degree of freedom in choosing the intraframe subband encoder within the motion-compensated prediction loop.

The configuration shown in Figure 8 is an example of a subband coder with ***out-band*** motion compensation. Here, out-band compensation refers to the domain in which the motion compensation takes place, namely in the picture domain: ***first*** the video frames are ***compensated*** for motion, and ***next*** the result is ***spatially decomposed*** into subbands. A consequence of placing the subband decomposition and reconstruction entirely inside the motion compensation loop is that any kind of scalability becomes virtually impossible.

In Section 4 we discuss two spatially scalable variants of the above out-band motion-compensated subband coder, namely an out-band compensation scheme with multiple motion compensation loops, and an in-band compensation scheme in which the motion compensation is carried out directly in the subband domain.

2.3 Interframe Coder Using 3-D Subband Decomposition

The interframe subband coding concept is not essentially different from the well-established MC/DCT approach. Not surprisingly, the performance and limitations of the two are comparable. A more promising approach to video is

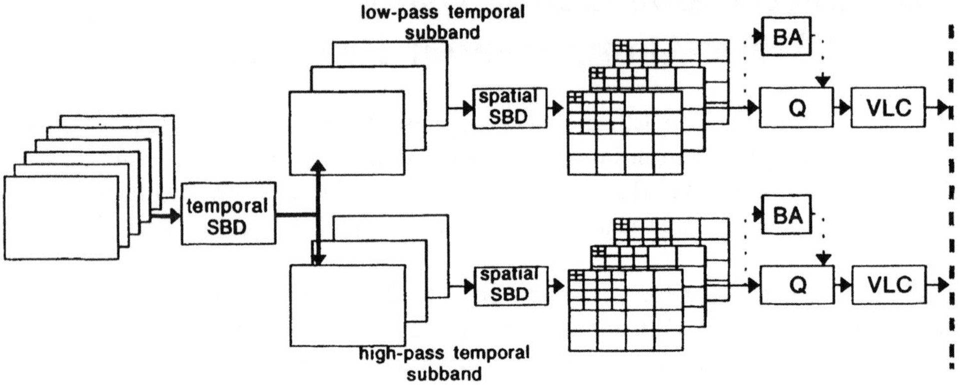

Figure 9 Basic configuration for a 3-D subband coder.

the extension of intraframe (spatial) subband coding to spatio-temporal or 3-D subband coding [57, 221, 249, 318, 392].

In 3-D subband coding, the video signal is not only decomposed into subbands in the spatial domain, but also in the direction of the temporal axis. In this way, extra compression can be achieved by employing the reduced visibility of coding artifacts in highpass temporal subbands. Further, temporal scalability becomes possible using temporal subband decompositions.

Figure 9 illustrates the basic configuration for a 3-D subband coder. Usually the spatio-temporal filter banks are taken to be separable, i.e. the spatial and temporal subband decompositions are independent, and are carried out by concatenating the temporal analysis filter bank and the spatial analysis filter bank. The temporal analysis and synthesis filter banks can be chosen freely, but to minimize the computational burden of the temporal processing, they are often based on simple two-tap filters. After filtering, temporal decimation by a factor of two follows. If we denote the k-th frame by $x_k[n_1, n_2]$, then the lowpass (tL) and highpass (tH) temporal subbands are computed as:

$$\begin{aligned} X_k^{tL}[n_1,n_2] &= \frac{1}{2}\left(x_{2k+1}[n_1,n_2] + x_{2k}[n_1,n_2]\right), \\ X_k^{tH}[n_1,n_2] &= \frac{1}{2}\left(x_{2k+1}[n_1,n_2] - x_{2k}[n_1,n_2]\right). \end{aligned} \tag{8.2}$$

Basically this means that two new sequences of half frame rate are created, where one sequence is nothing else than the average of successive frame pairs, and the other one is the difference of the frame pairs. Each of these (infinite length) sequences comprises a temporal subband. To achieve potentially higher compression, the lowpass temporal band can be decomposed further by concatenating two-band analysis filter banks in a tree-structure.

The lowpass and highpass temporal subbands are subsequently decomposed into spatial subbands. In this way we obtain several spatio-temporal subbands, which need to be encoded under the supervision of a single bit allocation algorithm. If we fully extend the concept of spatial subband coding to the temporal dimension, this would require the buffering of an extremely large, if not infinite, number of frames to build up the 3-D subbands. A numerical example can illustrate this. Assume that we have a sequence that consists of 100 frames, 576 lines per frame, and 720 pixels per line. A total of 8 spatio-temporal subbands is created by concatenating a two-band temporal, a two-band horizontal, and a two-band vertical analysis filter bank. The dimensions of the subbands then become *50 frames* by 288 lines by 360 pixels/line.

As such temporal subband dimensions are unacceptable in any practical application, the common notion of "what a 3-D subband is" is different from the one above. Instead of "collecting" frames over time to build up the temporal subbands, the ***individual frames*** in the lowpass and highpass sequences are called ***temporal subbands.*** For instance $X_1^{tL}[n_1, n_2]$ and $X_1^{tH}[n_1, n_2]$ are considered two temporal subbands, while in fact they are only two ***samples*** of the (infinite length) temporal subbands $X_k^{tL}[n_1, n_2]$ and $X_k^{tH}[n_1, n_2]$, respectively. As a result of this simplification, a 3-D subband coder degenerates to an intra-dualframe subband coder: one intraframe coder for the sum of a frame pair and another for their difference.

A drawback of the above way of temporal filtering is that the effects of motion are entirely ignored, yielding only moderate efficiency in employing temporal redundancies. We will discuss how motion compensation can be incorporated into the temporal subband decomposition in Section 5. At the same time we will indicate how temporal scalability can be implemented. Finally, we will discuss how a special form of spatio-temporal scalability, namely progressive-interlaced scalability, can be achieved by making use of non-separable 3-D subband decompositions.

3 SPATIALLY SCALABLE INTRAFRAME SUBBAND CODING

In this section we first discuss spatially scalable intraframe coding using quantization error feedback [58], and we show how the layering mechanism can take place. Next, we pay attention to the topic of multirate quantization. Finally, non-scalable and scalable intraframe coding are compared in terms of SNR performance and error robustness. Other forms of scalable coding using multirate quantization are described in [57, 366, 392, 391].

3.1 Scalable Intraframe Subband Coder

The scalable counterpart of the non-scalable intraframe coder shown in Figure 6 is shown in Figure 10. The video frames are decomposed into 31 subbands (using, in our particular numerical examples, the Johnston QMF16b filter [216]). With these 31 subbands, three spatial scales are created, namely

- the lowest spatial scale L^1 consisting of subbands 1–7,
- a middle spatial scale L^2 consisting of subbands 1–19,
- the highest spatial scale L^3 that includes all 31 subbands.

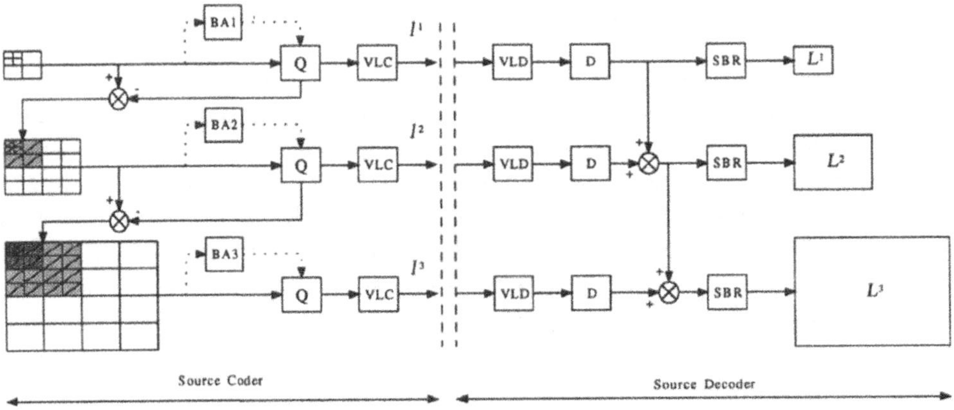

Figure 10 Scalable intraframe subband encoder and decoder.

The first transmission layer, denoted by l^1, carries the encoded subbands 1–7. These subbands are encoded under the supervision of the forward bit allocation BA1, which optimally distributes the available bit rate for scale L^1 among the subbands 1–7. The baseband is DPCM encoded spatially and subbands 2–7 are PCM encoded. As is shown in Figure 10, the quantization errors made in layer l^1 are *fed back* to the quantizers in the layer directly following (l^2 in this case). In this way the quantization errors of layer l^1 cannot propagate into the reconstruction of the higher spatial scale reconstructions L^2 and L^3, independent of the selected bit rate for scale L^1.

The second layer l^2 is needed for a decoder that wants to proceed from the lowest spatial scale L^1 to the next spatial scale L^2. This layer contains the following encoded subbands: (i) subbands 8–19, and (ii) subbands 1–7, which contain the *quantization errors of layer* l^1 (indicated by the shaded subbands in Figure 10). As in layer l^1, a bit allocation BA2 distributes the available bit rate for layer l^2 among the subbands 1–19, which are now all PCM encoded. Note that the quantizers for subbands 1–7 operate on quantization errors; the statistical properties thereof are quite different from subbands 8–19. Further, whereas the baseband is DPCM encoded in layer l^1, its DPCM quantization errors are PCM encoded in subsequent layers.

To reconstruct the video signal at full spatial resolution (highest spatial scale L^3) a decoder should also receive and reconstruct the third layer. Layer l^3 is created the same way as layer l^2. The bit allocation BA3 distributes the available bit rate among the subbands 20–31 and subbands 1–19, which now contain the quantization errors of layer l^2. Note that subbands 1–7 in fact contain quantization errors of the re-quantized quantization errors of layer l^1.

3.2 Multirate Quantization

In a scalable subband coder, higher layers not only carry additional highpass subbands, but also refinement information on lower frequency subbands that have already been transmitted in previous layers. Lower layers contain coarsely quantized subband coefficients which are used for the reconstruction of low spatial scale video signals. For the reconstruction of higher spatial scales, refinement information of these subband coefficients is transmitted in higher layers. Since some subband coefficients are encoded successively in stages, there are several bit rate contributions to the overall rate of a given subband, which is called *multirate quantization.* We consider two specific forms of multirate (scalar) quantization, namely *multistage* and *embedded* quantization.

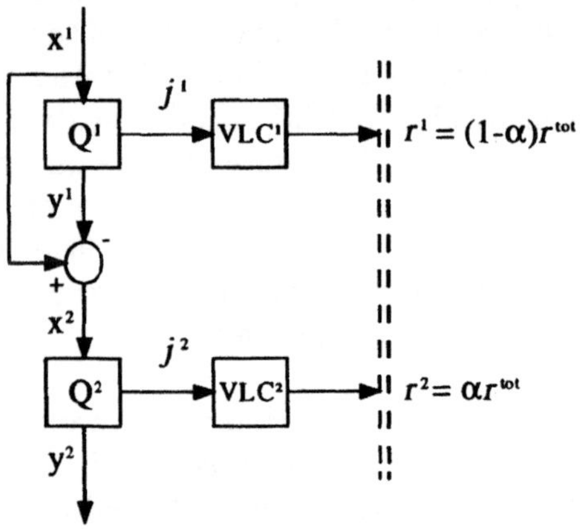

Figure 11 Multirate system using two-stage quantization.

In general, multirate quantization (MRQ) is *less efficient* than single-rate quantization at the same overall bit rate. This implies that in any scalable coder the highest scale reconstruction is of lower quality than in a comparable non-scalable coder. More specifically, the video frames encoded by the non-scalable coder in Figure 6 have a higher SNR quality than those reconstructed at scale L^3 of the scalable coder in Figure 10. Besides the increase in complexity of the overall system configuration, this is another price to be paid for obtaining scalability. However, with proper design of multirate quantizers the performance loss can be minimized [60, 286].

Figure 11 illustrates the most simple form of MRQ, namely *two-stage quantization.* The first stage quantizer, Q^1, quantizes the original subband coefficients, denoted by x^1. The selected quantizer representation levels, j^1, are losslessly VLC encoded, yielding a bit rate of r^1 bpp. The second stage quantizer, denoted by Q^2, quantizes the quantization errors of Q^1. After VLC encoding of the selected quantizer representation levels, the resulting bit rate is r^2 bpp. The output produced by VLC^1 is, for instance, carried by layer l^1, while the output of VLC^2 is transmitted in layer l^2.

In a practical coding system, the bit rates r^1 and r^2 are controlled by the bit allocations of the related layers. Here we evaluate the performance of the two-stage quantizer by varying the ratio between the bit rates while keeping the total bit rate constant. In Figure 11 this is specified by two parameters, namely

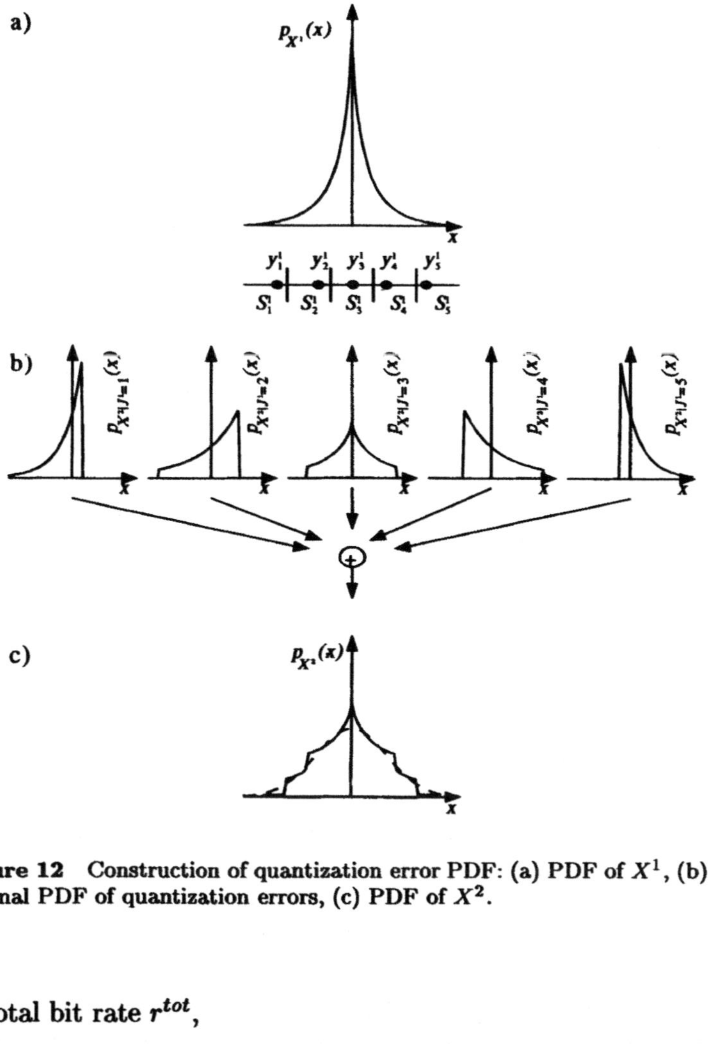

Figure 12 Construction of quantization error PDF: (a) PDF of X^1, (b) conditional PDF of quantization errors, (c) PDF of X^2.

- the total bit rate r^{tot},
- a fraction α, which divides the total bit rate between the two quantizers. In practical systems the lower layers usually have the lowest bit rate, i.e. $\alpha > 0.5$ is the most relevant case.

The subbands are assumed to have a generalized Gaussian probability density function (PDF) $p_{X^1}(x)$ with shape parameter $c = 0.75$ [439]. The first stage quantizer Q^1 is a UTQ, optimized for this PDF, followed by an optimal VLC coder [143]. The design of the second stage quantizer Q^2 requires the PDF of the quantization errors of the first stage quantizer. This PDF $p_{X^2}(x)$ is

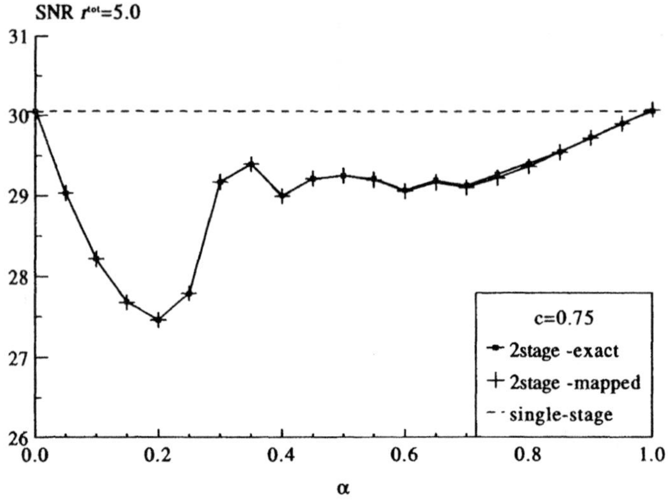

Figure 13 Performance of two-stage quantization.

analytically given as follows [213]:

$$p_{X^2}(x) = \sum_{i=1}^{K^1} p_{X^2|J^1=i}(x) P(J^1 = i), \tag{8.3}$$

where $p_{X^2|J^1=i}(x)$ is the conditional PDF of the quantization errors, K^1 the number of representation levels of Q^1, and $P(J^1 = i)$ the probability of occurrence of representation level i. Figure 12 illustrates the construction of the PDF of the quantization errors using (8.3).

The performance of two different two-stage quantizers, varying slightly in implementation details, are shown in Figure 13 for the case of $r^{tot} = 5$ bpp. The performance is expressed using the SNR performance measure defined in (8.1). Also shown is the performance of single-stage quantization using a UTQ at 5 bpp (dashed line). It can be seen that the two-stage quantizer performs poorer than the single-stage quantizer. For $r^1 < 3$ bpp (i.e. $\alpha > 0.4$), the performance degradation is approximately 1.0 dB, while for $r^1 > 3$ bpp the maximum degradation is 2.5 dB. This behavior is typical for fixed-rate multistage quantization.

An important cause for the performance degradation in two-stage quantization is that once the quantizers Q^1 and Q^2 have been designed, the second stage *quantizer* and *VLC coder* do not make use of knowledge on how x^1 was encoded

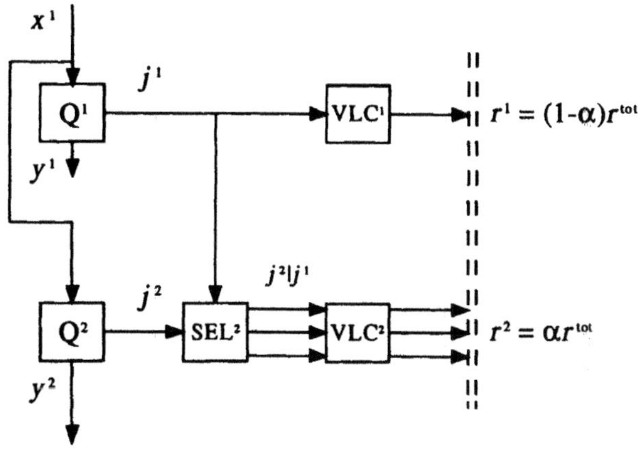

Figure 14 Multirate system using embedded quantization.

in the first stage, i.e. the two stages are independent. Use of this knowledge can, however, improve the coding performance of the second stage, and therefore the overall coding performance of the two-stage quantizer. In *conditional entropy coding* and *conditional quantization* the *VLC encoding* VLC^2 and the characteristics of the *quantizer* Q^2 are made dependent on the output J^1 of the first stage quantizer. The justification of these approaches are the dissimilar shapes of the conditional PDFs $p_{X^2|J^1=i}(x)$ shown in Figure 12.

The potential for conditional entropy coding and quantization is clearly demonstrated in [60], although the resulting quantizers do not consistently outperform two-stage coding. However, a more severe drawback of the concept of conditional entropy coding and quantization is that the complexity grows rapidly out of hand in a practical subband coding system.

A related, but far more attractive approach to MRQ is *embedded quantization* [60, 392]. This approach is illustrated in Figure 14. The applied quantizers have *embedded decision intervals*, which means that the decision levels of the coarser quantizer Q^1 coincide with some of the decision levels of the finer quantizer Q^2 (Figure 15). As a consequence, the refinement of the quantization of X^1 by the second stage quantizer requires only that the encoder indicates which subdecision interval has been selected by Q^2 within the decision interval selected by Q^1.

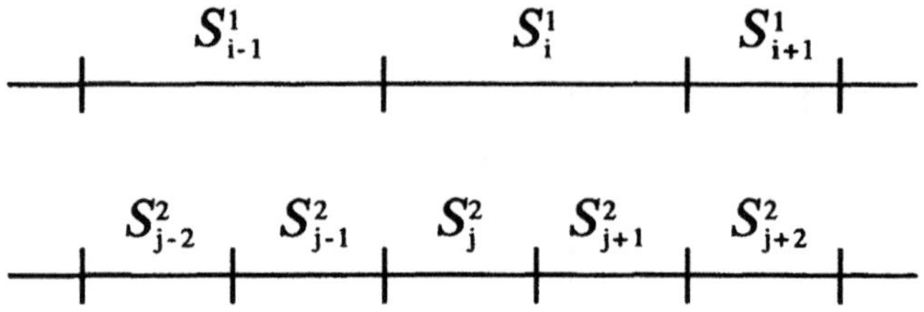

Figure 15 Example of an embedded quantizer. The decision intervals S^2 are entirely embedded in the decision intervals S^1.

The finest quantizer is first designed to be an optimal UTQ. This assures that the overall performance of the two-stage quantizer is equal to the performance of single-stage UTQ quantization. The coarse quantizer Q^1 is then recursively designed from the finest quantizer using a pruning algorithm. The pruning algorithm repeatedly combines those decision intervals that decrease the output rate most for a given increase in distortion. The price to be paid for this is a small performance loss in the first quantization stage since Q^1 is no longer a UTQ.

The first stage performance of embedded quantization is illustrated in Figure 16 for the case of $r^{tot} = 5$ bpp. Indeed a small loss of 0.1 dB in performance in the first stage needs to be tolerated. Since in practice also the second quantization stage may be designed by pruning a high bit rate UTQ, similar small losses can be expected for the overall performance of multistage embedded quantizers compared to similar single-stage quantizers. As a final remark we point out that two-stage quantization can be regarded as a bottom-up or coarse-to-fine approach, while embedded quantization can be regarded as a top-down or fine-to-coarse approach. Because of this difference, the implementation of embedded quantization is different from two-stage quantization, namely it does not explicitly include quantization error feedback. The principle of information refinement by a multirate approach, however, is the same in both approaches.

3.3 Performance Evaluation

To conclude this section we return to the scalable intraframe subband coder of Figure 10 and compare its performance and error robustness with a similar non-scalable subband coder. The experiments have been carried out using 40 frames of the (progressively scanned) "mobile" sequence.

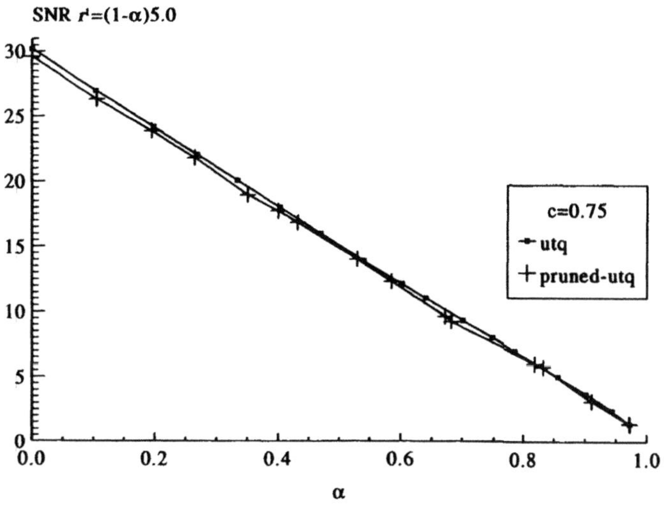

Figure 16 Performance of first quantization stage in embedded quantization.

The scalable coder supports three spatial scales with the bit rates 8 Mbs, 2 Mbs, and 0.5 Mbs, respectively. The quantization of the subbands is carried out using multistage quantization or embedded quantization. Figure 17 shows the obtained SNRs for the highest spatial scale. The non-scalable scheme outperforms the scalable ones by 0.2 dB for embedded quantization, and by 0.7 dB for multistage quantization.

To illustrate the error robustness of layered transmission, we have optimized the scalable and non-scalable coders for transmission of the layers over an ATM network [60]. However, no error concealment or forward error correction methods have been used to reduce the effect of cell losses in the ATM network. The non-scalable/non-layered coder has been subjected to cell-loss ratios (CLR) of 10^{-6} and 10^{-3} (Figure 18). The scalable/layered coder has been exposed to a CLR of $10^{-8}, 10^{-6}, 10^{-6}$ and $10^{-8}, 10^{-3}, 10^{-3}$ for layers l^1, l^2, l^3, respectively (Figure 19). As can be seen from the figures, the non-layered system is severely affected by ATM cell losses, while the layered system is much more robust. Subjective evaluations of the reconstructed sequences confirm these numerical results because errors in the layered system are hardly visible, as opposed to the annoying artifacts caused by cell losses in the non-layered scheme.

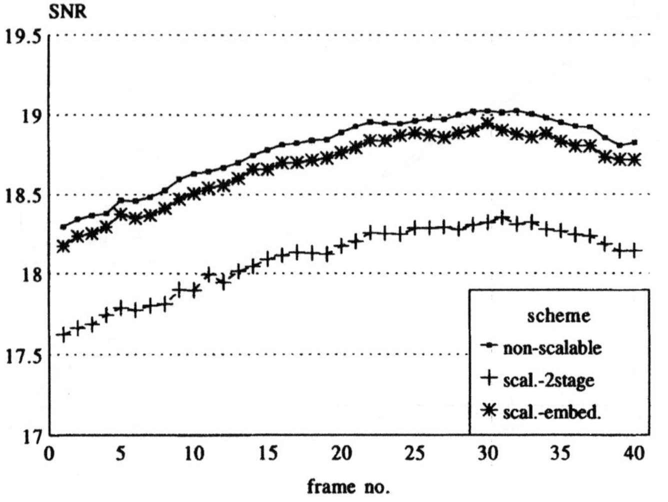

Figure 17 Coding performance for non-scalable and two scalable intraframe subband coders.

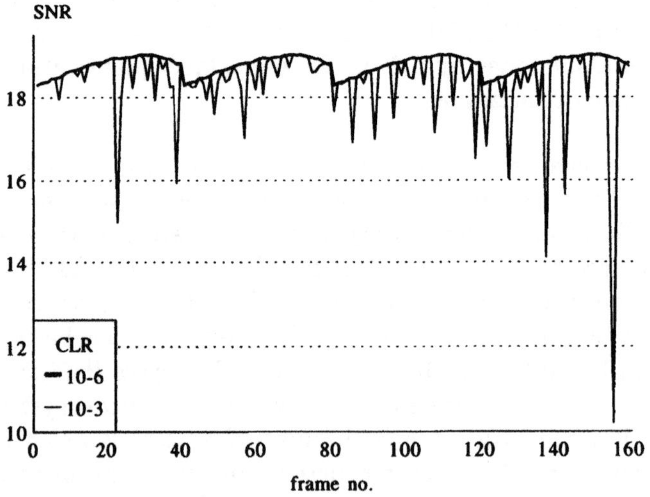

Figure 18 Effects of cell losses on a non-layered transmission scheme (sequence of 40 frames has been repeated four times with different cell-loss patterns).

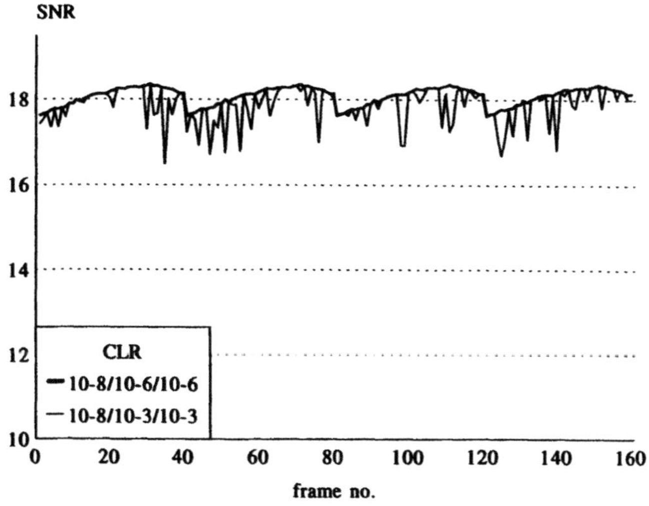

Figure 19 Effects of cell losses on a layered transmission scheme (sequence of 40 frames has been repeated four times with different cell-loss patterns).

4 SPATIALLY SCALABLE MOTION-COMPENSATED INTERFRAME SUBBAND CODING

The out-band motion-compensated subband coder in Figure 8 cannot be used directly for scalable video. This is because the subband decomposition and reconstruction are placed entirely inside the motion-compensated prediction loop. The motion-compensated prediction in the encoder is therefore always based upon the highest spatial scale. Because lower scale decoders cannot carry out the same motion-compensated prediction as the encoder since they do not have available the highest spatial scale video signal, the reconstruction of video at lower spatial scales will suffer from *drift*.

Two ideas have been employed for designing spatially scalable interframe subband coders, namely: (i) to use only those subbands in the encoder's motion-compensated prediction which the decoder also has available when reconstructing video at a certain spatial scale, and; (ii) to carry out the motion compensation in the subband domain directly (in-band compensation). Both methods will be discussed in this section. Temporal scalability is, however, generally not possible using the interframe subband coders discussed here.

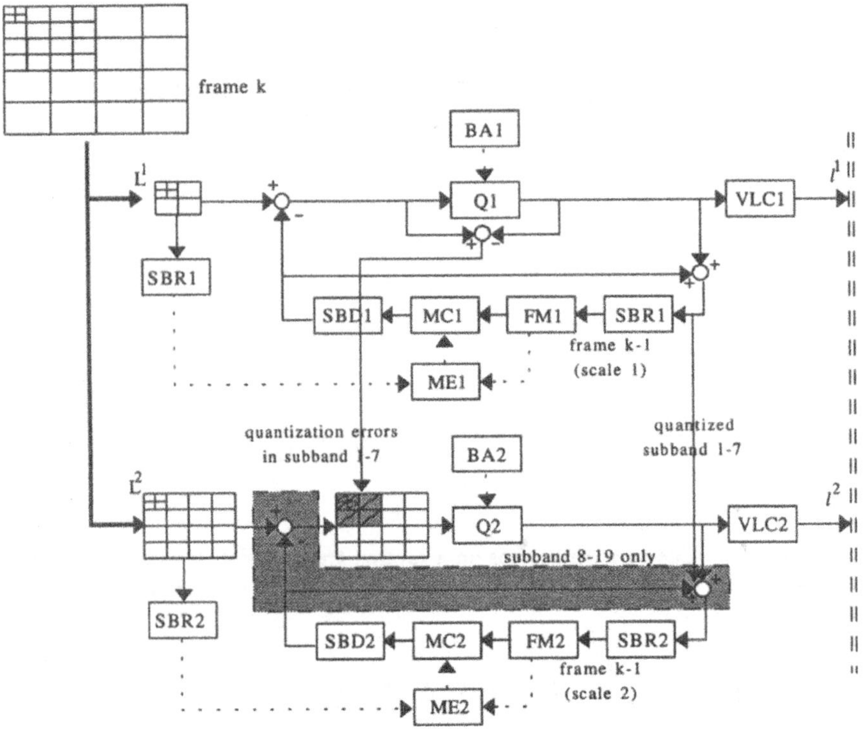

Figure 20 Scalable interframe subband coder with out-band motion compensation.

4.1 Out-Band Motion Compensation

Figure 20 shows the structure of a typical scalable out-band motion compensation subband system with three spatial scales [215, 287, 286, 424, 423, 455]. For the purpose of simplicity, only the encoding of the first two scales is shown in the figure. To encode the video frame k, first the full resolution frame is spatially decomposed into 31 subbands. The same three spatial scales are used as in the previous section.

The lowest spatial scale L^1 is obtained by reconstructing subbands 1–7 (SBR1). Frame memory FM1 contains the reconstructed frame at spatial scale L^1 at time $k-1$. Motion estimation is carried out between these two frames (ME1), followed by the motion compensation (MC1) and a 7-band subband decomposition (SBD1). The temporal differences between the subbands 1–7 of the frame k and the subbanded[1] motion-compensated prediction are quantized by Q1 under the supervision of bit allocation BA1, and losslessly VLC encoded. This establishes the information carried by layer l^1. The encoder also includes an embedded decoder for subbands 1–7 as the encoding of scale L^1 of frame $k+1$ requires the decoded result at scale L^1 of frame k. Observe that although temporal differences are taken in the subband domain, the crucial process of motion compensation takes place in the picture domain, prior to a subband decomposition. Thus Figure 20 is indeed an out-band compensation scheme.

The spatial scale L^2 is defined by subbands 1–19. Frame memory FM2 contains the reconstructed frame $k-1$ at this spatial scale. Motion estimation is carried out on the spatially scaled original frame (subband reconstruction SBR2 operating on subbands 1–19) and the frame in FM2. Next, a motion-compensated prediction is formed, which is decomposed into 19 subbands (SBD2). The prediction error subbands 8–19 which are additional with respect to scale L^1, are input to the quantizer Q2 and are carried by layer l^2. Information on the motion-compensated prediction errors in subbands 1–7 has already been transmitted in layer l^1. Therefore, Q2 merely refines the subbands 1–7 by re-quantizing the feedback *quantization errors* of layer l^1. Observe that the motion-compensated prediction of subbands 1–7 computed for scale L^2 is *not* used at all. For this reason, the motion compensated prediction loop in Figure 20 applies *only* to subbands 8–19. The reconstruction of subbands 1–7 in scale L^2 is carried out on the basis of scale L^1 and the requantized quantization errors of L^1. This quantization refinement information is carried in layer l^2. The described scalable interframe subband coder avoids drift in the decoder at all spatial scales, because subbands are predicted using only that temporal information which a decoder, working at a specific spatial scale, has available.

As Figure 20 shows, a motion estimation loop has to be maintained for each supported scale. A simplification is to dispense with the motion estimation and compensation of the higher frequency spatial subbands, since the gain in compression due to motion compensation of these subbands is relatively small [287]. Alternatively, the required motion vectors may be obtained by downscaling a dense vector field estimated at the highest scale, or by refining a coarse motion field estimated at lower spatial scales [215, 287]. The latter method essentially takes advantage of the multiresolution nature of subband coders to efficiently implement a hierarchical motion estimation algorithm.

---

[1] A "subbanded" signal is a signal that has been decomposed into subbands.

Figure 20 is comparable to the scalable intraframe coder discussed in Section 3 in the sense that the propagation of quantization errors into higher spatial scales is prevented by MRQ. Thus the multirate quantization theory discussed in Section 3.2 applies to scalable interframe coders as well [286].

4.2 In-Band Motion Compensation

A drawback of out-band motion compensation is that the complexity of the scalable system may become very large, mainly due to the multiple motion estimation loops. This drawback can be avoided by performing motion compensation directly in the subband domain. An example of such an *in-band motion compensation* scheme is illustrated in Figure 21 [59, 161, 314, 433, 453]. Again only the encoding of the first two scales is shown. Note, however, that the system contains only a single motion estimation loop.

First, a motion vector field is estimated between the reconstructed previous frame at the highest scale L^3 and the current frame (not shown in Figure 21). Next, the current frame is decomposed into 31 subbands. The same scale definitions are used as before. Each *subband* is now temporally predicted using this motion information and the corresponding reconstructed subband of the previous frame. Note that although the motion estimation may be carried out in the picture domain, the motion-compensation is carried out directly in the subband domain.

In layer l^1, the resulting motion-compensated differences of subbands 1–7 are quantized under the supervision of bit allocation BA1. The resulting quantization errors are fed back to the next layer. In layer l^2, subbands 8–19 are compensated for motion in the subband domain, and encoded together with the quantization errors of L^1 under the supervision of bit allocation BA2. In this way, subbands 1–7 are encoded in a multirate fashion. Drift cannot occur, since at each spatial scale, prediction takes place using only those subbands that belong to that scale.

While the complexity of the in-band compensation system is much lower than that of the out-band compensation system, its performance is significantly poorer. The reason for this is that although the correlation between consecutive frames is high, the temporal correlation between the subbands is generally much lower. In [59, 61, 287] it is shown that even if two frames are perfectly shifted versions of each other, their subbands are *not* because of the aliasing components in the subbands. Motion compensation carried out directly on subbands

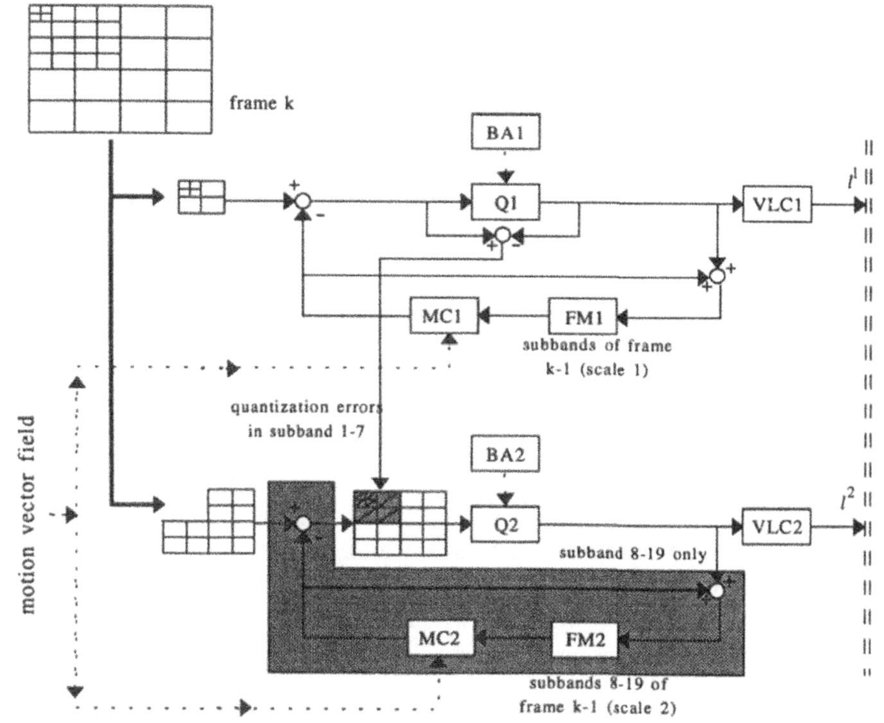

Figure 21 Scalable interframe coder with in-band motion compensation.

is therefore not successful in compensating the subband aliasing terms, leading to severe performance losses. The only way to correctly compensate the aliasing for motion is to reconstruct the subbands prior to motion compensation. This is, however, nothing else than out-band compensation.

Attempts have been taken to alleviate the above performance limitation [59, 433]. Often this either highly increases the complexity of the encoder and decoder, or leads to a coding concept that is somewhere half way between in-band and out-band compensation. This class of ***interpolated in-band coders*** seems to be an attractive alternative to purely out-band or in-band compensation spatially scalable interframe subband coders.

5 SPATIO-TEMPORALLY SCALABLE 3-D SUBBAND CODING

For the interworking of different services, not only spatial scalability is of importance, but also temporal scalability [326, 413, 432]. This may refer to compatibility between video formats that have different frame rates, but also to compatibility between interlaced and progressive video formats. In this section we discuss how both kinds of scalability can be achieved using 3-D subband decompositions.

5.1 Motion-Compensated Temporal Subband Decomposition

The 3-D subband coder in Figure 9 is only suitable for video with low motion activity. If motion occurs, the correlation along the temporal filter direction becomes significantly lower. For this reason, motion-adaptive 3-D subband coding schemes have been developed, which essentially apply temporal subband decomposition only in regions of low motion activity [319, 318].

A subband decomposition method that fully takes into account motion has been proposed in [306]. Figure 22 illustrates the analysis stage in a *motion compensated temporal subband decomposition.* Instead of operating the analysis filters along the temporal axis, the video signal is filtered along the motion trajectory. Of course, first a motion vector field needs to be computed between successive pairs of frames [305]. Although motion-compensated subband decompositions can be carried out with arbitrary length analysis filters, two-tap filters are commonly used to restrict the computational complexity and delay. The motion-compensated lowpass and highpass temporal bands are then given by:

$$
\begin{aligned}
X_k^{tL}[n_1, n_2] &= \frac{1}{2}\left(x_{2k+1}[n_1 - d_h[n_1, n_2], n_2 - d_v[n_1, n_2]] + x_{2k}[n_1, n_2]\right), \\
X_k^{tH}[n_1, n_2] &= \frac{1}{2}\left(x_{2k+1}[n_1 - d_h[n_1, n_2], n_2 - d_v[n_1, n_2]] - x_{2k}[n_1, n_2]\right), \quad (8.4)
\end{aligned}
$$

where $(d_h[n_1, n_2], d_v[n_1, n_2])$ is the (backward) motion vector computed at the coordinate $[n_1, n_2]$ between frame $2k+1$ and $2k$. Observe that $X_k^{tH}[n_1, n_2]$ is quite similar to the motion-compensated *prediction error* in an interframe coder. The subband synthesis also takes place in a motion-compensated fashion.

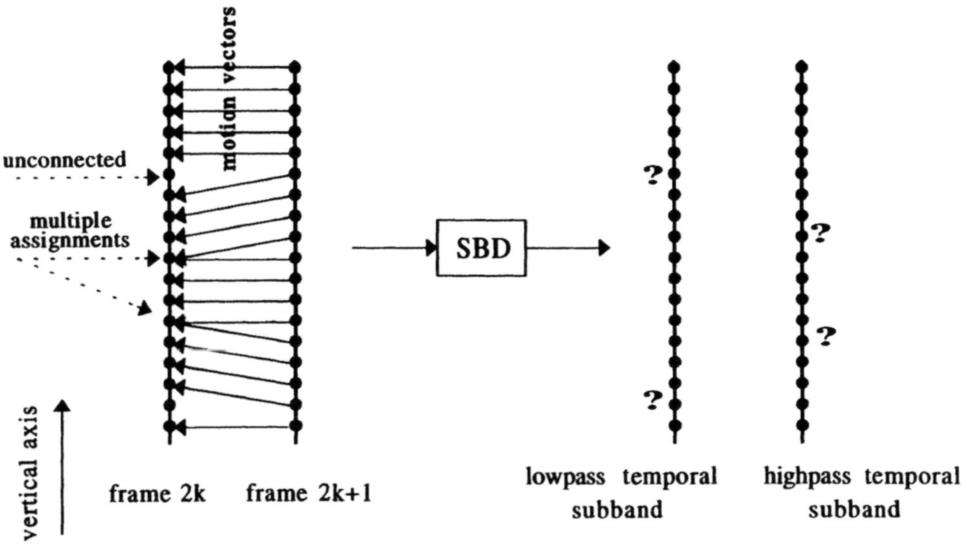

Figure 22 Principle of motion-compensated subband decomposition. Only the vertical and temporal axis of the sequence are shown.

The motion-compensated filter bank defined by (8.4) does not generally come with the perfect reconstruction property. The cause of losing perfect reconstruction are discontinuities in the motion vector field. Wherever variations or discontinuities in the motion vector field occur, several motion vectors may point from frame $2k+1$ to the *same location* in frame $x_{2k}[n_1, n_2]$ ("multiple assignments" in frame $2k$). At the same time, some coordinates in $x_{2k}[n_1, n_2]$ may *not be referenced to* at all ("unconnected" in frame $2k$). Both cases are illustrated in Figure 22.

The consequence of motion field discontinuities is that the subband analysis stage (8.4) is no longer reversible. One way to recover the perfect reconstruction property of the motion-compensated filter bank is the following. All multiple motion vector assignments are reconsidered, and all assignments but one are removed. Locations in frame $2k$ and $2k+1$ are now uniquely connected or not connected at all. For the uniquely connected locations, the lowpass and highpass components can be computed using (8.4). The unconnected locations in frame $2k$ and $2k+1$, however, cannot be passed through an analysis filter. These values are therefore given a special treatment [306], for instance values at unconnected locations in frame $2k+1$ are directly copied into the corresponding

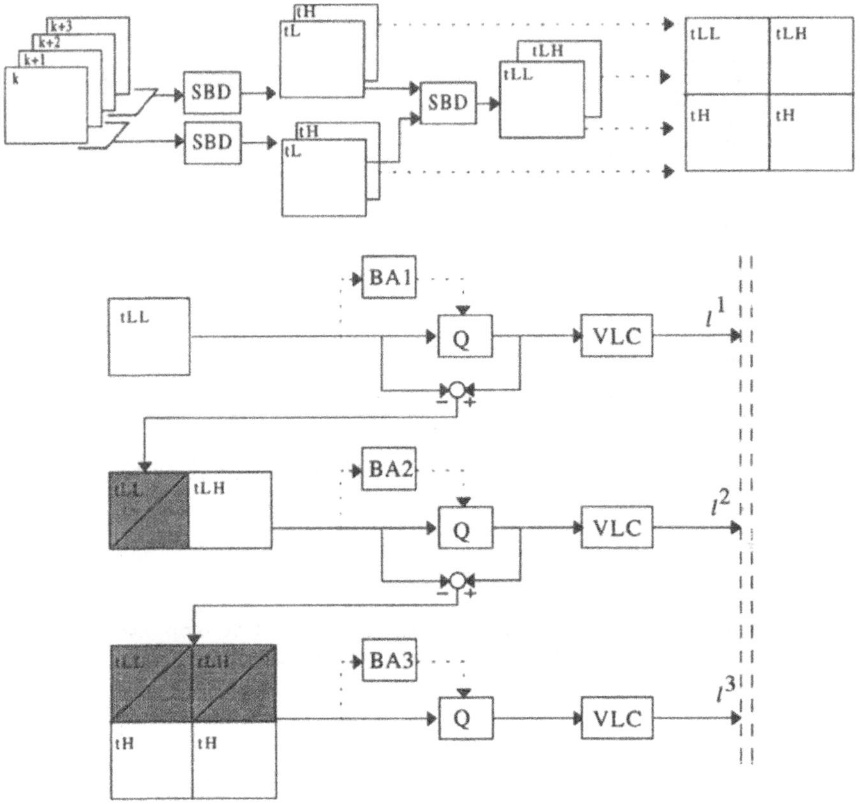

Figure 23 Temporal analysis filter bank and a subband coder with 3 temporal scales.

"holes" in the lowpass temporal subband, while the values at unconnected locations in frame $2k$ are directly copied into the corresponding "holes" in the highpass temporal subband. These 'holes" in the subbands are indicated by question marks in Figure 22.

Figure 23 shows the structure of a temporally scalable subband coder. This particular example allows for 3 temporal scales, with relative frame rate of 4, 2, and 1. The inputs of the temporal subband decomposition are 4 frames; the outputs are 2 highpass temporal subbands (tH), 1 low-highpass temporal subband (tLH), and 1 low-lowpass temporal subband (tLL). In each layer, the temporal subbands are encoded using an intraframe subband coder under the

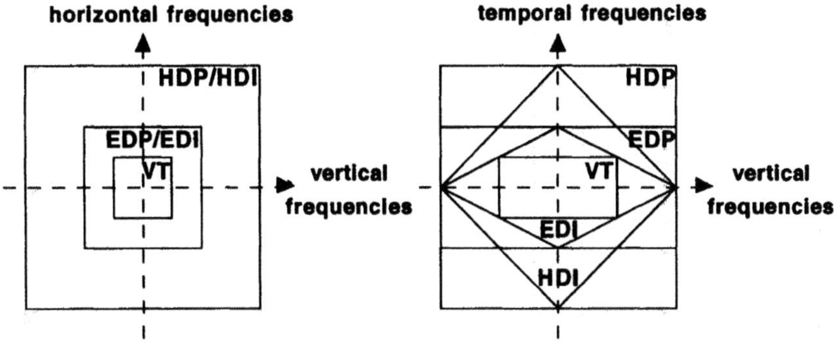

Figure 24 Spatio-temporal spectral support of different progressive and interlaced scales.

supervision of a bit allocation algorithm. At the temporal scale L^1, only the tLL subband is encoded. Quantization errors are fed back to layer l^2. For temporal scale L^2 the additional temporal subband tLH and the refinement information of temporal subband tLL are needed. Scale L^3 requires the refinement information of temporal subbands tLL and tLH, and the two additional highpass temporal subbands tH. From this discussion it should be clear that temporal scalability is the logical counterpart of spatial scalability in intraframe subband coding.

The temporally scalable system in Figure 23 is still rather incomplete. For instance, the frames in the subband tLL are temporally correlated. Should an interframe coder be used for this subband? Or, alternatively, should the number of temporal scales be increased to employ longer range temporal dependencies? Further, the question how to combine the spatial and temporal scalability discussed in this chapter in an optimal manner, is an issue of on-going research [392, 391].

5.2 Progressive-Interlace Scalable Subband Decomposition

All forms of scalability discussed so far were based on *progressively scanned* video formats. Many of today's video formats are, however, based on interlaced scanning. Clearly, there is a need for scalable systems that include progressive and interlaced video formats, such as the ones listed in Table 1 [64, 413, 432].

Table 1 Relevant progressive and interlaced video formats.

Video format	Scanning parameters	Sampling parameters	
		Luminance Y	Chrominance UV
HDP	1250/50/progressive	108 MHz/1440	54 MHz/720
HDI	1250/50/interlace	54 MHz/1440	27 MHz/720
EDP	625/50/progressive	27 MHz/720	13.5 MHz/360
EDI	625/50/interlace	13.5 MHz/720	6.25 MHz/360
VT	312.5/25/progressive	3.375 MHz/360	1.6875 MHz/180

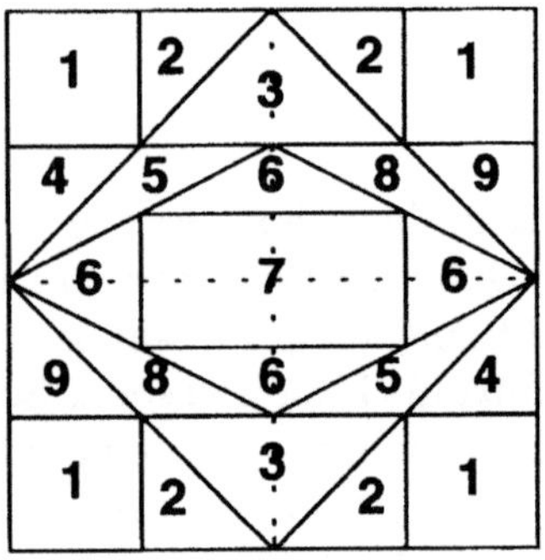

Figure 25 Vertical-temporal subband decomposition needed for progressive-interlace scalability.

Figure 24 shows the relationship between supports of the spatial frequencies and vertical-temporal frequencies of the formats in Table 1. Since the supports of the vertical-temporal spectra of interlaced formats are not rectangular, progressive-interlace scalability requires non-rectangular, i.e. non-separable subband decompositions [57].

The decomposition of arbitrary signals into subbands using multidimensional filter banks with non-separable subsampling pattern are detailed in [37, 431]. The basic building blocks that we need here are 2-D diamond-shaped and directional analysis filters. With these non-separable filters and their corresponding downsampling patterns, a vertical-temporal subband decomposition can be carried out into the 9 vertical-temporal subbands shown in Figure 25. An example of a complete non-separable vertical-temporal analysis filter bank is shown in Figure 26. By appropriately selecting subbands for the synthesis process, the

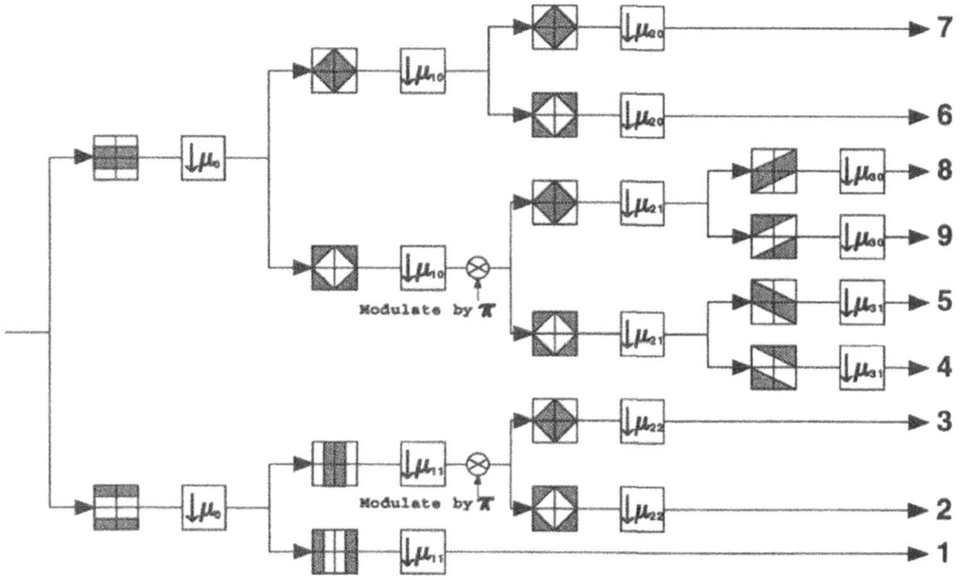

Figure 26 Vertical-temporal subband analysis filter bank for HDP format.

correct spectral support for the formats in Table 1 can be reconstructed. For instance, to recover the EDI vertical-temporal spectral support, we would synthesize only subbands 6 and 7, while for the EDP support, subbands 4 to 9 are needed.

A complication of the above described subband decomposition is that it still cannot fully support a progressive-interlaced scalable coding system. This is because the relation between the HDI and EDP vertical-temporal spectral supports is ambiguous. None of the two is a subset of the other. In addition the lower scale vertical-temporal sampling lattices are not necessarily a subset of sampling lattices at higher spatio-temporal scales. This is for instance the case for the scaling from EDI to HDI and HDP [57]. As a consequence, there are restrictions on the number of different interlaced formats that can be present in a progressive-interlace scalable subband coding system. A full evaluation of the described progressive-interlace scalable subband coding system is given in [57].

9

AUDIO CODING WITH FILTER BANKS

James D. Johnston

Signal Processing Research Department,
AT&T Bell Laboratories, Murray Hill, NJ, USA

1 INTRODUCTION

Recently, there has been a burst of work in coding of audio signals. Digital audio signals are typically signals with an analog bandwidth of 20 Hz to 20 kHz and amplitude resolution of 16 bits or more. Digital audio coders are not, for the most part, lossless coders in the information-theoretic sense. In fact they are quite lossy, using knowledge of the human auditory system in order to remove parts of the signal that the human auditory system cannot distinguish. Because they use knowledge of the perception process as their guide for lossy coding, these coders are commonly referred to as "perceptual coders," and the part of the signal they remove is referred to as "irrelevant."

2 THE HUMAN AUDITORY SYSTEM

The Human Auditory System (HAS) has been extensively studied for many years, and there have been some repeated discoveries. Most notable is the concept of the "critical band," a frequency range over which masking behavior is approximately invariant [358], which seems to relate to the somewhat nonlinear mechanical filter implemented in the human cochlea. As it turns out, the cochlea can be modeled as a filter bank, followed by a detector. In terms of the associated time-frequency representation, the frequency scale where each interval of one unit corresponds to one critical band is known as the "Bark" scale. Table 1 shows the relationship of this scale to the standard frequency scale in Hertz. It has been determined that this Bark scale also corresponds to physical distance on the basilar membrane of the human cochlea (inner

Band number (Hz)	Lower edge (Hz)	Center (Hz)	Upper edge (Hz)
1	0	50	100
2	100	150	200
3	200	250	300
4	300	350	400
5	400	450	510
6	510	570	630
7	630	700	770
8	770	840	920
9	920	1000	1080
10	1080	1170	1270
11	1270	1370	1480
12	1480	1600	1720
13	1720	1850	2000
14	2000	2150	2320
15	2320	2500	2700
16	2700	2900	3150
17	3150	3400	3700
18	3700	4000	4400
19	4400	4800	5300
20	5300	5800	6400
21	6400	7000	7700
22	7700	8500	9500
23	9500	10500	12000
24	12000	13500	15500
25	15500	19500	

Table 1 Critical band center and edge frequencies. Source: [358].

ear). Some care should be exercised in interpreting the cochlea as a filter bank, because the center frequencies of the analysis filters are not fixed, but continuous. Thus the distance of one Bark is approximately the bandwidth of the filter at a given location on the basilar membrane. The cochlear filter response differs by position on the basilar membrane in a continuous fashion, and there can be arbitrarily many such filter sets generated.

Another repeated result concerning the HAS is the concept of "masking" where one signal (the masker) can change or eliminate the loudness of another signal (the masked signal, or probe). There are many measurements of masking, depending on the exact structure and content of masker and probe, but two classic sets are the results of tones masking noise [358, 147] and noise masking a tone [180]. These results are particularly interesting because they show quite different masking levels for tone maskers as opposed to noise maskers with as much as 25 dB difference in the masking level provided by a sine wave as opposed to narrow-band noise of the same energy.

It is this concept of "masking" that provides the substantial compression in an audio perceptual coder. In effect, one can reduce the signal-to-noise ratio inside a critical band from 96-120 dB to 5-40 dB, depending on the structure of the audio signal, and reduce the bit rate required to code that part of the signal accordingly. As this masking takes place within the above-mentioned "critical bands," the characteristics of the auditory filter bank becomes important for several reasons, the foremost being that of the time response and frequency resolution of the auditory filter bank.

In auditory experiments, a signal (called a masker) is provided to the listener. Another signal (called the probe) is then provided at a frequency and time relative to the masker. Importantly, it has been shown that non-simultaneous masking of the probe is substantial for short delays of the probe following masker, which is called forward masking. However, there is very little, if any, masking for the converse situation of masker following probe, which is called backward masking. In part, this is understandable by the time-response implied by the auditory filter bank, but there are also clearly another issues involved in forward masking. There have been several tests of audio coders where backward masking has been insufficiently considered, leading to a situation where a "pre-echo" (a noise signal that precedes the masker) is audible.

3 THE AUDITORY FILTER BANK

The auditory filter bank has been modeled by several different researchers [358, 147]. Here, we will use examples given in [12]. Figure 1 shows the log-log frequency response of two cochlear filters, each shown on the log frequency scale (Hz), not on the Bark scale. A Bark frequency vs. amplitude plot is shown in Figure 2. Interestingly, in the Bark domain, the filter shape and bandwidth appear to be reasonably similar along the Bark or the "distance" scale, [172] suggesting that the scale is useful in understanding masking, time response of the auditory system, and so on.

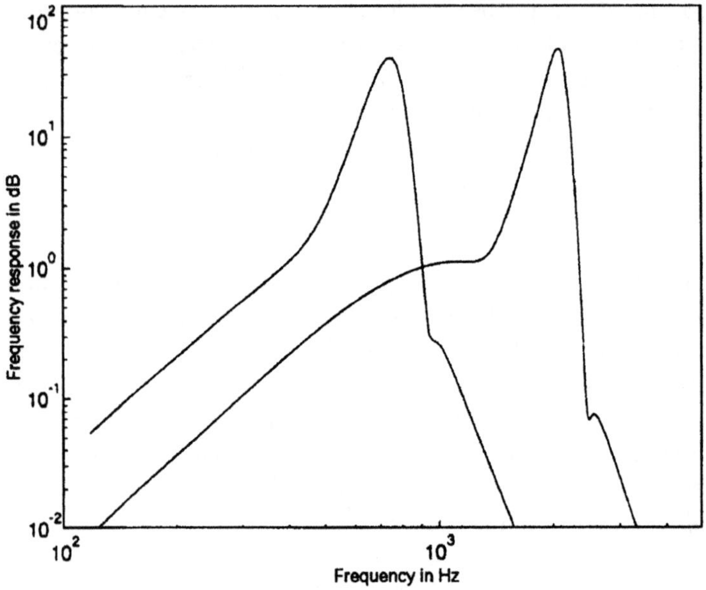

Figure 1 Log amplitude vs. frequency plot of two example cochlear filters due to [12].

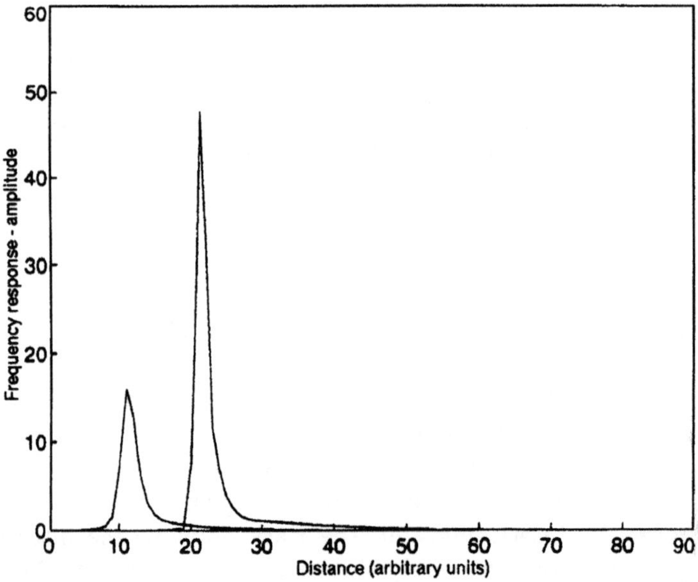

Figure 2 Amplitude vs. Bark frequency plot of the same two cochlear filters.

3.1 Channel Bandwidth(s)

The bandwidths corresponding to one critical band are shown in Table 1. While several mappings of this scale have been proposed, most implementations either use the scale as provided by Scharf, or the close approximation

$$B = 13.0 \tan^{-1}\left(\frac{0.76f}{1000}\right) + 3.5 \tan^{-1}\left(\left(\frac{f}{7500}\right)^2\right) \text{ Barks,}$$

where f is frequency in Hz and B is bark frequency. Effectively, either of these mappings corresponds to approximately uniform filter bands at low frequencies, and approximately $\frac{1}{3}$ octave critical bands at higher frequencies. Example time domain impulse responses for the filters shown in Figure 1 may be seen in Figure 3. Note that the time duration of the impulse response of the 750 Hz filter is approximately 19 ms, while the time duration of the impulse response of the 2050 Hz filter is approximately 10 ms. Also note that the bandwidth of the 750 Hz filter is approximately 140 Hz, and the bandwidth of the 2050 Hz filter is approximately 290 Hz, suggesting that the durations of the impulse responses of the cochlear filters can be reasonably approximated by assuming that the filters are minimum phase. Some researchers [12] suggest that the minimum-phase assumption is reasonable.

We must point out that the basilar membrane of the cochlea is clearly a non-linear detector. There are other forms of processing being done as part of the HAS functionality, as evidenced by masking of a delayed probe by a masker to an extent not directly attributable to the cochlear filter bank, variable spreading with frequency, and other effects noted in the literature [401, 402, 278, 72].

3.2 Temporal Implications of the Cochlear Filter

As is evident in Table 1, the bandwidth of a critical band varies by approximately 40 to 1 over the audio frequency band, suggesting that the time-response of the auditory system will vary by at least 40 to 1 as well. The various unanticipated audible effects and coding-gain problems resulting from the variation in time response are major issues in audio coder filter bank design.

As mentioned above, a well documented example of a time domain effect is the "pre-echo" noted in coders with long impulse response filter banks, or so-called

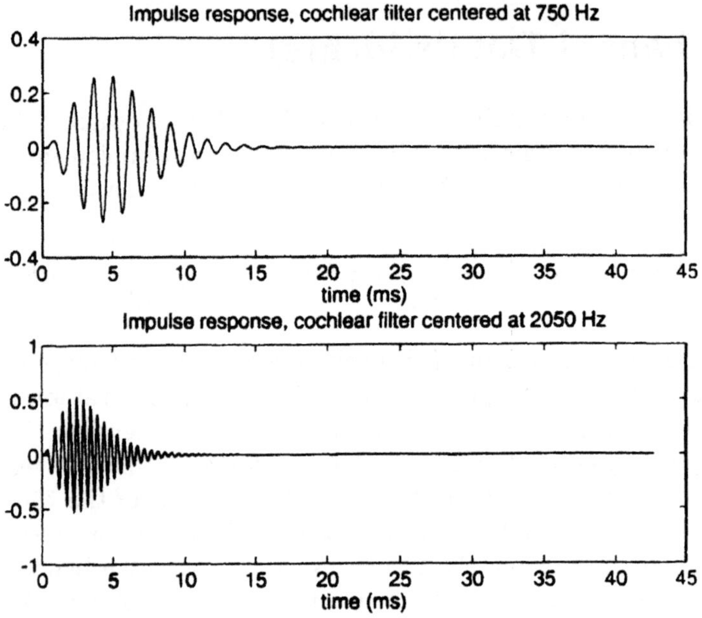

Figure 3 Impulse response (amplitude vs. time) for the two cochlear filters of Figures 1 and 2.

"transform coders." Early coders with high frequency resolution demonstrated this "pre-echo" quite clearly. An example of this effect can be seen in Figure 4, where a pre-echo has been deliberately demonstrated by removing the pre-echo control facilities from a current coder.

As an aside, no coder currently proposed for audio coding has, strictly speaking, a "block transform" as its effective filter bank. All current coder proposals use a critically sampled, overlapped filter bank, that, regardless of its implementation as a transform or as an erstwhile modulated filter bank, has critically sampled outputs and does not obey Parseval's theorem at each subsampling interval. Hence all coders in use today are properly known as subband coders, albeit some of the more sophisticated ones implement modulated overlap-add filter banks with a transform or transform-like structure at the heart of the system.

The new MPEG-I standard for encoded audio and video signals has three audio layers. Layers 1 and 2 use a filter bank with a much lower frequency resolution of 750 Hz at a 48 kHz sampling frequency, and which has little or no innate difficulty with time-domain vs. quantization artifacts. However, as the next two sections describe, these coders have great difficulty with signals whose interactions with the HAS require good frequency resolution, good coding gain, or both.

3.3 Frequency-Domain Implications of the Cochlear Filter

The 40 to 1 variation in bandwidth can also create difficulties in the frequency domain. Layers 1 and 2 of MPEG-I, which use a 32-band nearly exact reconstruction subband filter bank for analysis, have demonstrated some difficulties in controlling the spread of noise, especially at low frequencies. As an interesting problem, examine the necessary signal-to-noise ratio required in the DC to 750 Hz band of that coder for a sine wave presented at 750 Hz. Referring to Figure 1 for an estimation of the frequency response of the cochlear filter centered at 750 Hz, and to Figure 5 for the approximate masking spectrum of a sine wave centered at 750 Hz, it is clear that in order to maintain masked noise at 50 Hz, a signal to noise ratio of more than 70 dB is required. This is one example of where accommodating the shortest time response of the HAS penalizes the coder designer in parts of the frequency spectrum where the HAS has good frequency resolution. There are other issues and difficulties that arise

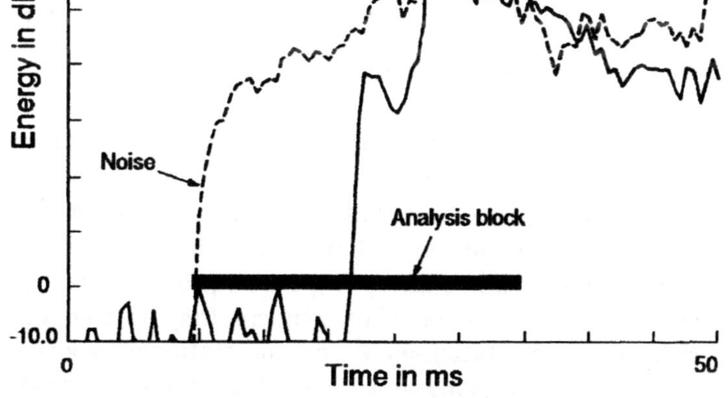

Figure 4 Example of "pre-echo" phenomenon.

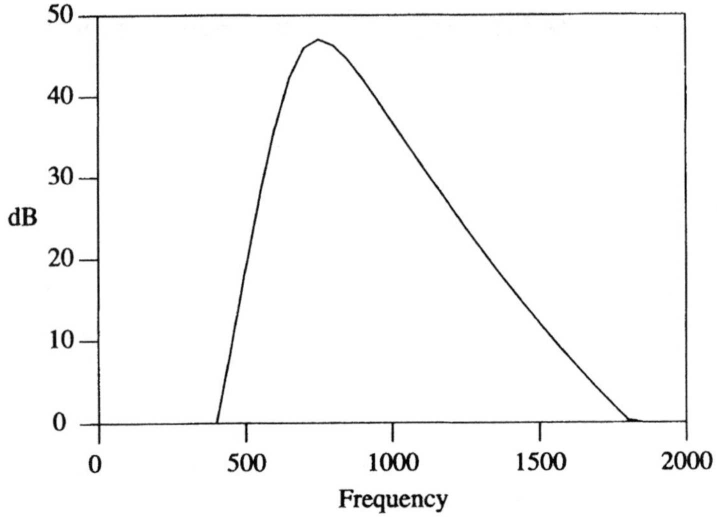

Figure 5 Approximate masking spectrum for a 750 Hz tone at a level of $+ = 3000$, compared to a reference level of 0 db $= 1$.

from the need to apply a psychoacoustic metric with good frequency resolution to a filter bank with low frequency resolution and aliasing. For instance, in layer 3 of MPEG-I, which uses a hybrid filter bank to increase the frequency resolution, it is necessary to actively cancel the inter-band aliasing in the layer 1 and 2 filter bank, in order to both ensure coding gain and prevent uncancelled aliasing from interfering with the basic audio quality of the coder.

3.4 Results of the Bandwidth Variation in the Cochlear Filter

As the two previous sections have shown, no uniform filter bank can, without some careful consideration to its application, provide a good coder filter bank from the viewpoint of the HAS. Toward this end, several authors [63, 370] have tried filter banks that have attempted to mimic the cochlear filter bank to some extent. While these coders have, in general, provided good performance on signals requiring good time-control of artifacts, they have shown difficulties in dealing with classes of stationary or pseudo-stationary signals that have many frequency components (sometimes called harmonics or partials) requiring encoding. These difficulties are seen in the next section to arise from a lack of coding gain, i.e. a lack of diagonalization of the input signal in the filter bank.

As a result of these issues, it is clear that there is no one "correct" choice of filter bank for signals being presented for human audition. In fact, the best choice of filter bank depend not only on human audition, but also on the signal being coded.

4 THE AUDIO SIGNAL

Audio signals are typically digitally supplied in the form of 16 bit or 20 bit PCM words, with one word/sample/channel at a sampling rate of 44.1 kHz or 48 kHz, depending on origin. Contained in the PCM signal is a time waveform that can be generated by anything the producer sees fit, from spoken voice to percussion instruments to synthesized signals that could not be created by normal acoustical methods. In this universe of signals, it is very difficult to know or assure the presence or absence of any particular kind of source or source model, very much unlike the usual situation in speech coding, where a production model is often valid. In this sense, audio coding is much more like general image coding, where the source of the input signal and the applicable model can vary substantially and capriciously over time and signal source. Some very basic measures of a selection of audio materials are shown in Table 2. This table shows, for a variety of music signals, statistics of the LPC gain [327] for a 16 pole predictor, including the mean gain in dB, the standard deviation in dB, and the maximum and minimum in dB. The table also shows the mean and maximum LPC distance for a predictor delayed relative to the signal by 256 samples, as well as statistics of spectral flatness measure (SFM) [250] for a 512 point von Hann-windowed FFT. LPC distance is a measure in dB of misfit in the predictor. It is the ratio of gain between the ideal (time aligned) and delayed predictor, and the SFM, which is a measure of the amount of redundancy available in the signal at that time scale.

From a quick examination of Table 2, it is seen that the mean prediction gain, even for a predictor of large order, is not generally enough to provide good predictive coding of a 16 bit signal with any substantial reduction in bit rate. Further, it is evident from the Max Distance to Mean Distance ratio that signal models remain constant for long periods of time, and then change suddenly, requiring a good tracking of the signal by the predictor. This, by itself, rules out most backward-adaptive predictor strategies because of their settling delay.

Examination of the SFM for these signals shows that the amount of easily extractable signal redundancy is not sufficient to provide substantial bit rate

Signal	Mean gain (dB)	Sigma (dB)	Max (dB)	Min (dB)	Max distance (dB)	Mean distance (dB)	Mean SFM (dB)
Narrowband noise	40.5	5.2	59.2	25.0	19.1	1.00	43.5
Bassoon	40.1	3.8	23.8	28.8	10.9	0.68	45.6
Viola	30.7	3.7	43.5	20.9	11.9	0.92	34.1
Castanets	19.9	4.5	38.8	4.9	17.2	1.68	22.9
Solo piano	36.9	3.9	51.1	23.3	12.1	1.03	39.6
Cabasa	5.4	0.5	7.6	3.7	1.6	0.50	8.7
Female operatic	32.6	4.3	44.6	14.5	10.7	0.77	36.9
Trumpet concerto	36.6	2.8	47.1	28.2	6.7	0.64	39.7
Zarathustra	33.6	4.9	50.6	12.7	10.6	0.65	37.7
Metal percussion	32.1	4.2	35.1	8.2	26.0	0.84	29.2
Suzanne Vega	23.4	5.7	36.4	3.7	18.0	1.40	27.3

Table 2 LPC statistics for 16 poles, compact disc sources.

reduction, in either a lossy or non-lossy sense, even given a 512-point FFT for coding in the transform domain, when using a source-modeling coder or other coder that exploits the characteristics or redundancies of the signal source. This is why perceptual coders are a necessity. One must remove the irrelevancy in the signal as well as the redundancy in the signal in order to achieve a substantial bit rate reduction.

4.1 Redundancy

Table 2 effective shows the redundancy in the signal for one set of signals at one FFT size. Figure 6 shows SFM vs. FFT size for a large selection of easy and difficult audio materials, all sampled at either 44.1 kHz or 48 kHz. As is clear, even easier audio signals have insufficient redundancy to encode a signal transparently by purely source coding techniques, unless small compression ratios are considered. In Figure 6, the amount of redundancy varies considerably as a function of audio signal and window length, but with the redundancy growing as the time window becomes longer. This shows that for signals that do not have substantial time-domain perceptual requirements, a long filter bank is essential to coder efficiency. Unfortunately, for signals where the time-domain perceptual requirements are important, it may be impossible to use a filter bank of greater than critical-band resolution, or alternately of 32 line uniform resolution at a sampling rate of 48 kHz, without introducing a protection method that requires substantial bit rate peaks at some instants.

Figures 7a and b show the SFM for two signals, castanets and harpsichord respectively. A second plot on each figure also shows the gain realizable for that signal and time scale when a time artifact protection method called Over-Coding Protection (OCP) is employed in order to prevent time-domain artifacts. OCP is a protection method based on a particular psychoacoustical model used in the AT&T Perceptual Audio Coder [218]. These plots show that the time structure of a signal may in some cases require a short filter bank in order to efficiently code a signal, and that other signals may require a high time resolution in order to provide efficient coding.

As is clear from the two illustrations, the gain in the harpsichord signal is not substantially changed by the time-domain structure of the signal, while the gain in the castanet signal is completely overwhelmed by the need to control time-domain artifacts. In fact, these two signals are in some sense diametric opposites as far as coding is concerned. In addition to the time-domain issues, the harpsichord is substantially "tonal", requiring a greater local SNR in order to be perceptually transparent, while the castanets are mostly atonal, requiring a smaller local SNR for transparency, assuming that the time domain issues are correctly handled. This could be restated as saying that the harpsichord has little irrelevancy, while the castanets have substantial irrelevancy. As a result of both the time domain and filter bank gain effects, a longer window is necessary for good compression of the harpsichord signal, while it is absolutely undesirable for the castanets.

Many authors, including [399, 370], have proposed using critical band filter structures for audio coding. This kind of coder provides good results in the case of signals such as castanets, where time-domain artifacts predominate. Unfortunately, the lack of coding gain available from such banks prevents their use in efficient coders that must encode signals such as harpsichords, where a recovery of the signal redundancy is compelling due to the lesser irrelevancy available in the signal.

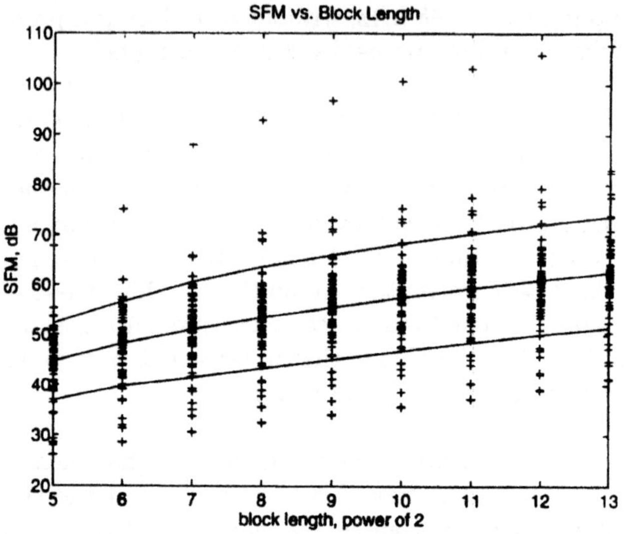

Figure 6 Scatter plot of SFM vs. block length for 44100 and 48000 kHz sampled signals. Each '+' represents an observation for one example signal at 1 block length. The middle curve is the mean, and the upper and lower curves $\pm 1\sigma$.

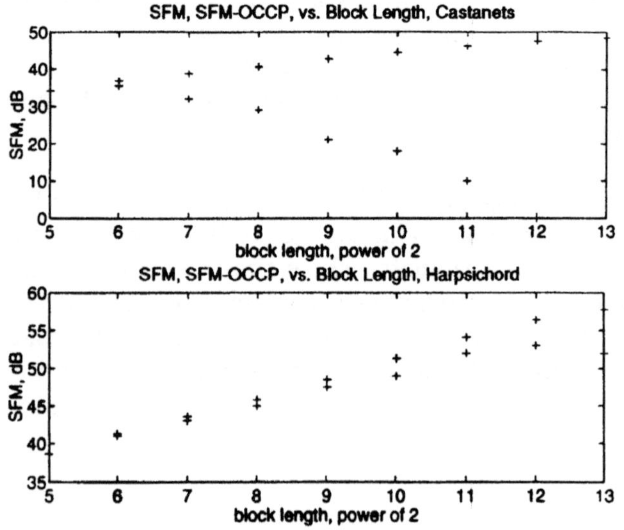

Figure 7 Plot of SFM and SFM–OCP for two signals, harpsichord and castanets. The upper '+' in each plot is the SFM for the given signal and block length, the lower is the SFM–OCP for that signal at that block length.

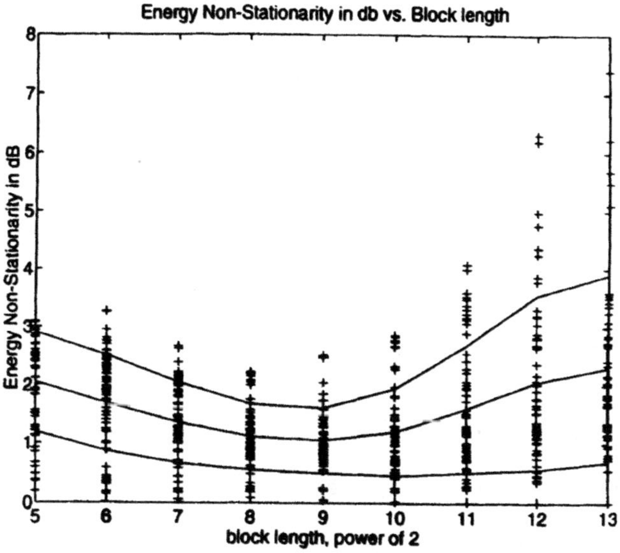

Figure 8 Scatter plot of energy non-stationarity in dB for the same signals as in Figure 6. Again, each '+' represents one observation, and the curves mean and mean $\pm\sigma$.

4.2 Stationarity

In Figure 8 we show data points, marked "+," as a function of block length that represent the non-stationarity in energy. The mean is shown by the solid line in the middle. The lines above and below denote $\pm\sigma$ of the mean. The non-stationarity in energy is calculated as abs $\log_{10}\left(\frac{E_t}{E_{t-1}}\right)$, where the abs represents the absolute value function. In all cases including those below, any results related to silent or near-silent blocks are discarded. Figure 9 shows the maximum value for each signal at each block length. Figure 10 shows the mean change in spectral energy, which is to say the same measure as in Figure 8, but applied individually to each frequency in the power spectrum of the DFT of the given block length. Figure 11 shows the maximum value for each signal at each blocklength.

As Figures 6, 8–11 show, there is a strong variation among audio signals with respect to stationarity. As Figure 7 also shows, the variation is increased substantially by the need for pre-echo protection at any given block length. An interesting observation is that on the average, audio signals start to show increasing non-stationarity beyond 512 to 1024 samples in window length. More interesting is that the amount of gain in most signals continues to grow faster than the increase in non-stationarity until the range of 1024 to 2048 point transforms. This shows that until the filter bank reaches a total impulse duration of approximately 40 ms, the increase in coding gain on the average exceeds the loss due to non-stationarity. This suggests that the default block length for coding should be somewhere between 1024 and 2048 samples. This particular calculation does not take into account all of the issues involved in pre-echo prevention, but does approximately reflect the safety factor that one must use to prevent the changing masking levels from revealing coding noise.

5 INTERACTIONS BETWEEN AUDITORY FILTER BANK AND THE AUDIO SIGNAL

Several interactions between the HAS and the audio coder filter bank, such as pre-echo, unmasked aliasing, and insufficient frequency resolution have already been mentioned. The following is an incomplete list of the interactions between the HAS and the coding filter bank:

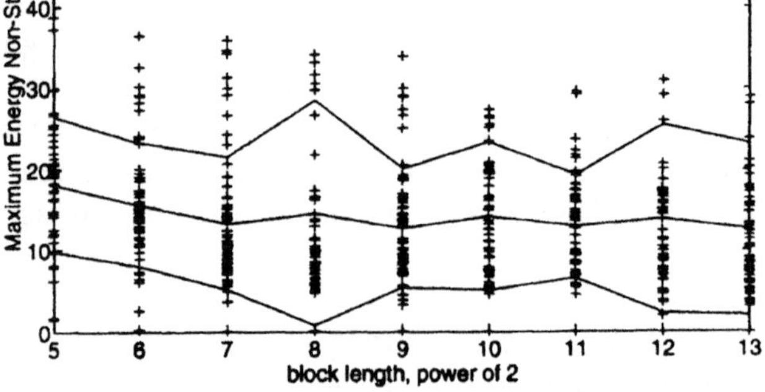

Figure 9 Maximum, calculated over all non-silent blocks of each signal, of the non-stationarity in Figure 8.

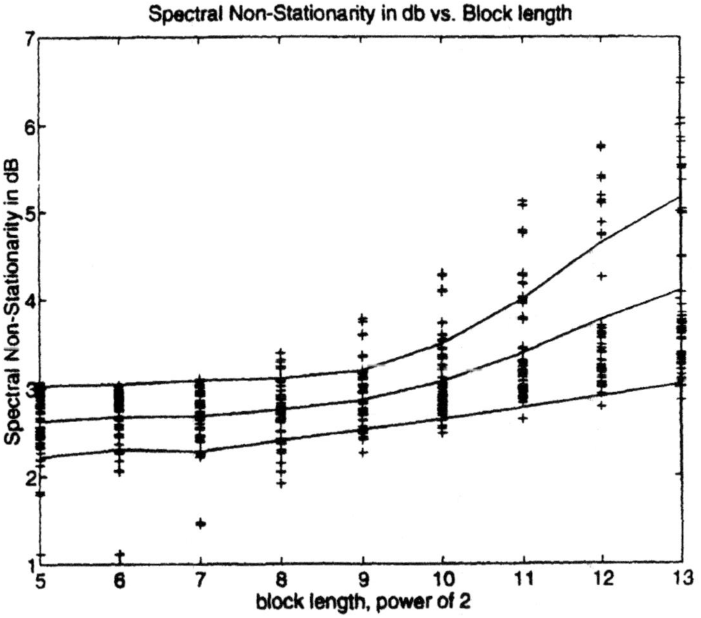

Figure 10 Scatter plot of spectral non-stationarity in dB for the signals in Figure 6, presented in the same format as Fig's 6, 8, 9.

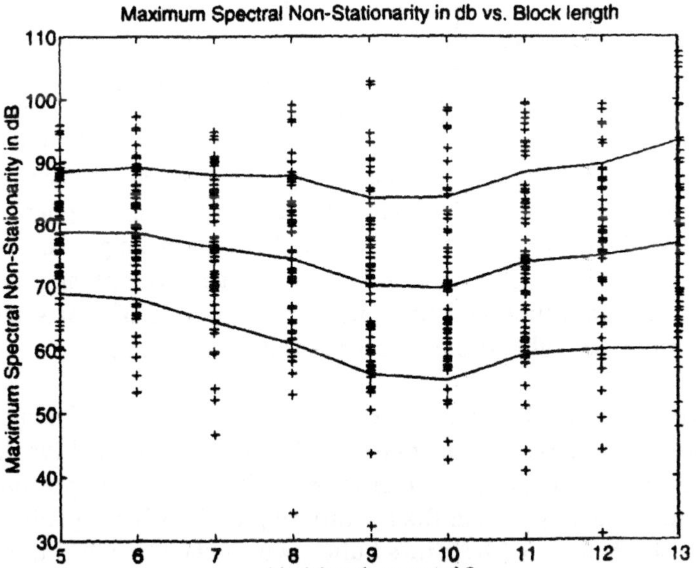

Figure 11 Maximum of the data in Figure 10 for each signal and block length.

- Time Domain Effects
  - Pre-Echo
  - Stereo Imaging artifacts
- Frequency Domain Artifacts
  - Unmasked Aliasing
  - Large overcoding requirements due to filter bank bandwidth
- Coding Gain Problems
  - Lack of sufficient coding gain
  - Loss of coding gain due to pre-echo or non-stationarity issues

The issues of unmasked aliasing and insufficient frequency resolution, i.e. frequency domain artifacts, have already been discussed in sufficient detail. Discussion of the other issues follows in this section.

5.1 Time Domain Effects

The primary time-domain effects resulting from the interaction of the coding filter bank and the audio signal are pre-echo and stereo imaging time domain issues. Figure 4 shows graphically what constitutes a "pre-echo." All filter banks will spread the noise injected in the frequency domain across the entire impulse response of the filter bank. In most filter banks, and all filter banks that have an overall pure delay (linear phase) response, this will imply that the noise will precede the signal by some amount of time, leading to the phenomenon of a "pre-echo." While some pre-echo is acceptable, the masking levels required for pre-echo protection in a practical coder suggest that the phenomenon is very limited, both in amplitude and time, i.e. the amplitude of a pre-echo must be low, and it must not substantially precede the masking signal.

In the processing of preventing pre-echo in coders with good frequency resolution, it is necessary to control the masking threshold calculated for the coder by adding "pre-echo protection," as mentioned in Figure 7. The primary result of this protection is large peaks in the bit rate required for high quality coding. In some coders this is handled by a rate buffer, where the large demand is supplied from a bit rate buffer, and the bits are then taken from less demanding parts of the signal. In other coders, the actual filter bank is switched into a mode with less frequency resolution, so that the amount of "protection" is reduced, and the bit rate peak reduced. Some of these coders [218] still use a buffer, others such as MPEG-I, layer 3, do not use a large bit buffer.

Stereo Imaging

The masking results described above are established for a masker and probe applied to one ear. While a coherent discussion of the complexities and ambiguities of stereo (two signals, each applied to a loudspeaker, with the listener in the reverberant field) or binaural (headphone) listening is well beyond this text, and is in fact poorly understood, several notable effects, such as spatial imaging and Binaural Masking Level Depression (BLMD) surface [278].

There seem to be several cues that the HAS uses to create a stereo image. Time onset of the envelope of the signal inside a given critical bandwidth has been shown to be important in creating the spatial image of a non-stationary audio signal, under both stereo and binaural conditions. In addition, the inter-aural delay (phase) of the signal itself, stationary or not, is shown to be important to imaging at low frequencies, certainly under 500 Hz and to some extent up to several kHz. It is only for stationary signals above approximately 4kHz when the envelope of the signal is approximately constant, that the total energy in a critical bandwidth seems to provide the spatial clue. At high frequencies, where the quantization noise in a coder may be near the energy of the signal, the length of the impulse response of a given filter bank and the levels of quantization permitted by the coder interact very strongly in controlling time-onset issues. As such, the issue of stereo imaging goes beyond the filter bank domain, into the domain of the psychoacoustic masking model and coder structure, but it is clear that long impulse responses require a more careful examination of onset time issues.

The issue of inter-aural delay also surfaces in masking, as the phenomenon of Binary Masking Level Depression. This effect is well-described in [278]. Effectively what BMLD permits is an imaging of the probe and the masker such that the probe may be audible, for some choices of probe and masker, at as much as a 20 dB lower level than it would be in the monaural case. This is true if and only if either the masker or the probe is uncorrelated or anti-correlated in the two ears, and the other of masker or probe is strongly correlated. As a result, a monaural coding result may be satisfactory, while unmasked noise or stereo imaging problems may arise in the case of a stereo coder from distortions of the time properties of the signal.

5.2 Effects of Filter Bank "Rate Gain"

As suggested before, extracting the irrelevancy due to HAS masking is only part of the coding battle. After the irrelevancy is extracted, the remaining redundancy must also be extracted for efficient compression. Effectively, the irrelevancy determines how much signal can be thrown away, and the redundancy determines how much can be recreated at the decoder without cost. Clearly, the useful gain from a filter bank will vary with signal and filter bank. As Figure 6 shows, there are substantial differences in the coding gain available from the filter banks, and as Figure 7 shows, even more differences in the amount of redundancy that can safely be extracted from a given signal with a particular filter bank impulse response.

Figure 7 shows that the differences between the auditory filter bank and the coding filter bank can result in substantial loss in the gain available from the filter bank. Effectively, a filter bank that has too long an impulse response, i.e. high frequency resolution, will usually be prevented from doing efficient coding because of the compensation that must be used to prevent time domain artifacts from becoming audible, and a filter bank with too little frequency resolution will lack the necessary rate coding gain to successfully code audio signals with a high demand for SNR as a function of frequency. The castanets and harpsichord signals are good examples of a pair of signals that exhibit the need for good time and good frequency resolution respectively.

The tradeoff in filter bank choice is quite evident. A long (good frequency resolution) filter bank will have enormous problems with pre-echo and overcoding requirements, and a short (good time resolution) filter bank will have insufficient redundancy extraction for good coding. This suggests strongly that a fixed filter bank is not appropriate for audio coding. It would, in the best of all possible worlds, perhaps be possible to take a properly calculated perceptual threshold, examine the time-frequency tesselation of that threshold for parts of the time-frequency plane that have the same threshold, and then generate a filter bank that would fulfill that tesselation, under the optimization constraint of minimum coded bit rate. Currently, this sort of highly adaptable filter bank does not exist.

5.3 Current Solutions

There are several different current solutions to the filter bank problem. Musicam [400] and the lower layers of the MPEG-1 audio coder use a low resolution filter bank and abandon attempts at high compression rates. PAC [218] and other coders use a long, high resolution filter bank in the form of a modified discrete cosine transform (MDCT), and switch the length of the filter bank when signal conditions require it. Sinha [370] and other researchers propose a wavelet-based filter bank that mimics the response of the auditory system, but that does not provide the necessary redundancy extraction for difficult signals.

In all of these cases, the coders are "subband coders," i.e. they break the signal down into frequency-controlled subbands, with a filter bank that provides a 1:1 ratio of input to filtered signals, and that provides, in the end, either exact or nearly-exact reconstruction of an unquantized signal. A common misnomer is to refer to subband coders with a few (such as 2, 4, 32) subbands as a "subband" coder, and coders that use many subbands (such as 128, 256, 512, 576, 1024, 1152) as a "transform" coder. This usage is incorrect in a very basic sense. A transform is a specific construct that obeys, among other things, Parseval's theorem, on a block-by-block basis. Subband filters, regardless of their implementation as polyphase filter banks or MDCT's do not do this. Many forms of subband filters are energy conservant, but they are not energy conservant on a block-by-block basis, and hence should not be referred to as transforms. It is likely that this misidentification originated in the use of filter banks that used FFT-like structures for efficient implementation. As a matter of practicality, nearly all audio coder filter banks use a "transform-like" structure for ease and efficiency of implementation. A very good discussion of such structures for filter bank applications can be found in [254].

Because of the need to maintain either time-domain resolution or frequency-domain resolution, depending on the signal, some coders, including PAC [218], and ASPEC (the precursor of MPEG-1 layer 3) [217], adapt the filter bank in some sense in order to provide increased efficiency. In PAC, for instance, the "normal" filter bank that uses 1024 channels in frequency is replaced by a set of 8 128-channel filters when the coder measures that it is more efficient to use the better time resolution. This sort of block switching algorithm, while effective, does not provide the level of efficiency that will be necessary for very low rate audio coding with high quality.

5.4 Future Needs in Filter Banks

The primary need in audio coding is the ability to flexibly change the time-frequency tiling of the filter bank, in a way dictated by the joint understanding of the signal and the HAS, so that a joint minimum in bit rate can be reached for each kind of audio signal. For some audio signals, as shown above, this may require a 4096 or more band filter bank. For other signals it may require a 32 band bank, or a filter bank with filters on or near a critical-band scale.

The key in such a filter bank is the flexibility of time and frequency tiling, in the sense that frequency portions of the signal that are highly varying may require a time resolution matching that of the critical band, and portions that are highly stationary may require very, very high frequency resolution, both in the same "frame" of the coded signal.

Such a filter bank must be able to maintain continuity of both signal and frequency meaning, both during and between the switching of resolutions. In particular, a filter bank that switches between a critical band state and a high-resolution uniform state, but has as its intermediate step a filter bank set that consists of allpass like basis vectors will not provide a good solution, because the projection of the perceptual threshold onto such an analysis set will require extraordinary overcoding in order to avoid unmasked noise in the reconstructed output.

As well as maintaining continuity, the flexible filter bank must have both analysis and synthesis filters that have well-established frequency meaning. A filter bank that has a good reconstruction frequency separation, but very bad analysis separation, or vice versa, will cause either a substantial loss of redundancy extraction or irrelevancy due to the bad frequency separation of the analysis or synthesis.

In addition to everything else, a desirable property of such a filter bank would be 1:1 and onto performance, i.e. performance that is obtainable from an orthonormal filter bank set. While a nearly-exact-reconstruction filter bank can be used, the inexact nature of such a filter bank creates problems with level shift of the digital signal, as well as creating potentially audible distortions in either frequency shaping or signal clipping in the 16 bit PCM word.

Finally, such a filter bank must be easy to implement in an actual hardware realization. Audio coding will be used not only for "professional" purposes, but also, and perhaps mostly, for "consumer" applications, where the price of implementation is of paramount consideration.

6 CONCLUSIONS

The filter banks currently used in audio coding are well understood in the filter bank art. In general, they are filter banks of uniform resolution, provide exact reconstruction, and are easily implemented at sampling rates of 48 kHz/channel. These filter banks introduce specific difficulties into the audio coding field. Because of these difficulties, detailed above, the audio coders that provide state-of-the-art compression rates (i.e. compression on the order of 12:1 for difficult signals) must use two, or perhaps even more, frequency resolutions in order to avoid the time/frequency problems inherent in systems that must deal with the human auditory system. A new filter bank structure, flexible both in time and frequency, and switchable on demand, would eliminate specific and well-understood inefficiencies in coding algorithms, as well as eliminate problems of audible distortion in signals with outlier statistics.

10

WAVELETS AND FILTER BANKS IN SPREAD SPECTRUM COMMUNICATION SYSTEMS

Gary J. Saulnier, Michael J. Medley* and Pankaj K. Das

Electrical, Computer and Systems Engineering Department, Rensselaer Polytechnic Institute, Troy, New York 12180-3590

**Rome Laboratory, Griffiss Air Force Base, New York 13441*

1 INTRODUCTION AND BACKGROUND

In recent years, interest in spread spectrum communication techniques has expanded from the military into the commercial domain. The term "spread spectrum" refers to a class of digital modulation techniques that produces a transmitted frequency spectrum much larger than the bandwidth of the information being transmitted. For example, a data sequence having a rate of several kilobits per second (kbps) may be transmitted using a signal having a bandwidth of several MHz. While this may seem wasteful of bandwidth, the process of "spreading" and "despreading" the spectrum produces some important, desirable properties. It is interesting to note that when used in a multiuser environment, spread spectrum techniques can actually be more bandwidth efficient than narrow-band techniques.

While there are many types of spread spectrum modulation, they all share a number of desirable features. Among these features are interference rejection, low power spectral density, high-resolution ranging capability, code-division multiplexing and multiple access, and resistance to multi-path fading. The first three items are the primary reasons why spread spectrum systems have found applications in military environments. The combination of low power spectral density and interference rejection makes spread spectrum signals difficult to detect and jam. The ranging property has made spread spectrum signals useful for target tracking and radar. The last two items have fueled the recent growth in spread spectrum applications in the commercial environment. The urban mobile radio channel is characterized by severe multi-path fading, which hinders the effectiveness of narrow-band communications systems [405, 389].

This multi-path environment produces a *frequency-selective* channel whose frequency response may contain deep nulls. Due to their large bandwidth, spread spectrum signals provide *frequency diversity* which makes them less susceptible to this fading. Code division multiplexing and multiple-access (CDM/CDMA) allow many spread spectrum signals to occupy the same bandwidth at the same time, making it possible for receivers to benefit from the frequency diversity while maintaining an efficient use of the frequency spectrum.

Clearly, the properties that are of primary interest to military systems can also be beneficial to commercial systems. For instance, commercial systems may be subjected to unintentional interference or may *overlay* [273], i.e. share bandwidth, with narrow-band services. In this case, both the interference rejection and low power spectral density properties are important because they minimize the mutual interference between the narrow-band and spread spectrum signals. The interference rejection property makes the spread spectrum signal relatively immune to the interference caused by the narrow-band signal, while the low power spectral density property helps the spread spectrum signal avoid interfering with the narrow-band signals.

There are many different types of spread spectrum modulation techniques. The remainder of this section will discuss the two most common types, Direct Sequence (DS) and Frequency Hopping (FH) spread spectrum techniques, and provide some insight into the origins of the properties cited above. A later section will discuss how wavelet and filter bank concepts and techniques can be applied to enhance the performance of DS spread spectrum systems in the presence of jamming. Additionally, the generation of new wavelet-based spread spectrum signaling waveforms, which emphasize desired signal properties, will be discussed.

1.1 Direct Sequence and Frequency Hopped Signals

DS systems modulate the data with a pseudo-noise (PN) sequence having a rate that is much higher than the data rate. Maximal sequences (m-sequences) [127], which consist of a series of 0's and 1's, are often used for this application because of their ease of generation and good randomness properties. These sequences are *shift register* sequences, which are generated by a shift register with a feedback network consisting of *modulo*-2 adders. For a shift register of length n, the m-sequence that is generated is of length $2^n - 1$. Each sample in the sequence is called a *chip*, meaning that an m-sequence is $2^n - 1$ chips

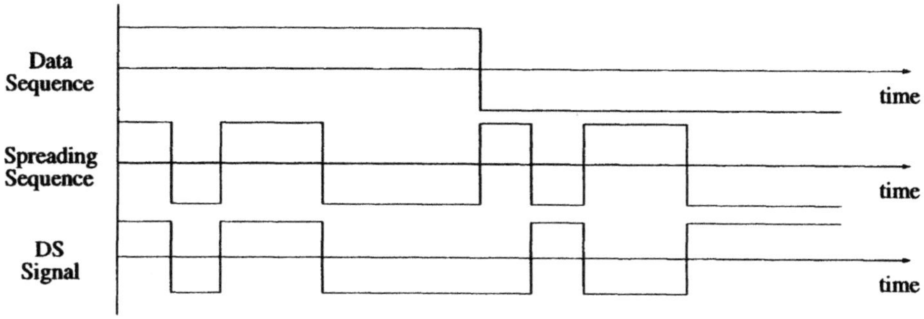

Figure 1 Direct sequence spread spectrum signal generation. The spreading sequence modulates the data sequence to produce the DS signal.

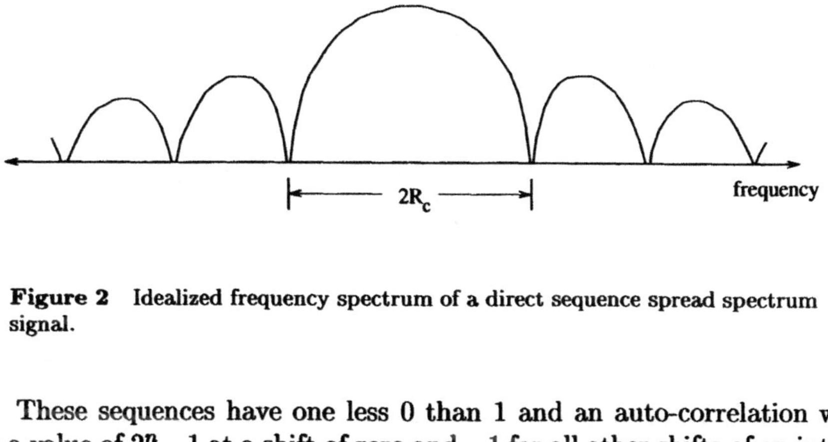

Figure 2 Idealized frequency spectrum of a direct sequence spread spectrum signal.

long. These sequences have one less 0 than 1 and an auto-correlation which takes a value of $2^n - 1$ at a shift of zero and -1 for all other shifts of an integral number of chips. For large n, the balance of 0's and 1's results in a very small DC component when the 1's and 0's are converted to ± 1's for modulating the data. The DS signal is produced by multiplying the data bits (± 1's) by the m-sequence, producing a high rate sequence of ± 1's as shown in Figure 1. The rate of the PN sequence is called the *chip rate* of the system, R_c. The resulting frequency spectrum has a $\sin(x)/x$ shape with a main lobe bandwidth of $2R_c$. A plot of the magnitude-squared frequency response is shown in Figure 2. If desired, the chips can be shaped using, for example, Nyquist pulse shaping in order to suppress the side lobes and limit the bandwidth of the signal [456].

FH systems create a broadband signal by moving the carrier of a narrow-band signal in a pseudo-random manner over a wide range of frequencies. Often, the narrow-band signal is produced using Frequency-Shift Keying (FSK) and

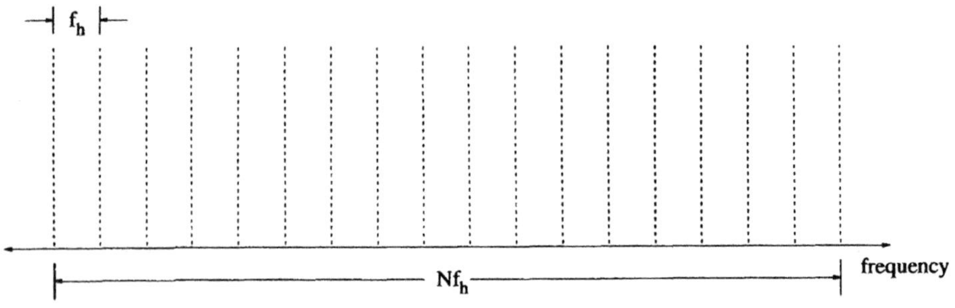

Figure 3 Idealized frequency spectrum of a frequency-hopped spread spectrum signal.

the carrier frequency is hopped through a pattern of equally-spaced carrier frequencies as directed by a PN sequence. In Fast FH systems (FFH), the rate at which the carrier frequency hops will be an integer multiple of the data symbol duration. As a result, the same symbol may be sequentially transmitted on a number of carrier frequencies. In Slow FH (SFH) systems, an integer number of data symbols are transmitted during each hop. Figure 3 shows the idealized frequency spectrum of a FH signal in which the N hop frequencies are spaced at intervals of f_h Hz.

1.2 Processing Gain and Jamming Margin

In both the DS and FH systems, the receiver must know the PN sequence used to spread the data signal in order to despread the signal and recover the data. A DS receiver correlates the received signal with a locally-generated copy of the PN sequence that is synchronized with the spreading sequence in the received signal. Similarly, in a FH receiver, the correlation is performed using a locally-generated copy of the hopping carrier. Correlation is performed by multiplying the received signal by a local copy of the spreading sequence or hopping signal and lowpass filtering. The multiplication process collapses the data signal down to its original bandwidth while spreading any additive noise or interference to a wide bandwidth. The lowpass filtering operation is then able to recover the data signal and reject a large fraction of the spread noise and/or interference. At the receiver, the ratio of the spread signal-to-noise ratio after despreading, SNR_O, to the input signal-to-noise ratio, SNR_I, is defined as the *processing gain*, G_p, i.e.

$$G_p = \frac{\text{SNR}_O}{\text{SNR}_I}. \tag{10.1}$$

The processing gain can be approximated by taking the ratio of the spread bandwidth to the data information bandwidth (rate), meaning that

$$G_p \approx \frac{BW_{spread}}{R_{info}}. \tag{10.2}$$

Processing gain forms the basis for the interference rejection and low power spectral density properties of spread spectrum signals. For a given information rate, a larger spread bandwidth means that the transmitted signal energy is spread over a larger bandwidth, meaning that the receiver can operate with a lower value of SNR_I. The system performance is ultimately a function of the value of SNR_O, which determines the bit-error-rate (BER) experienced by the communications link. The range of SNR_I for which the receiver can provide the required performance is determined by the *jamming margin*,

$$M_J = \frac{G_p}{\text{SNR}_O * L_{system}}, \tag{10.3}$$

where L_{system} accounts for any losses due to receiver implementation. For example, consider that the required SNR_O is 10 dB, L_{system} is 1.5 dB and G_p is 30 dB. In this case, M_J is 18.5 dB, meaning that the required SNR_O will be attained with a spread signal-to-noise ratio of 18.5 dB or greater.

1.3 Code-Division Multiple Access

Code-Division Multiple Access (CDMA) is a technique in which users share the channel by employing different spreading codes. In a CDMA system, all the users occupy the same frequency band at the same time. This is in contrast to Frequency-Division Multiple Access (FDMA) and Time-Division Multiple Access (TDMA) schemes in which the signals occupy the channel at disjoint frequencies or times, respectively. In a CDMA system, each transmitter has a different spreading code that is known by the intended receiver. For effective operation, the cross-correlation between codes used by different users should be minimized so that the despreading process will reject most of the *multi-user interference.*

CDMA systems can be implemented using either FH or DS techniques, though DS is by far the most popular choice. One reason for this is that when multiple users occupy the same hop frequency at the same time, the transmitted data will likely be lost by all users. In a SFH system, these ***hits*** will result in the loss of received symbols and some form of channel coding must be employed to overcome the losses. In a FFH system, the frequency diversity introduced by sending the same data symbol on multiple hops helps perform the same function as coding in a SFH system. In essence, the problem is that individual signals occupy only a small portion of the total bandwidth at a given time, resulting in the amount of interference between user signals varying greatly with time, i.e. the correlation between signals is high when hits occur and zero otherwise.

By contrast, DS signals occupy the full bandwidth at all times, meaning that the correlation and, hence, the interference between users, is relatively uniform from symbol to symbol. DS/CDMA systems, however, encounter problems when signals from interfering users arrive at a receiver having a power level that is significantly higher than the signal intended for that receiver. To illustrate the problem, consider a case where two transmitters use spreading codes having a cross-correlation of -20 dB (normalized to a peak auto-correlation of 0 dB). If both signals arrive at a particular receiver having the same power, the desired signal will have a power level that is 20 dB above the interfering signal after despreading, i.e. the signal-to-interference ratio (SIR) will be 20 dB. However, if the desired signal arrives with a power level that is 25 dB below that of the interfering signal, the SIR after despreading will be -5 dB and acceptable BER performance will not be possible. Since this second situation is likely to occur when the interfering user is closer to the receiver than the desired user, it is often referred to as the *near-far problem.* The way to mitigate the near-far problem is to use spreading codes with very low cross-correlation coupled with transmitter power control.

While m-sequences are well-suited for single-user spread spectrum communications, there is an insufficient number of m-sequences having a given length that have low cross-correlation for all relative shifts of the codes. Since m-sequences are poorly suited for use in a CDMA system, there has been a considerable amount of work on the design of sets of sequences with better cross-correlation properties. The most well-known of these sequences are Gold codes [168, 190]. For a positive integer value of n not divisible by 4, there exists a set of $2^n + 1$ Gold codes, each of length $2^n - 1$, which have maximum cross-correlation values of $2^{(n+1)/2} + 1$ for n odd and $2^{(n+2)/2} + 1$ for n even. Since the maximum auto-correlation value for any single code is $2^n - 1$ the maximum *normalized* cross-correlation values are

$$|R(k)| \leq \begin{cases} \frac{2^{(n+1)/2}+1}{2^n-1} & n \text{ odd} \\ \frac{2^{(n+2)/2}+1}{2^n-1} & n \text{ even and not divisible by 4.} \end{cases} \qquad (10.4)$$

Additionally, the magnitude of the cross-correlation between the codes takes only two values. For n odd, the magnitude of the normalized cross-correlation equals $1/(2^n-1)$ approximately 50% of the time, and the maximum value given in Equation (10.4) approximately 50% of the time. For n even, the magnitude of the normalized cross-correlation equals $1/(2^n-1)$ approximately 75% of the time, and the maximum value given in Equation (10.4) the remainder of the time.

1.4 Communications and Intercept Receivers

In this section, we are primarily concerned with two types of receivers, *communications receivers* and *intercept receivers.* Communications receivers are used to demodulate and decode the received waveform; they are typically assumed to have knowledge of the spreading sequence or, in the case of FH, the hopping pattern. In contrast, the intercept receiver is an "enemy" receiver that monitors the channel with the goal of detecting the presence of a signal. Once a signal is detected, it may direct a jammer to try to disrupt the communications channel or may work to extract information from the signal. Unlike the communications receiver, the intercept receiver does not have any *a priori* knowledge about the parameters of the transmitted signal, e.g. spreading sequence, chip rate and data rate.

A block diagram of a time domain DS receiver is shown in Figure 4a. The receiver multiplies the received signal, $r[n]$ by a locally-generated copy of the spreading sequence and integrates the result over a data bit period. The locally generated spreading sequence must be *synchronized*, i.e. brought into time alignment, with the spreading sequence in the received signal. Techniques for achieving synchronization will not be considered here but are covered in detail in [190, 127].

An alternative receiver structure that performs despreading without synchronization is the *sliding correlator* shown in Figure 4b. The sliding correlator is simply an FIR filter having tap values that equal the spreading sequence and its output is the complete cross-correlation waveform between the input signal and the reference spreading code. Detection is performed by sampling the correlation waveform at the appropriate time and making a decision. This

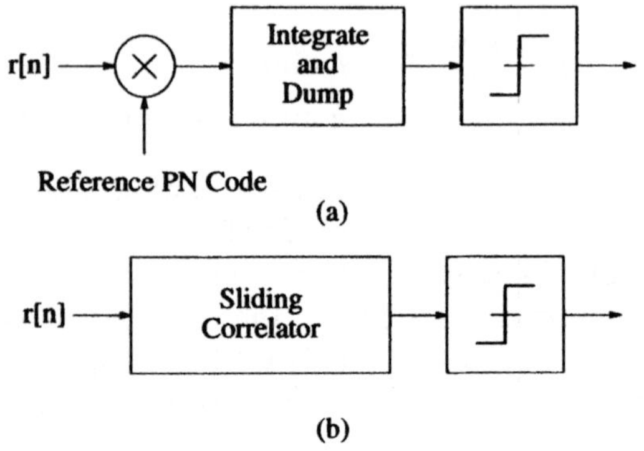

Figure 4 Time domain DS receivers.

approach is generally more computationally intensive than the synchronous approach shown in Figure 4a.

It is well-known that time domain convolution and correlation can be performed as a multiplication in the frequency domain [307]. In this case, the Fourier Transform of the received signal is multiplied by that of the time-reversed reference spreading sequence and the result is processed by an inverse Fourier transform to produce the correlation waveform. Since the frequency domain multiplication performs a circular convolution, data reconstruction algorithms such as the overlap-and-save or overlap-and-add techniques [307] may be necessary to produce the desired linear convolution. The output of the inverse transform is the same as that from the sliding correlator described above.

The DFT is a block transform that operates on segments of the input signal. When the blocks of the input signal are selected such that they are time-aligned with the data bits, i.e. each block contains only one complete data bit, it is possible to perform the data detection in the transform domain. In fact, it has been shown [259, 261] that unambiguous information regarding the polarity of the received data bit is present in the transform domain and may be used to make the bit decision directly in that domain. To take advantage of this information, it is necessary to correlate the transform of the received signal with that of the known reference PN code as shown in Figure 5. This correlation process is expressed as

$$D[k] = \sum_{i=0}^{N-1} R[i,k]C[i,k] \tag{10.5}$$

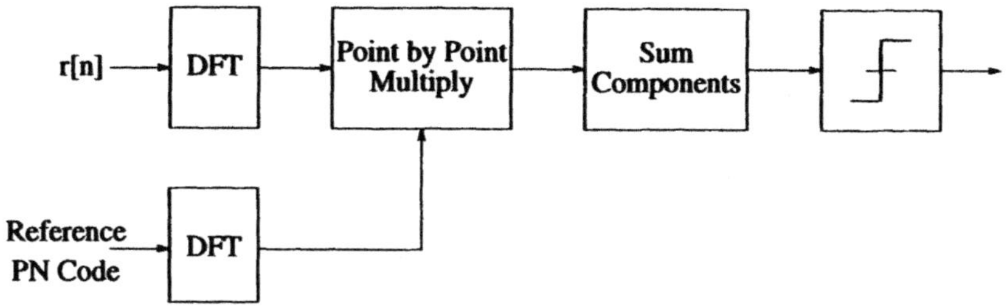

Figure 5 Frequency domain DS receiver with frequency domain detection.

where $D[k]$ is the correlation value for the kth data symbol, $R[i,k]$ is the ith component of the transform of the kth data symbol, $C[i,k]$ is the ith component of the transform of the spreading sequence used for the kth data symbol and N is the order of the transform. To make a data symbol decision, $D[k]$ is passed through a decision device. Since the correlator output is immediately compared to a preset threshold to make the bit decision, an inverse transform is not required. Hence, in the receiver of Figure 5, all bit decisions are made in the transform domain and an inverse transform need never be performed.

Unlike communications receivers, intercept receivers simply try to detect the presence of a signal in the channel. The classical technique for detecting the presence of a DS signal is to use a total power radiometer, which consists of a bandpass filter, a squaring device and an integrator. The detected power of the received signal is compared to a threshold which is set using information about the receiver and channel noise. A decision on the presence of a DS signal is made when the measured detected power exceeds this threshold. The probability that the threshold is exceeded given a DS signal is present is the probability of detection, P_d, while the probability that the threshold is exceeded in the absence of the signal is the probability of false alarm, P_{fa}. The performance of an intercept receiver can be characterized by plotting P_d versus P_{fa}, producing a receiver operating characteristic (ROC) for the detector.

An intercept receiver can be also be implemented in the transform domain. In this case, the total power in the transform domain is determined by squaring and summing the transform components as shown in Figure 6. As in the time domain case, the result of the squaring and summing operations is compared to a threshold in order to make a decision on whether a signal is present or not.

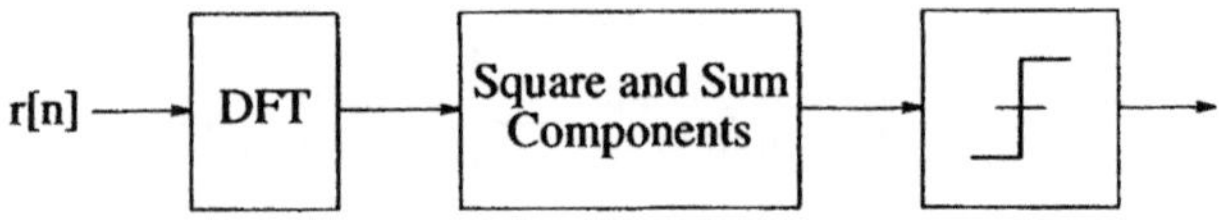

Figure 6 A frequency domain total power radiometer.

1.5 Interference Suppression

One of the important properties of spread spectrum signaling is that it inherently provides some protection against interference; in fact, any level of interference protection can be obtained by designing the signal with sufficient processing gain. The price for greater protection, however, is an increase in the bandwidth of the transmitted signal for a given data bandwidth. Practical considerations such as transmitter/receiver complexity and available frequency spectrum can serve to limit the reasonably attainable processing gain. As a result, it is beneficial to apply signal processing techniques to augment the processing gain of the spread spectrum signal itself, allowing greater interference protection without an increase in bandwidth. In general, these *interference suppression* techniques discriminate between the desired spread spectrum signal and the interference and work to suppress the interference [269]. The processing can be performed in the time domain (e.g. adaptive transversal filtering) [355, 205, 194, 226, 244, 206] the spatial domain (e.g. adaptive array antennas)[277], or the transform domain [270, 272, 344, 271, 159]. The main focus here will be on transform domain processing techniques using wavelet transforms and filter banks.

When a signal is transformed or mapped to a different "space" and processed, the signal processing is said to have been done in the transform domain, or, in other words, that one is using transform domain processing. Note that this mapping should be unique and unambiguous, and that an inverse mapping or transformation, which can return the signal to the time domain, should exist. The most widely-used continuous-time transform is the Fourier Transform, but there are many others of importance such as the Fresnel, Hartley, Mellin, and Hilbert Transforms, to name but a few. In communications and radar applications, particularly ones using spread spectrum techniques, transform domain processing can be utilized to suppress undesired interference and, consequently, improve performance. Here, the basic idea is to choose a transform such that the jammer or the undesired signal is nearly an impulse function in the transform domain, while the desired signal is transformed to a waveform that is very

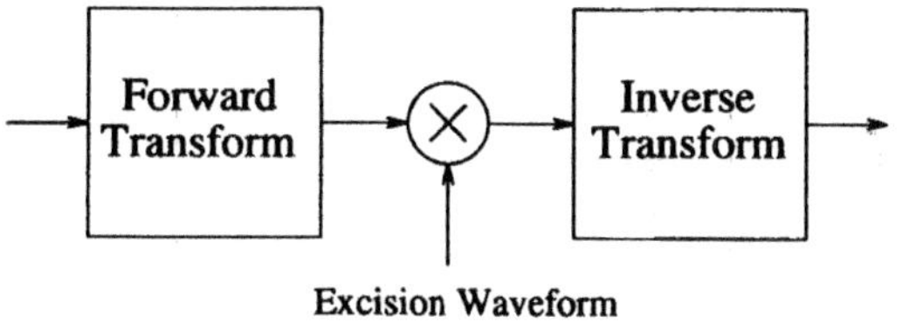

Figure 7 Block diagram of a transform domain excision system.

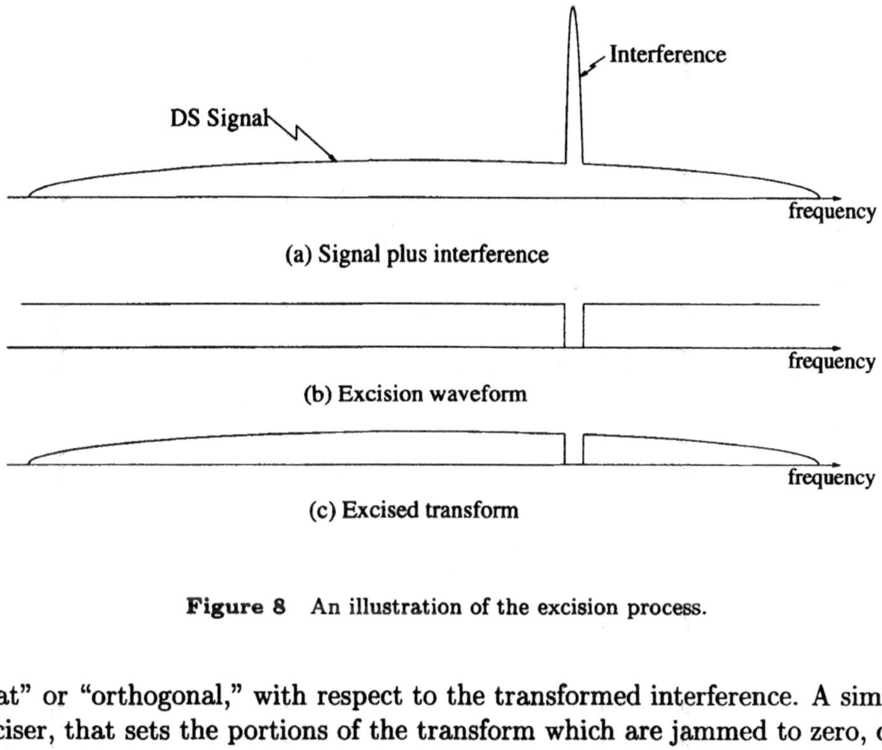

Figure 8 An illustration of the excision process.

"flat" or "orthogonal," with respect to the transformed interference. A simple exciser, that sets the portions of the transform which are jammed to zero, can then remove the interferer without removing a significant amount of desired signal energy. An inverse transform then produces the nearly interference-free desired signal [270, 272, 344, 271, 159]. Figure 7 is a block diagram of a transform domain excision system. The excision waveform usually takes only the values of zero and unity, resulting in the complete removal of portions of the transform that are determined to be "jammed." Figure 8 illustrates the excision process.

Most of the research and development related to the application of transform domain signal processing has been restricted to Fourier transform-based applications. In fact, many practical implementations utilize "short-time" Fourier transforms, which observe only a segment of the input signal. In particular, discrete-time implementations typically use the Fast Fourier Transform (FFT), which is a computationally efficient implementation of the DFT.

Under ideal circumstances, the transform domain representation of the interfering signal should appear as an impulse function. Implementing the short-time Fourier transform, however, requires the use of windowing functions to localize the input data in time. These windowing operations yield frequency domain representations that are characterized by undesired side lobes. As a result, when time domain windows are used, an excessive amount of interference energy may be detected and represented in neighboring spectral bins. The amount of energy contained in these side lobes is a direct function of the windowing function used, e.g. a rectangular window yields very large side lobes due to its $\sin(x)/x$ frequency response whereas Hamming, Bartlett and other windows weight the data samples in such a manner that the size of the transform domain side lobes is significantly reduced at the cost of widening the main lobe [307]. In an excision-based system, it is desirable to represent the interfering signal in as few transform domain bins as possible in order to preserve the maximum amount of desired signal energy while removing the maximum amount of interference energy. Hence, it is advantageous to localize the interference energy to the main lobe of the frequency response and to keep the interference energy contained in the transform domain side lobes as small as possible. While using non-rectangular windows reduces the size of the side lobes, it requires the processing of overlapping segments of the input signal in order to get accurate reconstruction of the time signal, thus greatly increasing the computational requirements [65]. The *overlap-addition* method described in [327] is an example of the type of processing that is required.

2 TRANSFORM DOMAIN PROCESSING USING WAVELETS AND FILTER BANKS

Recently, the idea of using the discrete wavelet transform (DWT) and related subband filter banks to suppress undesired interference in spread spectrum communications systems has been introduced [259, 263, 396]. Whereas conventional block transforms such as the DFT partition the time-frequency plane into fixed cells with relatively high frequency resolution and poor time domain resolution, the discrete wavelet transform, which is typically implemented using subband filter bank structures, allows the time-frequency plane to be partitioned into time-frequency cells with varying time and frequency support. In addition, subband transform basis functions are often not restricted in length as are fixed length transforms such as the DFT and Discrete Cosine Transform (DCT). Indeed, subband filters may be designed with arbitrarily long impulse responses whereas the length of a traditional block transform impulse response is limited by the number of transform domain cells, or bins. Clearly, the length of the basis functions directly affects both spectral resolution and computational complexity; as a result, the flexibility in the impulse response duration offered by wavelets and filter banks may be used to improve transform domain resolution. Most importantly, subband filter banks and discrete-time approximations to the DWT may be implemented in a manner consistent with the *perfect reconstruction* (PR) property, thus making it possible to recover the time domain signal without distortion. The primary objective of the work presented in this section is to utilize the time and frequency localization afforded by the DWT and multi-rate filter banks to obtain improved receiver performance in the presence of various types of interference.

There have been a number of approaches taken to applying wavelets and filter banks to suppressing interference in a DS spread spectrum signal. The goal of this work is to utilize the energy compaction properties of the wavelets and filter banks to overcome the limitation of the windowing effects associated with the short-time Fourier transform. Additional investigations have applied the wavelet transform to suppressing pulsed interference. Here, the time-localization afforded by the wavelet transform makes it possible to suppress more interference while removing less DS signal energy. In both the sinusoidal and pulsed interference cases, the type of filter bank or wavelets utilized has an impact on both performance and implementation complexity.

The sections below summarize some of the work performed to date using wavelets and filter banks to suppress interference. Since there is no "standard" wavelet, i.e. there is an infinite number of possible mother wavelets, there is an unlimited number of possible wavelet implementations that can be used to suppress interference. Only a few of the possible variations have been tried and they will be discussed here. Virtually all the work described is ongoing and the results presented provide only a preliminary indication of the potential benefits to be gained through the application of the techniques. All the implementations are discrete-time systems that utilize a uniformly-sampled input signal. As a result, the comparisons will be made to transform domain systems that utilize FFTs.

2.1 Filter Banks Based on Daubechies' Wavelets

One approach to suppressing interference is to use a binary subband tree structure that is constructed using hierarchical stages of two-channel paraunitary quadrature-mirror filter (QMF) analysis filter banks. The analysis portion of the overall tree structure consists of cascaded two-channel units which may form a full, dyadic or irregular subband tree. Clearly, different subband filters yield different values for the corresponding QMF bank filter coefficients. The analysis tree produces transform domain coefficients that can be processed by an exciser to remove any coefficients that are determined to be primarily interference. After excision is performed, a synthesis filter bank, constructed from complementary two-channel paraunitary QMF synthesis filters may be used to perform the inverse subband transform. The constraints on these filters, as well as their relationship to the appropriate analysis filters are discussed in detail in references [7, 418]. It is interesting to note that, if no processing (i.e. excision) is performed on the spectral coefficients, the input signal may be perfectly reconstructed at the output of the processor.

The dyadic binary subband tree structure is equivalent to the DWT when the input signal is projected onto the wavelet transform scaling space at full resolution [7]. At each stage of subband decomposition, the QMF analysis filter banks split the input signal into lowpass and highpass frequency components and down-sample both filter outputs by a factor of two; subsequent processing is only performed on lowpass filter outputs. As a result, this transformation method partitions the signal spectrum into octave frequency bands, providing good frequency resolution at low frequencies while retaining good time resolution at high frequencies. Unlike conventional block transforms, which project the input signal onto a set of basis functions with fixed time and frequency support, this method of multi-resolution analysis transforms the input signal into the wavelet transform domain wherein the coefficients have various time and frequency support. Thus, in the discrete wavelet transform domain, high frequency time domain information, such as the location within the data record of transient noise or interference energy bursts, is retained while lower frequency information is simultaneously available.

If the interfering signal is narrow-band, the full binary subband tree structure can be quite effective in localizing a significant amount of jammer energy to a small number of transform domain bins. This method of subband decomposition recursively divides the input signal spectrum into separate lowpass and highpass components. Each resulting component is subsequently partitioned into finer resolution lowpass and highpass frequency bands until the desired level of frequency resolution is obtained. As a result, this method of signal analysis yields a uniform partition of the signal's frequency spectrum much like the DFT. However, unlike the DFT, the spectral side lobes generated by the filter bank structure are dependent on the QMF filters that are used. Thus, depending on the FIR filter coefficients, the side lobes may not be as large as those produced by non-windowed DFTs and, hence, fewer bins may need to be excised to remove a specific amount of interference energy. Consequently, given that the same number of bins are removed, subband decomposition of the received signal via hierarchical filter bank structures may yield improved receiver performance relative to that achieved using DFT-based techniques.

Communications Receiver Performance

In Figures 5 and 6, the DFT was used as the transformation of choice since it represents the conventional transform method used in many DS spread spectrum receivers. In general, however, spread spectrum receivers are not limited to Fourier-based operators. In fact, if the DFT is replaced by an arbitrary linear transform, the generic receivers of Figures 9 and 10, which use time and transform domain correlation, respectively, may be realized. Recently, performance results achieved by a DS communications receiver employing subband transform domain excision have demonstrated that the full and dyadic subband transforms are capable of improving receiver performance, relative to that achieved using the DFT, in the presence of both continuous and transient interference [259]. In this study, the level of performance improvement was measured in terms of the receiver BER and was shown to be a function of the QMF bank filters used, as well as the hierarchical subband tree configuration.

To evaluate the performance of the DS communications receivers of Figures 9 and 10, simulations were performed in which FIR filters with coefficients derived by Daubechies [113] were used in the paraunitary QMF bank substructures. In these simulations, a 63-chip m-sequence was used to modulate a binary input data stream, thus producing a DS signal that was sampled once per chip or, equivalently, 63 times per data bit. The resulting spread data signal was then transmitted over an Additive White Gaussian Noise (AWGN) channel with either additive narrow-band or pulsed interference. At the receiver, five

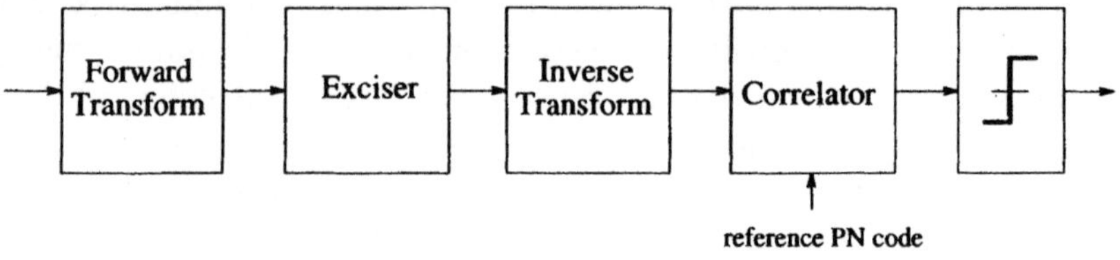

Figure 9 A DS communications receiver with subband transform domain excision.

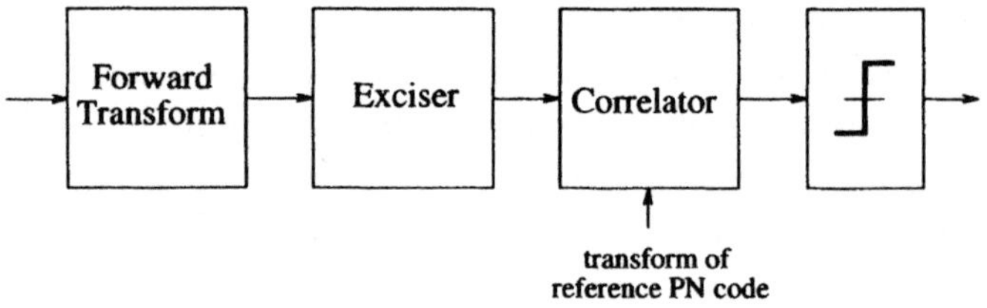

Figure 10 A DS communications receiver with subband transform domain excision and transform domain correlation [259].

hierarchical levels of the two-channel PR-QMF bank substructures were used to generate a 32-band filter bank which mapped the time domain input into the subband transform domain [322]. Full and dyadic subband tree structures using Daubechies' 4-tap FIR filters were used to analyze the received data signal in the presence of continuous and transient interference, respectively.

The resulting BER curves show the performance of the system as a function of the energy per data bit to one-sided noise power spectral density ratio, E_b/N_o. Since interference energy is not considered in the calculation of E_b/N_o, the theoretical BER is equivalent to the performance of binary phase-shift keying in AWGN and serves as a benchmark of DS receiver performance to which subsequent results will be compared. When interference is present, the interference or jammer power is defined by the ratio of jammer power to DS signal power and is denoted as JSR.

The performance of DS transform domain receivers using the FFT and full subband decomposition in the presence of a continuous sinusoidal interferer having a JSR of 30 dB and a normalized frequency of 0.127 Hz (i.e. frequency normalized to a chip rate of unity) is shown in Figure 11. When no interfer-

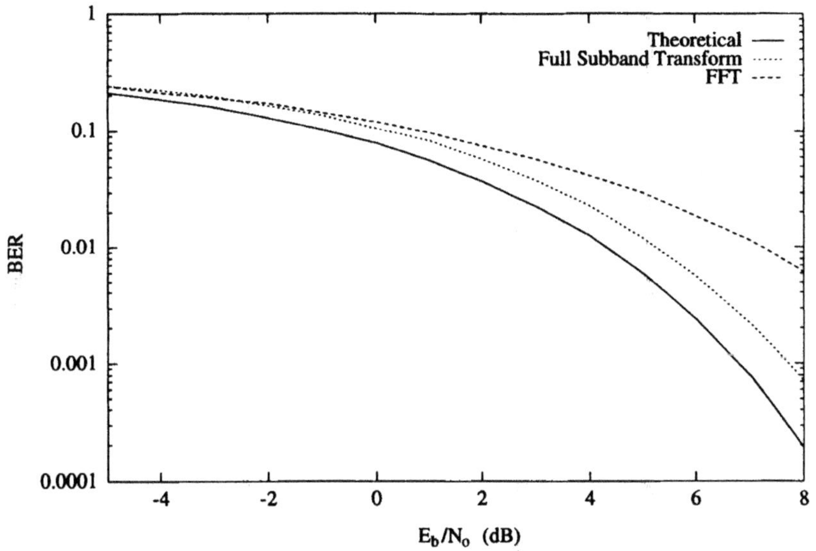

Figure 11 Full subband decomposition and FFT-based DS receiver performance in the presence of continuous sinusoidal interference with 8 bins excised (JSR = 30 dB, frequency = 0.127 Hz) [259].

ence suppression technique is employed, the resulting BER is approximately 0.5, regardless of the E_b/N_o. For the results shown here, however, the interference is mitigated by excising the eight transform domain bins with the largest magnitude and replacing these transform values with zero. As is evident from the figure, when these bins are removed the subband transform-based receiver yields a lower BER than the comparable FFT-based implementation. Since the 4-tap FIR filters derived by Daubechies were not specifically designed for interference suppression, receiver performance may be further improved by utilizing QMF banks with filters that are optimized for such applications.

Although transform domain excision using either full subband decomposition or FFT-based receivers may significantly improve receiver performance in the presence of continuous sinusoidal interference, their excision capability in the presence of a transient, or pulsed, interference source is far less effective. Such interference sources are typically characterized as relatively wide-band interferers which make the uniform spectral bands associated with the full binary subband tree and the FFT impractical in a transform domain excision system. As an alternative, transformation techniques that yield joint time-frequency resolution, such as the dyadic subband decomposition, may be more appropri-

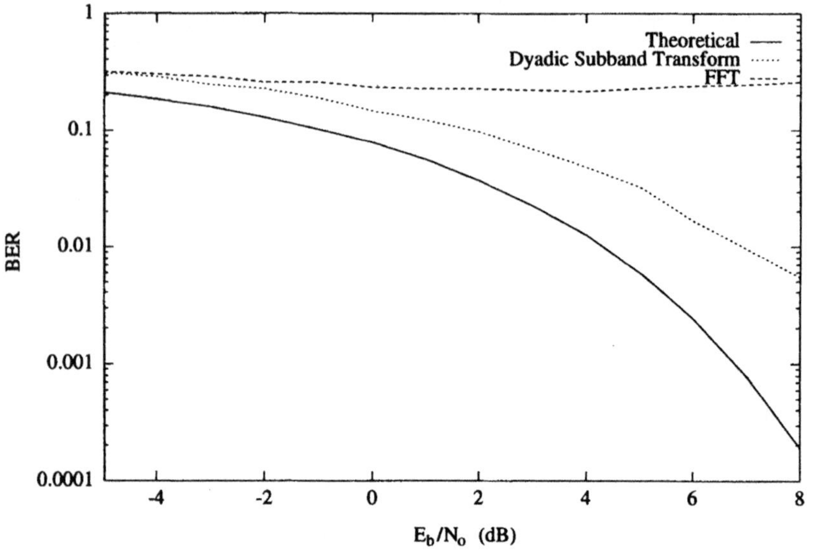

Figure 12 Dyadic subband decomposition and FFT-based DS receiver performance in the presence of transient interference with 8 bins excised [259].

ate. For comparison, Figure 12 illustrates the relative performance of the DS receiver using the FFT and dyadic subband transform methods in the presence of a pulsed white noise jammer when eight of the 64 transform domain bins are removed. In this case, the transient interference is represented by a white noise source with a duty cycle of 5% relative to a period of 70 samples, thus ensuring that the pulse duration is much less than that of the data bit while the pulse period is slightly larger. In addition, the noise variance has been specified to yield an average jammer-to-signal ratio of 23.3 dB.

Despite the apparent improvement in receiver performance achieved using the dyadic subband transform, neither of these transformation techniques has shown the ability to significantly mitigate transient interference in a practical setting. In fact, since the number of subbands affected by the pulsed interferer increases directly with the pulse duty cycle, several subbands may be affected by the jammer and, hence, excision in the transform domain may be completely inappropriate. As a result, it may be necessary to utilize a transform domain technique related to the dyadic filter bank and DWT implementations which provides an optimal joint time-frequency representation of the received signal. In fact, it may even be necessary to suppress the transient interference directly in the time domain if the pulse duration is sufficiently short. Clearly, one must first determine the domain in which excision is most appropriate, if at all. This issue has been addressed through the development of an adaptive time-frequency exciser [396] and will be presented in Section 2.2.

Intercept Receiver Performance

As discussed, Figures 11 and 12 illustrate that, when using an appropriate transformation, much of the undesired interference energy may be localized to a few transform domain bins and subsequently excised; as a result, reliable communications in the presence of such a source may be maintained or improved. This idea of transform domain processing is also useful to those wishing to detect spread spectrum signals using intercept receivers. In fact, it has recently been illustrated [260] that subband transform domain excision may also be used to improve the performance of traditional radiometric detectors, or intercept receivers.

As in the DS communications receiver applications, filter banks using the Daubechies' coefficients have also been applied to interference suppression in an intercept receiver [260]. As before, each binary data bit is modulated using a 63-chip PN sequence and the resulting signal, sampled once per chip, is summed with AWGN and interference. In this case, performance results are presented as ROCs which provide P_d and P_{fa} for a given signal-to-noise ratio (SNR) and JSR, where SNR is defined as the ratio of signal power to AWGN power. The threshold level decreases moving from left to right in the ROC, resulting in an increase in P_d and a corresponding decrease in P_{fa}.

The performance of an intercept receiver with excision using the full binary tree filter bank with 32 subbands is shown in Figure 13. For this figure the ratio of the signal power to AWGN power is -7.0 dB. The narrow-band interference is a sinusoidal jammer with a normalized frequency of 0.0635 Hz and a JSR of 20 dB; excision is performed by removing the four or eight transform domain bins that have the largest magnitude. It is clear from the figure that when no excision is performed, the probabilities of false alarm and detection are approximately equal, thus producing an unreliable and, in fact, "worst-case" determination of signal presence. Essentially, the probability that the threshold is passed is the same whether the signal is present or not because the jammer power is dominating the received signal. As frequency bins are excised P_d and P_{fa} increase and decrease, respectively, yielding a more reliable indication of the presence of the DS signal. The outermost curve represents the system's

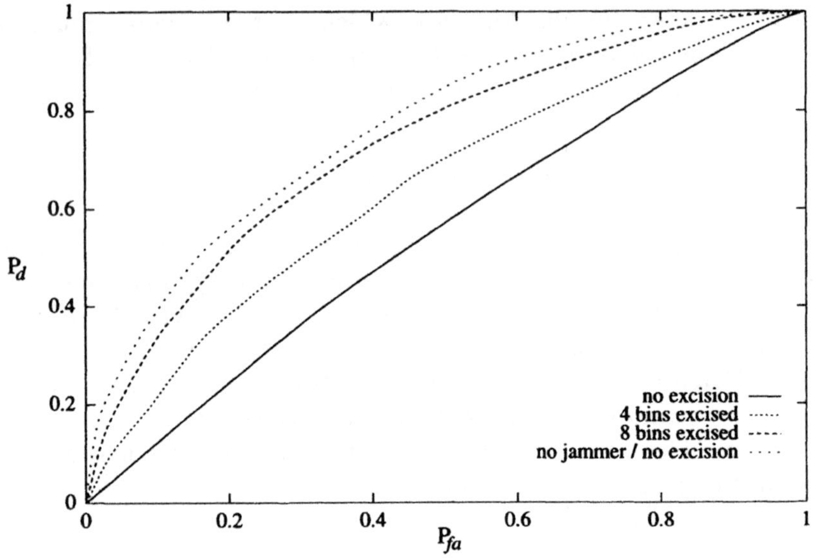

Figure 13 ROC using full subband decomposition and varying number of bins excised in the presence of continuous sinusoidal interference (JSR = 20 dB, frequency = 0.0635 Hz) [260].

performance when no interference is present and serves as a benchmark for performance comparison.

It is important to keep in mind that in most transform domain excision applications, including this one, the performance benefits accompanying excision are not without limit. In particular, if an excessive number of transform domain bins are removed, receiver performance will be significantly degraded since the excision of any bin removes both signal and interference energy. In fact, after a certain point, excising additional bins will remove more signal energy than interference energy, thereby worsening the overall system performance.

Figure 14 shows the receiver performance using a 32 subband transform with eight subbands removed in the presence of a 20 dB sinusoidal jammer. For this system, normalized jammer frequencies of 0.0635, 0.127 and 0.254 Hz were tested. As expected, performance varies with jammer frequency due to aliasing effects within the filter bank and the characteristics of the spreading sequence [355]. As with the communications receiver, aliasing and interference

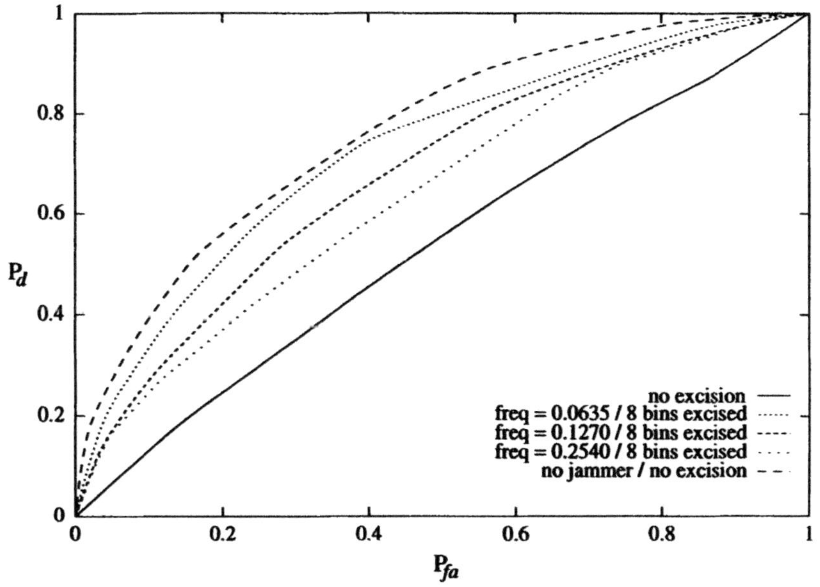

Figure 14 ROC using full subband decomposition and varying sinusoidal jammer frequency with 8 bins excised [260].

energy compaction within the subbands directly affect the subband transform's ability to isolate the undesired signal energy.

2.2 Adaptive Filter Bank Structures

The approaches to interference suppression that have been considered up until this point utilize fixed filter bank structures and an exciser which simply replaces any transform bins which are considered to be primarily interference with zero. While these systems are clearly effective at improving BER performance when interference is present, it seems likely that greater gains can be obtained by using processing systems that *adaptively* respond to the type of interference present and work to optimize some aspect of system performance. This section will discuss some of the approaches to making such an adaptive system and present some preliminary results obtained for one particular implementation.

There are a number of ways in which the filter bank structure can adapt to the interference. One approach is to vary the tap weights of the filters in the system with the goal of making the interference approximate an impulse function in the transform domain. This process is common in filter bank design and is formulated as the maximization of energy compaction. If taken from a transform point of view, the process is one of adapting the basis functions of the transform such that one basis function is essentially *matched* to the interference while, at the same time, other properties, such as orthonormality and perfect reconstruction, are maintained.

Another approach to adaptivity is to start following the full subband tree composed of two-channel paraunitary QMF filter banks but, instead of implementing the whole tree, make jammed/unjammed decisions at each level of the tree. When a signal at a particular level is found to be unjammed, that signal is not decomposed further but, rather, is passed directly to the synthesis section of the processor. Conversely, any signal that is found to be jammed is decomposed further with the goal of isolating the interference to a smaller band. With this technique the tree structure is determined by the interference and, by eliminating unnecessary decompositions, the computational load is reduced. Of course, the reduction in computational complexity is diminished somewhat by the requirement that jammed/unjammed decisions must be made at each level of the tree rather than once at the point of excision. If the interference were a single sinusoid at zero frequency, the resulting tree would be the dyadic subband. A single sinusoid at a different frequency would produce a tree structure that is equivalent to the dyadic structure in complexity but which would provide the greatest resolution in the region of the interference rather than at low frequencies. Only in the case of strong interference spaced throughout the frequency band, e.g. multiple tone jammers, would the resulting tree structure approach the complexity of the full subband tree.

A third approach to an adaptive system is to replace the simple two-level exciser with adaptable weights on each transform bin. While the exciser itself is adaptive, it is constrained to either keeping or completely removing a transform bin (weighting by zero or one) and the excision decision is generally not based on a true optimization of a performance parameter such as output signal-to-interference ratio or BER. A better system could be formed by adaptively weighting each bin of the transform using continuously-variable tap weights that are determined based on maximizing performance. One possible structure used to perform this optimization is a transform domain adaptive filter, such as the one shown in Figure 15. In this figure, the transform domain bins are multiplied by the adaptable weights, $\alpha_1, \alpha_2, \ldots \alpha_N$, where N is the order of the transform. An adaptive algorithm works to minimize the difference between the weighted transform and some *desired signal*, generally using one of many well-known approaches such as the Least-Mean-Square (LMS) algorithm, the

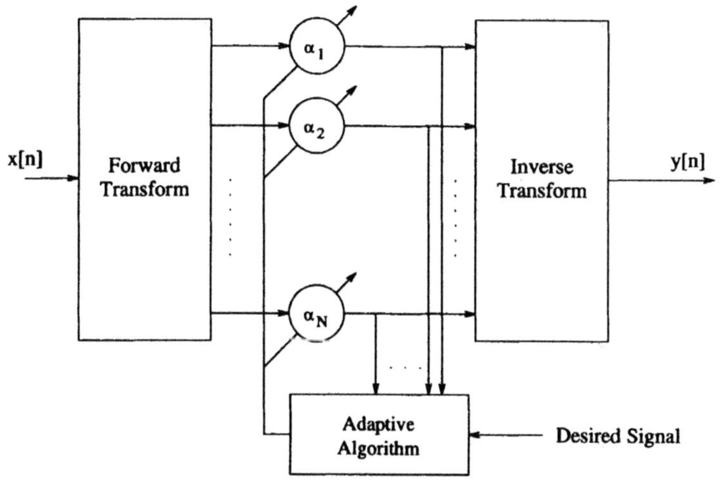

Figure 15 Transform domain adaptive filter.

Recursive Least-Squares (RLS) algorithm and their many variations [179]. Frequency domain adaptive filters incorporating these algorithms have been known for some time [280, 334], and the extension to the wavelet transform domain is straightforward [139]. Clearly, two keys to the performance of the system shown in Figure 15 are the adaptive algorithm and the desired signal. In many applications, the desired signal is the transform of some reference signal and the object of the adaptation is to minimize the difference between the weighted transform of the received signal, $x[n]$, and this desired signal.

For a DS spread spectrum receiver, it may be desirable to minimize the difference between the detected symbols and the output of the correlator. A system that does this and makes use of transform domain correlation is shown in Figure 16. In this system, as in the excision system depicted in Figure 10, no synthesis or reconstruction filter bank is required. The input signal $x[n]$ and the reference PN spreading sequence are both processed by forward transforms and the results are multiplied point-by-point. This multiplication performs the despreading operation. The products are then weighted by the adjustable taps and then summed in order to produce an estimate of the transmitted data bit $\tilde{d}[n]$ which is passed to the decision device to produce the recovered symbol $\hat{d}[n]$. The error signal that is used to drive the adaptive algorithm, $\epsilon[n]$, is the difference between $\hat{d}[n]$ and $\tilde{d}[n]$. If either the LMS or RLS adaptive algorithm is used, the tap weights will converge to minimize $\epsilon[n]$ in the mean-squared

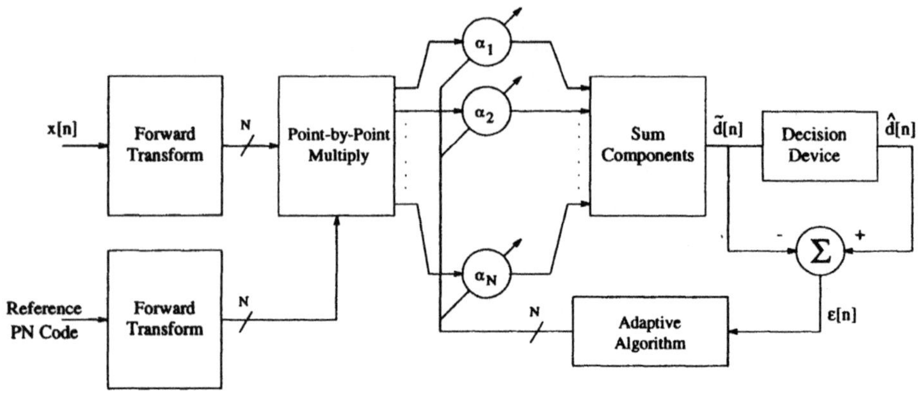

Figure 16 Adaptive transform domain DS receiver.

sense. For initial startup, a preamble, known to the receiver, can be used in place of $\hat{d}[n]$ to ***train*** the system, with a switch back to $\hat{d}[n]$ taking place after convergence.

While the adaptive system of Figure 16 can produce a set of optimal weights, it has the drawback that it may not be able to react as quickly to changes in the input as the simple exciser. Most adaptive algorithms are either iterative, in which case they require time to converge to the optimal solution, or require information about the statistics of the input signal, which the receiver must estimate when the channel has unknown (and changing) properties. Therefore, if the interference is changing rapidly, the algorithm may not be able to track these changes and will be ineffective at suppressing the interference. While some algorithms are better at tracking changing conditions than others, there is generally a tradeoff between the tracking capability and the *misadjustment noise*, which is an indication of the variation of the weight values around the optimal values after convergence.

Adaptive Time-Frequency Exciser

Since two-channel PR-QMF bank substructures are relatively simple to construct, it is not surprising that many M-band filter banks are implemented as hierarchical subband trees based on two-band prototype filter banks. Despite the relative ease with which these prototypes can be designed, a noticeable disadvantage of two-band structures becomes apparent when the signal energy is predominantly concentrated in a frequency range at or near $\omega = \pi/2$. In such a case, the binary spectral split can be inefficient. Hierarchical filter banks based on two- three- and M-band systems can be more effective. Recently, an efficient algorithm that exploits the frequency partitioning behavior of both two- and three-band filter banks has been introduced by Tazebay and Akansu [396].

The adaptive time-frequency (ATF) exciser presented in [396] is a transform domain excision algorithm that adaptively generates a signal dependent hierarchical subband tree using either two- or three-band prototype paraunitary FIR filter banks. This approach analyzes the received signal's energy distribution at each node during the hierarchical subband tree's development and determines whether to continue decomposition with additional filter bank stages or terminate the multi-level tree and process the received signal using the resulting filter bank structure. Since the interfering signal may vary in time, this process is repeated for each received data symbol, thus allowing adaptive reconfiguration of the subband filter bank. In the proposed algorithm, a subband node is decomposed only if its transform domain energy compaction measure exceeds the time domain compaction level and a predetermined decision threshold. As a result, unnecessary decomposition is avoided and, thus, transform complexity, in terms of the number of multiplications and additions, is reduced. In addition, since fewer subbands are decomposed, spectral leakage in the transform domain is minimized [396].

The ATF excision algorithm initially decides whether to process the received signal in the time domain, frequency domain or neither. This decision is based on the energy compaction measure associated with each domain which, for any M-band linear transform, may be expressed as

$$G = \frac{\sigma_x^2}{\left[\prod_{i=1}^{M} \sigma_i^2\right]^{1/M}} \tag{10.6}$$

where σ_x^2 and σ_i^2 denote the input variance and ith subband variance, respectively. This expression is closely related to the transform coding gain measure often utilized in waveform coding applications [213, 158] and serves as a measure of spectral flatness. In particular, low energy compaction values correspond to relatively smooth, or evenly distributed spectral power whereas high compaction levels indicate a significantly uneven distribution of power in the transform domain.

As suggested in Section 1.5, for interference suppression via excision to be effective, jammer energy must be localized in the transform domain. Such containment of the undesired signal energy precipitates a noticeably uneven

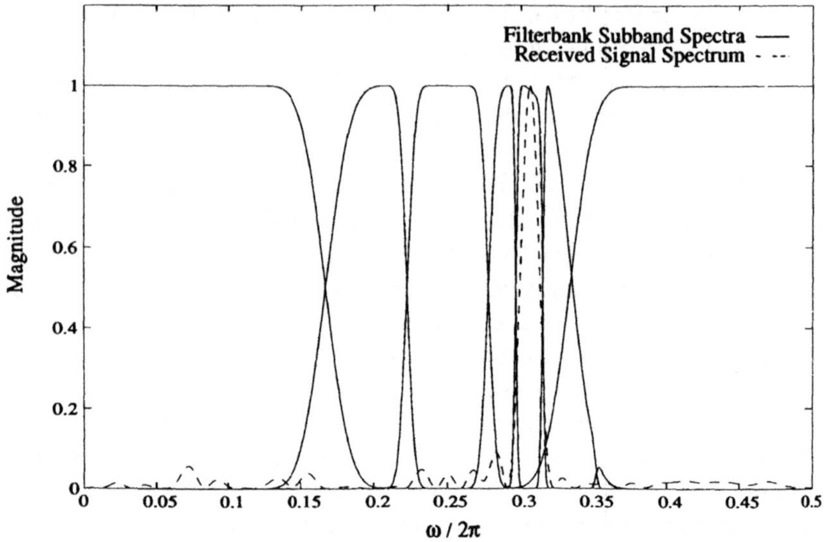

Figure 17 Spectral partition generated using the ATF excision algorithm in the presence of continuous sinusoidal interference (frequency = 0.3055 Hz) [396].

distribution of power throughout the transform domain. Therefore, it is desirable to perform excision in the domain where the interfering signal maximizes the energy compaction measure. Although block and subband transformation techniques are often utilized to decorrelate the received data samples and isolate the undesired signal energy, the excision process need not be limited to the transform domain. In fact, if the time domain samples are not significantly correlated and the input signal is of very short duration, no transformation may be necessary and excision may be performed directly in the time domain. To make this determination, the time and frequency domain energy compaction measures must be evaluated and compared to a constant energy compaction level that reflects the spectral flatness of the spread spectrum signal in the absence of narrow-band or transient interference. Clearly, the larger energy compaction value indicates the domain in which processing should be performed. If, however, neither level exceeds the preset threshold, the received signal is passed directly to the demodulator without excision.

One of the most promising attributes of the proposed ATF excision algorithm is its ability to perform signal dependent subband decomposition. In contrast

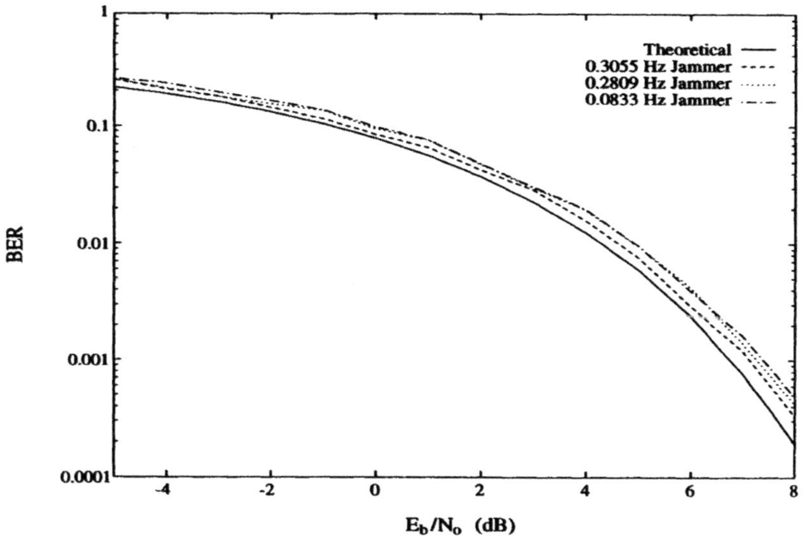

Figure 18 ATF BER performance in the presence of various frequency sinusoidal interference (JSR = 20 dB) [396].

to the fixed filter bank structures discussed in Section 2.1, the ATF excision algorithm adaptively reconfigures the hierarchical subband filter bank structure in an effort to minimize transform domain partitioning and improve spectral containment of the interfering signal. As an illustration, Figure 17 demonstrates the spectral partitioning capability of the ATF exciser in the presence of single-tone interference. The corresponding filter bank subbands are generated by recursively applying a three-channel prototype filter bank in a three-stage irregular, hierarchical subband tree. As is evident from the figure, the ATF excision algorithm divides the signal spectrum into seven unequal width subbands with relatively high spectral resolution near the frequency of interest.

As shown in Figure 17, the ATF exciser presents a robust transformation technique that tailors the subband frequency responses to the spectrum of the received signal. Consequently, the ATF excision algorithm is less sensitive to jammer frequency than fixed block and subband transforms. This is illustrated in Figure 18, which shows the robustness in BER performance achieved using the ATF excision algorithm in the presence of various frequency single-tone jammers.

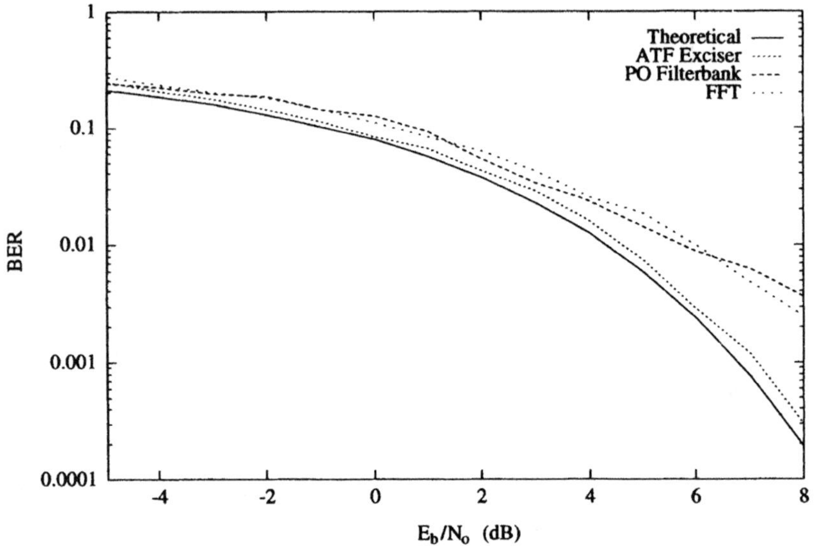

Figure 19 ATF BER performance in the presence of continuous sinusoidal interference (JSR = 20 dB, frequency = 0.3055 Hz) [396].

In [396], two-channel PR-QMF and three-channel paraunitary FIR filter banks were optimized with respect to the maximum energy compaction constraint. The resulting two- and three-band prototype filter banks, which utilized 16-tap and 56-tap FIR filters, respectively, were used in the ATF excision algorithm to generate the adaptive, irregular hierarchical subband tree. To illustrate the potential improvement in receiver performance, Figures 19 and 20 compare the BER results obtained using the ATF exciser, a 128-point FFT and a 64-band progressively optimized (PO) filter bank in the presence of continuous and transient interference, respectively. Unlike fixed transformation techniques, which may be optimally designed for a given input, this adaptive approach generates a signal dependent subband decomposition capable of tracking and suppressing time-varying interferers. This adaptive quality makes the adaptive time-frequency exciser a valuable interference suppression technique in many practical spread spectrum environments.

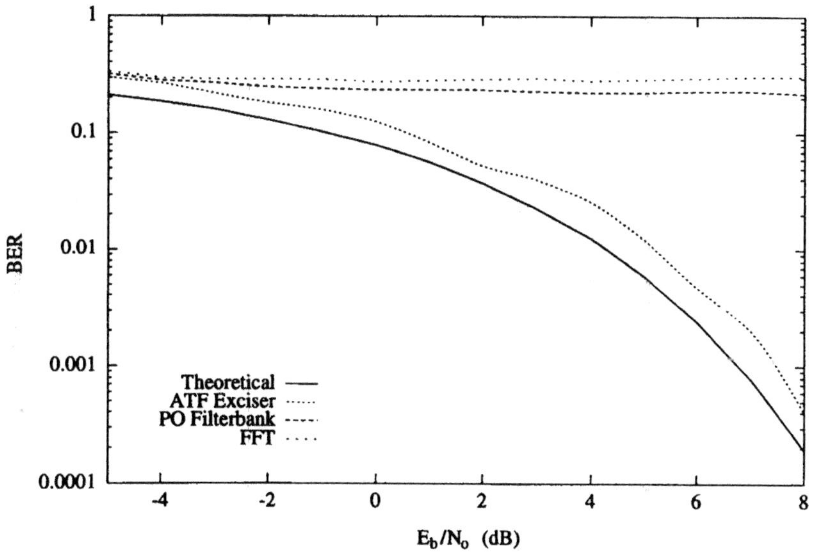

Figure 20 ATF BER performance in the presence of transient interference (duty cycle = 10%) [396].

3 DATA MODULATION USING FILTER BANKS

In addition to interference suppression, the recent interest in the DWT has spawned a growing interest in its application to multiple access and covert communications systems [449, 450, 325, 351, 89, 186, 239, 155]. Since wavelet and subband transforms may be designed to yield orthogonal basis functions, it is only natural to speculate as to whether or not such waveforms may be used to modulate or spread data signals. Since filter banks may also be optimized subject to various constraints, it is likewise natural to consider the development of practical objective functions and constraints that improve transmission reliability and covertness. Accordingly, several papers have recently explored these issues, thus resulting in a new class of "multi-rate modulation" communication systems.

In these multi-rate modulation systems, the transform basis functions are typically used as DS spreading codes to modulate the message data and are often generated by performing an inverse transform. The resulting spread waveform is then transmitted over the channel and, at the receiver, a forward transform is sufficient to correlate the received signal with the reference spreading codes. Since both the input vector and the transform basis functions are arbitrary system parameters, such an implementation facilitates a variety of possible waveform coding applications. In particular, the subband transform basis functions may be specifically designed to yield a composite waveform with spectral properties well-suited to the available transmission medium. Furthermore, these basis functions could additionally be designed to meet certain covertness or anti-interference criteria. These ideas, and their applications to spread spectrum and CDMA communications systems, are examined more closely in the following sections.

3.1 Communications in Fractal Noise Environments

One of the applications in wavelet-based coding [449] utilizes the transform basis functions as an efficient means to transmit data over an additive Gaussian fractal noise channel. In this context, a $1/f$ family of fractal noise processes is defined [225] as processes whose power spectra are of the form

$$S(f) \sim \frac{\sigma_x^2}{|f|^\gamma} \tag{10.7}$$

over a wide range of frequencies f, where $0 < \gamma < 2$ is a parameter typically near unity. In the proposed data modulation system [449], an inverse wavelet transform is used to synthesize a composite waveform from a block of pulse amplitude modulated (PAM) data values. In particular, if a complete orthogonal wavelet basis is defined in terms of its mother wavelet, $\psi(t)$, and its dilations by a factor m and translations by a factor n as

$$\psi_n^m(t) = 2^{m/2}\psi(2^m t - n), \tag{10.8}$$

a composite waveform can be generated by summing combinations of arbitrarily weighted basis functions. Accordingly, Wornell [449] has proposed using input data values to obtain the composite signal

$$c(t) = \sum_{\underline{M} \le m < \overline{M}} d_m[n]\psi_n^m(t) \tag{10.9}$$

where $d_m[n]$ represent the data values used to weight the mth channel basis functions and $\underline{M}$ and $\overline{M}-1$ represent the lowest and highest available channels, respectively. It is clear from the above expression that not all of the wavelet

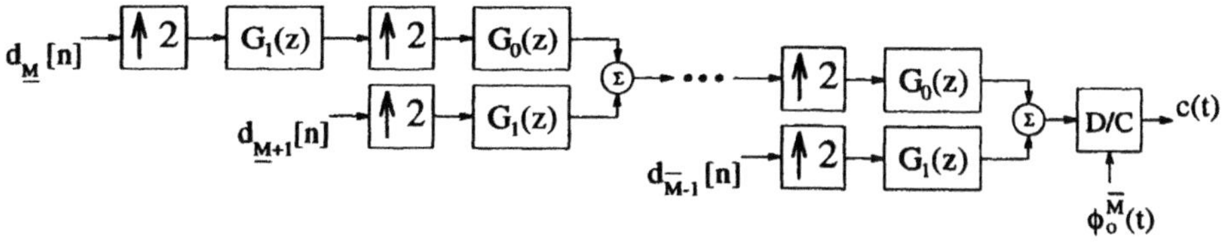

Figure 21 Composite waveform synthesis using a hierarchical dyadic subband tree [449].

basis functions may be utilized. In fact, whereas the highest usable frequency channel, denoted as the $\overline{M}-1$ channel, is limited by the total system bandwidth, the lowest available channel, the $\underline{M}$th channel, is limited by the amount of fractal noise present at lower frequencies [449]. Nonetheless, the synthesized waveform uniquely expresses the data block to be transmitted in terms of the orthogonal wavelet basis functions.

At the transmitter, the composite signal may be generated by exploiting the inverse discrete wavelet transform or equivalent synthesis filter bank structure as shown in Figure 21. Note that for the IDWT representation to be complete, the discrete-time filter bank output must be modulated onto the lowpass continuous-time scaling function.

As the subband index m varies from $\underline{M}$ to $\overline{M}$, the bandwidth and data rate of the corresponding frequency band are increased by a factor of two. Hence, the input data sequence must be segmented into appropriate length subsequences for each value of m. In terms of a discrete wavelet transform implementation, these data symbols may be viewed as DWT transform domain coefficients. As such, $d_m[n]$ denotes the $2^{m-\underline{M}}$ data symbols used to weight the wavelet basis functions associated with the mth octave frequency band. Thus, if it is assumed that each of the data symbols are generated by representing b_m data bits with symbols from an L_m-ary modulator, a total of $2^{m-\underline{M}}b_m$ bits will be allocated to each subband.

Since the basis functions form a complete orthonormal set, the original data may be demodulated at the receiver using an analysis filter bank derived from the synthesis filters used at the transmitter. In fact, in the absence of channel errors, perfect reconstruction of the transmitted data can easily be realized provided that the analysis and synthesis filter banks satisfy the paraunitary conditions expressed in Chapter 2. Hence, a DWT of the received data signal is sufficient to retrieve the transmitted data sequence.

In addition to its application to communications in a fractal noise environment, this data modulation technique has also been proposed as a novel means of transmitting data over a communications channel of simultaneously unknown duration and bandwidth [450]. The resulting modulation method has been termed "fractal modulation" and has been shown to achieve error rate values comparable to those obtained using minimum shift keying, MSK, with greater spectral efficiency [325].

In a broader M-band filter bank sense, the inverse discrete wavelet transform may be replaced with a generic M-band paraunitary synthesis filter bank. In such a case, the corresponding receiver could be realized simply as the complementary paraunitary analysis filter bank. Clearly, such a multi-rate subband modulation system could be designed to maintain the orthogonal basis set necessary to demodulate the received data signal while simultaneously offering a more robust design methodology. In particular, it may be feasible to use subband transforms characterized by basis functions with special properties, e.g. waveforms optimized with respect to energy compaction or inter-band aliasing constraints or, perhaps, low probability of intercept/detection, LPI/D, or M-ary simplex waveforms. In the following section, we will see that covert and multiple access communications systems have, in fact, recently considered multi-rate modulation schemes in an attempt to improve system performance.

3.2 Spread Spectrum and CDMA Applications

The basis functions corresponding to an arbitrary block transform may be designed to yield orthogonal sequences with support equal to the transform length. Under certain conditions (primarily, system synchronization), these functions can be used as alternatives to m-sequences and Gold codes in DS systems and CDMA applications [186]. In such applications, the transform basis functions are used to directly modulate the input binary data stream. The resulting signal is subsequently transmitted through the channel, and, at the receiver, a forward transform is performed to make the bit decision. In this system, unlike conventional DS systems, correlation with the spreading code is equivalent to correlation with the appropriate basis function; thus, only a forward transform, which inherently correlates the received signal with all transform basis functions, must be performed to recover the message signal.

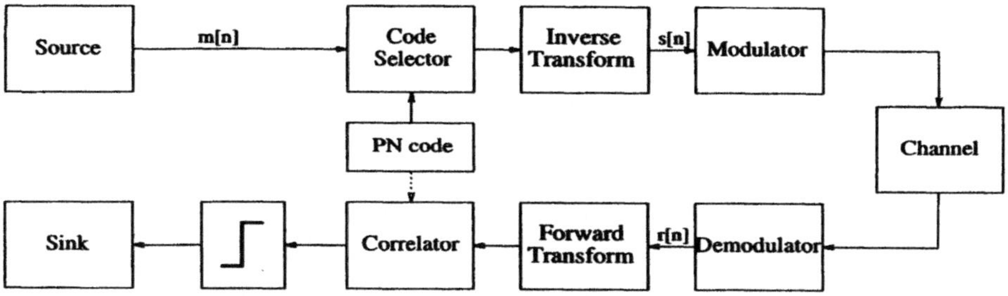

Figure 22 A subband transform-based spread spectrum communications system.

Hetling et al. [186] have suggested the filter bank-based spread spectrum system shown in Figure 22. In this system, the basis function to be used as the spreading waveform is determined by a PN sequence and is allowed to change with each incoming data bit. In particular, the selected spreading code is generated by performing an inverse transform on a unit impulse data vector. Note that the location of the impulse within the data vector, as indicated by the PN code, determines which basis function is used to spread the binary message signal, $m(n)$. The "spread" signal, $s(n)$, which appears as either a non-inverted or inverted replica of the basis function, depending on the data bit polarity, is subsequently modulated and transmitted over the channel.

At the receiver, the demodulated signal, $r(n)$, is mapped into the transform domain wherein an approximation to the originally transmitted unit impulse vector is reproduced and the bit decision is made. If the transformed signal yields a positive spike in the transform domain, a $+1$ data bit is assumed to have been sent; likewise, if the transform domain representation appears as a negative impulse, a -1 data bit is declared. If no noise or interference is present in the channel, perfect reconstruction of the impulse vector is possible; however, if either is present, the reproduced data vector will be corrupted and, depending on the undesired signal's power, only an approximation to the original vector may be reconstructed.

The unit impulse data vector used to generate the spreading code can be viewed as the transform domain representation of a particular signal, namely the basis function, whose energy is completely isolated to one transform domain bin. As a result, after the forward transform has been performed at the receiver, the energy content of the transform domain bin corresponding to the appropriate basis function is all that needs to be examined to make a bit decision. In short, if the appropriate bin value is positive (negative), a +1 (−1) data bit is declared. Such a bit decision is consistent with traditional spread spectrum correlators since each of the transform domain bins correspond to the correlation output of the received signal and the appropriate basis function.

Correct bit decisions in the absence of noise are clearly the result of the orthogonality of the wavelet transform basis functions. One can take further advantage of this orthogonality by transmitting two distinct messages simultaneously over the same channel using spreading codes corresponding to different basis functions. Such an approach naturally extends the application of the wavelet transform to communications systems in which code division multiple access (CDMA) is employed.

Since both the transmitter and receiver are assumed to know which spreading function is used, and, since these codes are known to be orthogonal, the system shown in Figure 22 may be used to implement a CDMA communications system. In such a system, each user is assigned a different spreading code, i.e. transform basis function, and allowed to transmit over the same channel. If all users transmit synchronously, the orthogonality of the spreading waveforms ensures that there is no interference between users.

A significant drawback regarding the proposed CDMA system, however, is that the orthogonality of the spreading codes cannot always be maintained. In particular, the previous comments regarding code orthogonality assumed that all transmissions were synchronized. In such instances where all users transmit synchronously, orthogonality of the spreading codes can indeed be assured. If transmissions occur asynchronously, however, these codes do not promise to be orthogonal over all possible time lags. As a result, orthogonality cannot be guaranteed in most applications. The conditions under which spreading code orthogonality can be maintained and how such wavelet-based codes may be applied to modified spread spectrum and CDMA communications systems are discussed in [186], [188] and [89]. In contrast to the filter bank approach suggested in [186] and [188], Cochran et al. [89] specifically address the issue of code orthogonality over all possible time lags of the wavelet basis functions and presents a means by which suitable wavelet bases with the desired cross-correlation characteristics may be designed.

It has been shown [186] that system BER performance in the presence of AWGN and no interference is identical to that achieved using binary phase-shift keying. Hence, in the absence of jammers and additional users, system performance using the wavelet-based spreading sequences is equivalent to that of traditional spread spectrum systems using PN and Gold codes. Through simulation, ad-

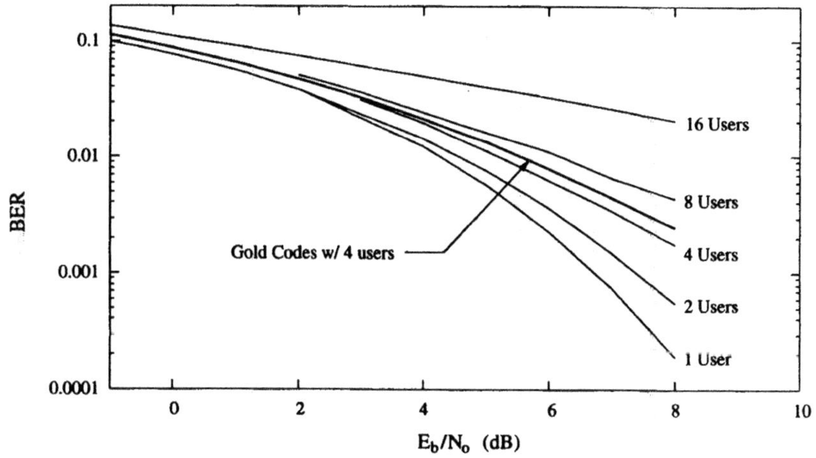

Figure 23 BER results for a CDMA system using spreading waveforms based on Daubechies wavelets [186].

ditional BER results demonstrating the potential feasibility of applying the wavelet-based coding concept to asynchronous multi-user applications have also been generated. In these simulations, the AWGN channel and signal jamming were implemented at baseband and the spreading waveform energy was normalized such that it nearly matches the energy of a 63-chip DS system. The orthogonal wavelet basis was generated using 4-tap FIR filters with wavelet coefficients derived by Daubechies [113]. The full binary tree implementation of the filter bank was used with five stages of filters resulting in 64 chips per message bit.

CDMA simulation results are shown in Figure 23. To generate these results, no restriction on the transmit start time was imposed on any of the additional users. In fact, each user on the channel was given a random start time which was uniformly distributed over the bit interval. Each data point on the figure is the average of the results from multiple simulations each with different random start times. Obviously, if all of the users are synchronized, the orthogonality of the spreading codes ensures that the resulting BER will be equivalent to the theoretical rate achieved using binary phase-shift keying, BPSK, in an AWGN environment. For reference, Figure 23 also contains a curve illustrating system performance when using 63-chip Gold codes [168] in a channel with four users. In such a case, the BER obtained using spreading codes based on Daubechies' 4-tap FIR filters is slightly better than that achieved using Gold codes.

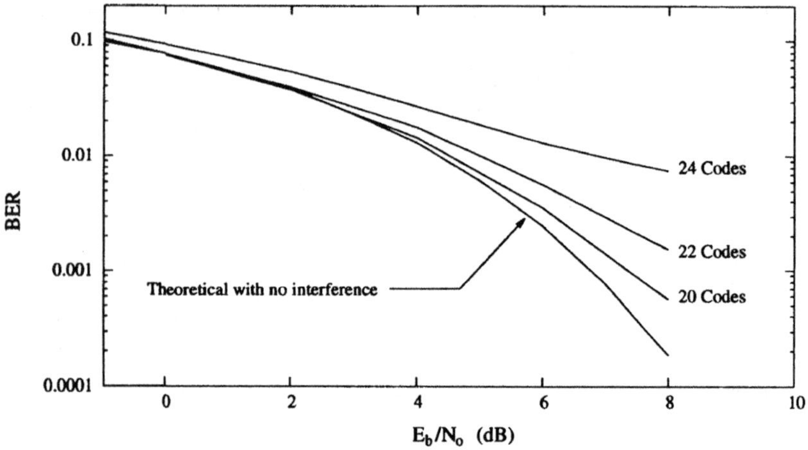

Figure 24 BER results obtained using spreading sequences optimized for anti-jam performance [186].

Using anti-jam waveform constraints, a new waveform design methodology that produces spreading codes which are particularly well-suited to communications in the presence of various interferers has also been introduced [187]. In fact, the BER simulation results shown in Figure 24 illustrate the potential use of such spreading sequences in a channel corrupted by narrow-band interference. To obtain these results, the subband filter bank basis functions were optimized with respect to an anti-jamming constraint to produce the maximum number of orthogonal spreading waveforms with minimal spectral content in the frequency range of interest, which, in this case, is 0.125 to 0.25 Hz, relative to a sampling frequency of unity.

More specifically, Figure 24 represents the system performance obtained in the presence of a 20 dB single-tone jammer at a normalized frequency of 0.21875 Hz. In accordance with the anti-jam design constraint, 4-tap FIR filters were designed and implemented in the hierarchical analysis and synthesis filter banks. Despite the existence of 32 possible spreading waveforms, which were obtained by extending the full binary subband tree structure to five levels, only 24 of the transform basis functions were sufficiently uncorrelated with the anticipated jammer; the remaining eight waveforms are highly correlated with the interfering signal and, thus, not utilized. The figure shows BER results obtained when the 20, 22 and 24 basis functions with the least correlation with the interference are used as the spreading sequence.

The anti-jam constraint was used to produce filter bank spreading codes that allow multiple users to communicate in the presence of narrow-band interference. In more general applications, similar constraints may be developed and used to derive an appropriate set of spreading functions that are well-suited to a given application. In fact, such constraints may be readily designed to yield covert LPI/D waveforms with spectral characteristics similar to those of the ambient noise. Clearly, such a set of basis functions could easily generate a composite signal that would be very difficult to detect by conventional means. In fact, using the wavelet transform, it has been shown [351] that a variety of specialized waveforms may be produced that significantly improve communications covertness.

4 CONCLUSIONS

Wavelets and filter banks are being applied to an ever-growing number of areas. In this chapter, the authors have attempted to summarize the work that has been performed to date on the application of filter banks and wavelets to spread spectrum communications. While we have tried to be comprehensive, the rapid growth in this area makes it likely that some work has been overlooked. In any event, the work that has been reported is, in general, quite preliminary in nature. We are certain that wavelets and filter banks will be applied even more effectively to problems in spread spectrum communications in the near future.

11

ACOUSTIC ECHO CANCELLATION USING SUBBAND ADAPTIVE FILTERING

Phillip L. De León II and Delores M. Etter

Department of Electrical and Computer Engineering, University of Colorado, Boulder, CO 80309, USA

1 INTRODUCTION

Adaptive filtering has been successfully used in canceling line echos in the telephone network since the 1960's [179, 443, 387]. In these cases, the adaptive filter is used to model the unknown echo path in a communications channel and then cancel the echo signals from the conversation. An adaptive filter is required because the echo path is not only unknown, but also because it is different for each network path. However, the acoustic echo cancellation problem is not as easily solved with conventional adaptive filtering systems due to the requirements for a much longer adaptive impulse response and for faster convergence since the acoustic echo path is time-varying during the communication transmission. In this chapter, we summarize key results from adaptive filter theory, and discuss the least mean square (LMS) and normalized least mean square (NLMS) algorithms used to adjust an adaptive filter. We then examine adaptive filtering in subbands as one method for acoustic echo cancellation. Simulations (using a measured room response with speech signals) demonstrate that the oversampled subband adaptive system provides improved performance for acoustic echo cancellation applications.

2 LINEAR FIR ADAPTIVE FILTERING

An adaptive system can be characterized by four parts: 1) the adaptive structure being modified, 2) the configuration which contains the adaptive structure, 3) the performance measure used to evaluate the current state of the adaptive structure, and 4) the algorithm used to modify the structure to improve its performance. In this discussion, we focus on the adaptive system used in acoustic echo cancellation.

2.1 The Adaptive Structure

An adaptive FIR filter $\hat{\mathbf{w}}$, shown in Figure 1, is one of the most commonly-used structures for adaptive systems. In an adaptive FIR filter, the filter coefficients (or weights) vary with time, and thus the impulse response and corresponding transfer function of the filter also vary with time.

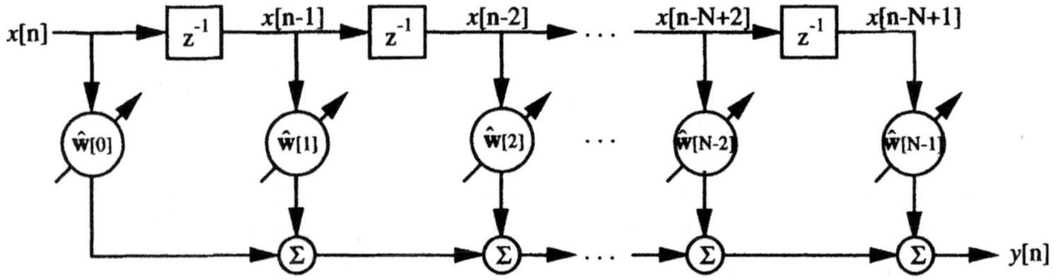

Figure 1 Adaptive FIR filter.

2.2 The Adaptive Configuration

An unknown system $\mathbf{w}$, may be changing over time, may be too complicated to compute directly, or may need to be computed in real-time. In these situations, a system identification or system modeling configuration is used as illustrated in Figure 2. In this configuration, an adaptive filter, $\hat{\mathbf{w}}$ attempts to model the unknown system using the input and the output of the unknown system, $\mathbf{w}$.

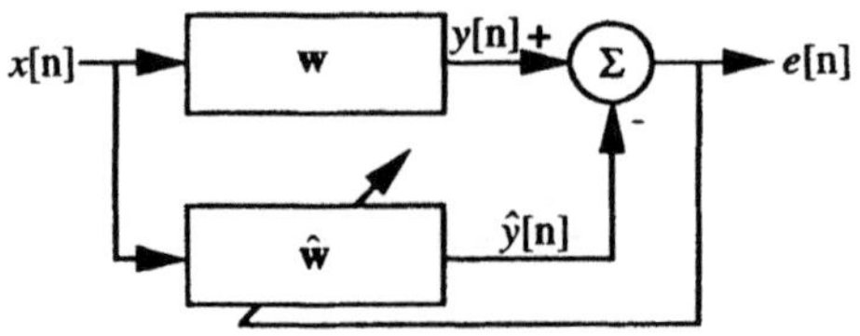

Figure 2 System identification configuration.

To see how the system identification configuration is used in the cancellation of acoustic echos, consider the hands-free teleconferencing system in Figure 3 which shows a speaker and a microphone at each end of the channel. Without echo cancellation systems, a filtered version of the signal from one side of the conversation (for example, originating from A) can pass through the network at B, and is then added to the signal originating from B; this causes an echo to be added to the signal received by A.

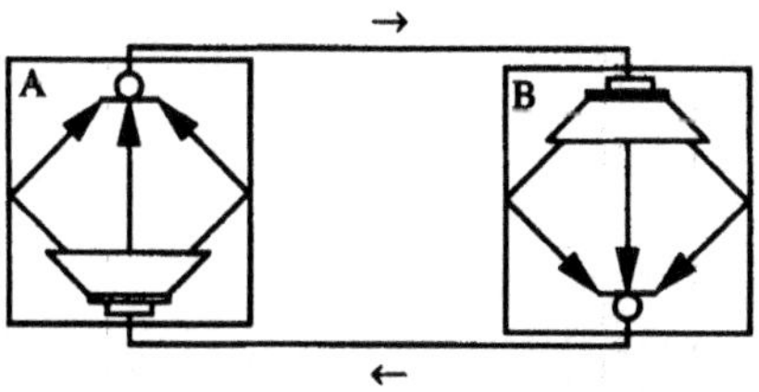

Figure 3 Teleconferencing system.

To eliminate this echo, an adaptive filter is introduced at each end of the teleconferencing system, in parallel to the room echo path, as shown for one end in Figure 4. In this system identification configuration, the adaptive filter, $\hat{\mathbf{w}}$ builds a model of the room echo path, $\mathbf{w}$ and the output from the model is subtracted from the output containing the echos, yielding a signal in the return path that has little or no echo.

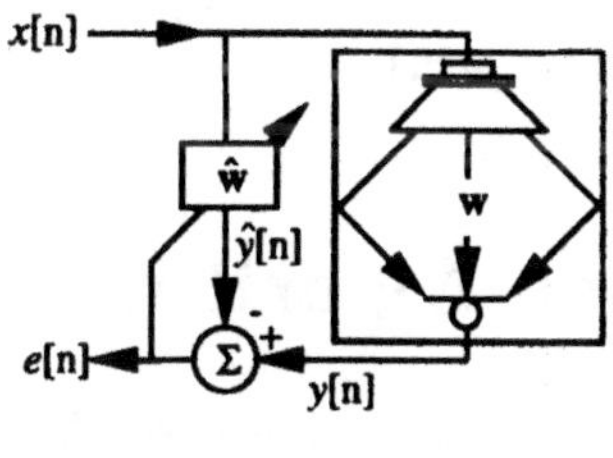

Figure 4 Acoustic echo canceler application.

2.3 The Performance Measure

The objective of the adaptive filter in Figure 2 is to modify an input signal, x such that the resulting signal estimates a desired signal, y. An error signal, e is computed by subtracting the output of the adaptive filter $\hat{y}$, from y, as shown in Figure 2. If the error signal is large, then $\hat{y}$ is not a good match to y; if the error signal is small, then the output of the filter is a good match to the desired signal. Thus, the error signal provides a ***measure*** of the ability of the adaptive filter to model the unknown system. The mean-squared error (MSE), ξ, is used as the performance measure for the adaptive filter (instead of just the error itself) because it provides a performance surface that is quadratic with respect to the filter coefficients. This performance surface can be shown to be a function of the correlation of the input signal, x and the cross-correlation of the input and output signals to the unknown system [179, 443]

$$\begin{aligned} \xi[n] &= E\left[e[n]e^*[n]\right] \\ &= E\left[\left(y[n]-\hat{y}[n]\right)\left(y[n]-\hat{y}[n]\right)^*\right] \\ &= E\left[|y[n]|^2\right] - \hat{\mathbf{w}}^H\mathbf{p} - \mathbf{p}^H\hat{\mathbf{w}} + \hat{\mathbf{w}}^H\mathbf{R}\mathbf{w} \end{aligned} \tag{11.1}$$

where $\hat{\mathbf{w}}$ is a column vector of the adaptive filter coefficients; $\mathbf{R}$ is the input correlation matrix, $E\left[\mathbf{x}[n]\mathbf{x}^H[n]\right]$; $\mathbf{p}$ is the cross-correlation vector, $E\left[\mathbf{x}[n]\,y[n]^H\right]$; and ^{H} refers to the conjugate transpose.

It can be shown [443] that the eigenvectors of the input correlation matrix define the principal axes of the error surface and that the corresponding eigenvalues give the second derivatives of the error surface with respect to the principal axes. The effect of these eigenvectors and eigenvalues on the adaptive process will become evident in the next section.

2.4 The Adaptive Algorithm

We now consider algorithms for adjusting the adaptive filter to its optimum, which minimizes the MSE. Since the MSE surface is quadratic, the minimum occurs at the point at which the gradient (derivative of the MSE) with respect to the filter coefficients is equal to zero. Therefore, taking the gradient of the MSE in Equation (11.1), we have

$$\nabla\left(\xi[n]\right) = 2\mathbf{R}\hat{\mathbf{w}} - 2\mathbf{p}. \tag{11.2}$$

Assuming that $\mathbf{R}$ is nonsingular, we can then set the gradient to zero and solve for the optimum filter coefficients

$$\hat{\mathbf{w}}_{\mathbf{opt}} = \mathbf{R}^{-1}\mathbf{p} \tag{11.3}$$

where $\mathbf{R}^{-1}$ is the inverse of $\mathbf{R}$. Thus, if $\mathbf{R}$ is known, we can solve directly for the optimum filter coefficients, using Equation (11.3) which is also called the Wiener filter solution.

In many applications, either $\mathbf{R}$ is not known or is time-varying, and thus a direct solution for $\hat{\mathbf{w}}_{\mathbf{opt}}$ may not be feasible or will not be optimal at all times. In these cases, an adaptive algorithm is desired that can start at any initial filter setting, and converge to $\hat{\mathbf{w}}_{\mathbf{opt}}$.

Most adaptive algorithms fall into two categories: gradient-based algorithms (search algorithms that use an estimate of the gradient to determine the direction of the minimum); and recursive least squares (RLS) algorithms (algorithms that use iterative techniques to determine the optimum filter coefficients at any point in time by computing $\mathbf{R}^{-1}[n]$, the inverse of $\mathbf{R}$ at time n). There are many variations of gradient-based algorithms, including steepest descent algorithms and random algorithms. There are also a number of variations of RLS algorithms, including the sequential regression algorithm (SER) and fast RLS algorithms that are able to efficiently estimate $\mathbf{R}^{-1}[n]$.

In the acoustic echo cancellation application, we use a steepest descent algorithm because it is simple and requires fewer computations per iteration than other algorithms. The general form of a steepest descent algorithm is the following

$$\hat{\mathbf{w}}[n+1] = \hat{\mathbf{w}}[n] - \frac{1}{2}\mu\nabla(\xi[n]) \tag{11.4}$$

where μ is a convergence parameter that controls the size of the step taken at each iteration along the performance surface. The factor $\frac{1}{2}$ is used merely for convenience. If you consider a simple quadratic function of one variable that has only positive function values, a point on the function with a positive gradient (or slope) will be to the right of the minimum, and thus the weight value should be decreased to move toward the optimum weight value. Similarly, a point on this quadratic function with a negative gradient will be to the left of the minimum, and the corresponding weight value should be increased to move toward the optimum weight value. Thus, in general, the weight increment should be in the direction of the negative gradient, as shown in Equation (11.4).

In real-time processing applications, we usually do not know $\mathbf{R}$ and $\mathbf{p}$, and in addition, they may be changing in time. Thus, we need to use an *estimate* of the gradient. The LMS algorithm [179, 443] uses an instantaneous estimate of the gradient to determine the following algorithm

$$\hat{\mathbf{w}}[n+1] = \hat{\mathbf{w}}[n] + \mu \mathbf{x}[n] e^{*}[n]. \tag{11.5}$$

Convergence of the LMS algorithm to the optimum weight values depends on the convergence factor, μ. Since the eigenvalues give the steepness of the performance surface, the bound for the step-size (or convergence parameter) is a function of the maximum eigenvalue of the correlation matrix, $\mathbf{R}$ [179, 443]

$$0 < \mu < \frac{2}{\lambda_{max}}. \tag{11.6}$$

Since the eigenvalues are generally not simple to estimate, a practical guideline for the upper bound of the convergence parameter can be developed that uses the fact that the trace of $\mathbf{R}$ is equal to the sum of its eigenvalues. Hence, the maximum eigenvalue is less than or equal to the trace of $\mathbf{R}$ which may be approximated by the filter length times the input signal power. Thus, an upper bound for μ is

$$0 < \mu < \frac{2}{N\,(\text{signal power})}. \tag{11.7}$$

The NLMS algorithm can be viewed as a modification of the LMS algorithm that gives it a time-varying step-size parameter that accounts for variations in the input signal power. The NLMS algorithm can also be developed as a solution to a constrained optimization problem [179] using the method of Lagrange multipliers. In either case, the final form of the NLMS algorithm is

$$\hat{\mathbf{w}}[n+1] = \hat{\mathbf{w}}[n] + \frac{\tilde{\mu}}{a + \|\mathbf{x}[n]\|^2}\mathbf{x}[n]\, e^*[n] \tag{11.8}$$

where a (used to avoid possible division by zero) is greater than 0, $\tilde{\mu}$ is between 0 and 2, and $\|\mathbf{x}[n]\|^2$ is the norm squared or power of the input vector $\mathbf{x}[n]$.

The next section of this chapter presents a subband adaptive filtering solution for acoustic echo cancellation, and compares it to the full-band solution.

3 THE SUBBAND ADAPTIVE FILTERING SYSTEM

The motivation for adaptive filtering in subbands stems from two well-known problems in full-band adaptive filtering. First, the convergence and tracking of a gradient-based adaptive filter can be very slow if the input correlation matrix is ill-conditioned or equivalently if the input signal has wide spectral dynamic range such as that found in speech [179, 171]. Second, very high-order adaptive filters are computationally expensive. One technique used to overcome these limitations is to decompose the signal into subbands and adaptively filter each subband signal using separate adaptive filters. The filter bank is the primary tool used to perform the subband decomposition. The earliest references to this technique used in acoustic echo cancellation applications were Furukawa in 1984; Itoh, Maruyama, Furuya, and Araseki in 1985; and Kellermann in 1985 [387, 209].

In the subband adaptive filter system for a system modeling configuration (Figure 5), the desired output signal, y, and the input signal, x, are split into M subband signals by analysis filters, $\mathbf{h}_0, \cdots, \mathbf{h}_{M-1}$, and downsampled by a factor of D. A bank of adaptive filters, $\hat{\mathbf{w}}_0, \cdots, \hat{\mathbf{w}}_{M-1}$, each adjust themselves (usually using the NLMS algorithm) so as to minimize their mean squared subband error, e_m, which as mentioned earlier, is taken as the square of the difference between the desired subband signal, y_m and the subband adaptive filter output, $\hat{y}_m$. The subband error signals may be used either directly (as in Figure 5) or indirectly to reconstruct a full-band output. Reconstruction of the full-band signal consists of upsampling by a factor of D and filtering with synthesis filters, $\mathbf{g}_0, \cdots, \mathbf{g}_{M-1}$.

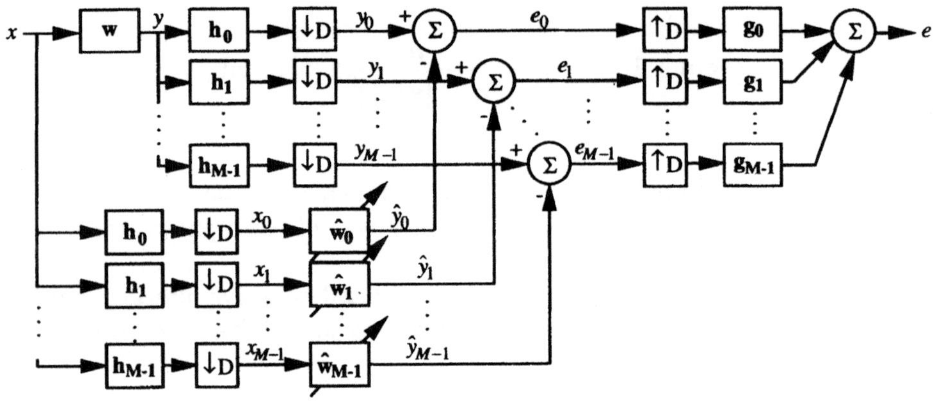

Figure 5 Subband adaptive filtering system (system modeling configuration).

The first benefit of this system results from the fact that the subband adaptive filters are shorter in length than the equivalent full-band adaptive filter (although the total number of adaptive FIR coefficients is usually the same) and operate at a downsampled rate. The lengths of the adaptive filters do not necessarily have to be equal and in fact there may be certain advantages in adjusting the length of each filter to better match the signal characteristics in that band [164]. The second benefit stems from the subband decomposition of the input and desired output signals by the analysis filters. By decomposing the signals, the adaptive filters operate on a smaller bandwidth and can be adjusted to take advantage of this fact. Furthermore, in a critically sampled ($D = M$), subband adaptive filter, the subband decomposition of the input may decrease the spectral dynamic range of the subband input, resulting in faster convergence for gradient-based algorithms. On the downside is the increased design complexity of the technique, the end-to-end delay associated with analysis and synthesis filtering, and the aliasing due to downsampling. In oversampled systems ($D < M$), there is the undesirable potential to increase the spectral dynamic range of the subband input which equivalently increases the eigenvalue disparity ($\lambda_{\max}/\lambda_{\min}$) and in turn, slows the convergence rate of the subband adaptive filter for gradient-based algorithms [279]. All these factors have warranted closer study and analysis in recent years.

4 COMPUTATIONAL COMPLEXITY OF THE SUBBAND ADAPTIVE FILTERING SYSTEM

To gain an understanding of the computational advantage the subband system has over the full-band system, we compute the computational complexity of the M/D oversampled, M-band subband adaptive filtering system (Figure 5), as measured by the number of real multiplications per input sample. This is computed in two parts: the complexity for subband filtering, $C_{\mathbf{subband},1}$ and the complexity for adaptive filtering $C_{\mathbf{subband},2}$.

Assume that x and y are real signals, that analysis and synthesis filtering is implemented with the polyphase uniform DFT filter bank [108], that the prototype analysis/synthesis filter is length L, and that M/D is an integer. Then $C_{\mathbf{subband},1}$ is computed as follows. There are a total of M polyphase filters, each of length L/D operating at a rate of $1/D$ in the filter bank thus requiring $\frac{LM}{D^2}$ real multiplications per input sample. This operation is performed three times: for the analysis filtering of x and y and for the synthesis filtering of $e_0, \cdots, e_{M-1}$. The M-point DFT and IDFT are implemented (assuming M is a power of 2) with a radix-2 FFT which requires approximately $\frac{M}{2}\log_2 M - M$ complex multiplications. For real data, the M-point IDFT can be realized with an $M/2$-point FFT and $M/2$ complex multiplications [104]. This results in $M \log_2 \frac{M}{2}$ real multiplications for the analysis filtering of x and y; a similar realization holds for the synthesis filtering of $e_0, \cdots, e_{M-1}$. Thus the total number of real multiplications for subband filtering per input sample is

$$C_{\mathbf{subband},1} = \frac{3LM}{D^2} + 3M \log_2 \frac{M}{2}. \tag{11.9}$$

Since the input and desired output signals are real, the DTFT is symmetric. Exploiting this symmetry will require processing of $\frac{M}{2} + 1$ of the subbands with subbands $\frac{M}{2} + 1, \cdots, M - 1$, taken as the respective complex conjugates of subbands $\frac{M}{2} - 1, \cdots, 1$. Furthermore, the uniform DFT bank will yield real signals in subbands 0 and $M/2$ and complex signals in the other subbands. Thus two adaptive filters will have real coefficients and $\frac{M}{2} - 1$ will have complex coefficients. Assume the length of the impulse response to be modeled is N, each adaptive filter is of length N/D operating at the downsampled rate, and the LMS algorithm (either real or complex) is used for the update. The total number of real multiplications for adaptive filtering per input sample is

$$C_{\text{subband},2} = \frac{2\left(\frac{2N}{D}+1\right)+4\left(\frac{M}{2}-1\right)\left(\frac{2N}{D}+1\right)}{D}$$
$$= \left(\frac{2N}{D}+1\right)\left(\frac{2M-2}{D}\right). \quad (11.10)$$

The complexity for the M/D oversampled, M-band subband adaptive filter system is then taken as the sum of Equations (11.9) and (11.10)

$$C_{\text{subband}} = C_{\text{subband},1} + C_{\text{subband},2}$$
$$= \frac{3LM+4MN-4N}{D^2} + \frac{2M-2}{D} + 3M\log_2\frac{M}{2}. \quad (11.11)$$

Figure 6 contains a plot of the normalized ($C_{\text{subband}}/C_{\text{full-band}}$) subband computational complexity with $L = 129$ and $N = 512$ versus the number of subbands, M, for critically sampled and $2\times$ oversampled systems.

It is clear from the plot (for the parameters used) that critically sampled subband systems with 4, 8, 16, or 32 subbands (in the derivation we assumed M is a power of 2) are computationally more efficient than the equivalent full-band system. Furthermore, $2\times$ oversampled subband systems are more efficient with 16 or 32 subbands.

5 EXPERIMENTAL SETUP

The experimental results presented in this chapter were conducted in software on a general purpose computer and use both full-band and subband adaptive system modeling configurations for comparison. The unknown system, $\mathbf{w}$, is an actual impulse response for a room, sampled at 8 kHz and truncated to 512 samples. In all experiments, $M = 4$ and efficient polyphase, uniform DFT filters are used for analysis and synthesis filtering [108]. The analysis filter prototype was designed using the Parks-McClellan algorithm and is of length 129. The synthesis filters were the same as the analysis filters. Magnitude responses for these filters are illustrated in Figure 7.

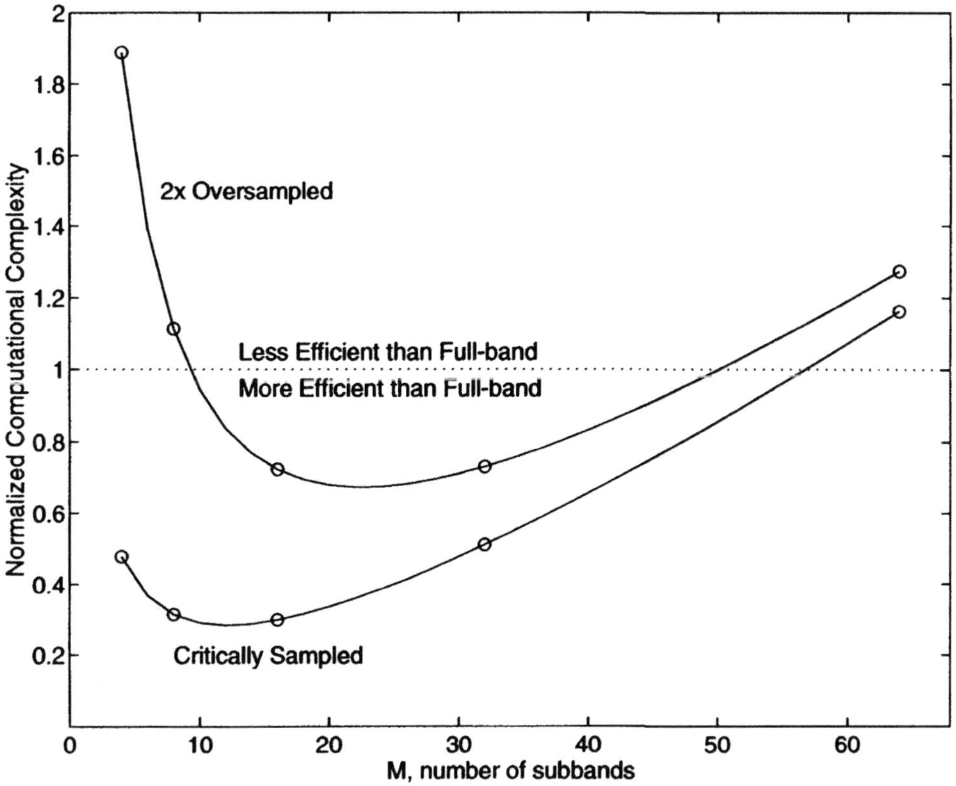

Figure 6 Normalized computational complexity for the subband adaptive filtering system.

The adaptive filters were all of length $\lceil \frac{512}{D} \rceil$ and use the NLMS algorithm for the update with $a = 0.01$ and $\tilde{\mu} = 1/3$. Adaptive filtering was done only in subbands 0, 1, and 2 since $y_3, \hat{y}_3$ were taken as $y_1^*, \hat{y}_1^*$ respectively, due to the complex conjugate symmetry of the uniform DFT filter bank. The adaptive filters in subbands 0 and 2 were real and in subband 1 was complex. Zero-mean, unit-variance white Gaussian noise input was used to measure the convergence of both the full-band and subband systems, and experiments were averaged over 100 simulations. MSE plots were smoothed with a 100 point moving average filter and 2× oversampled subband MSE plots were smoothed with a 50 point moving average filter. Power spectrum plots were obtained from the last 1024 points of the hanning-windowed error signal. In a separate experiment, speech input was also used to measure the performance. The speech signal was composed of the utterance "Zero One Two Three Four Five Six Seven Eight Nine", sampled at 8 kHz, and is shown in Figure 8.

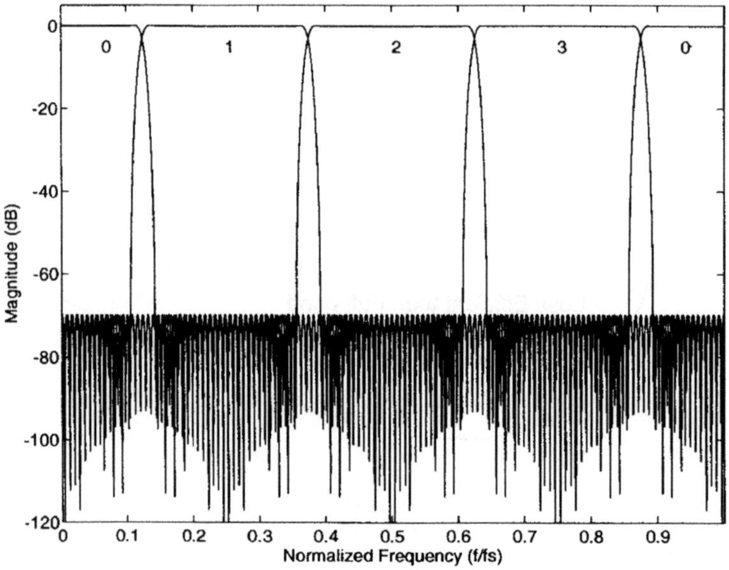

Figure 7 Polyphase, uniform DFT analysis and synthesis filters used in experiments.

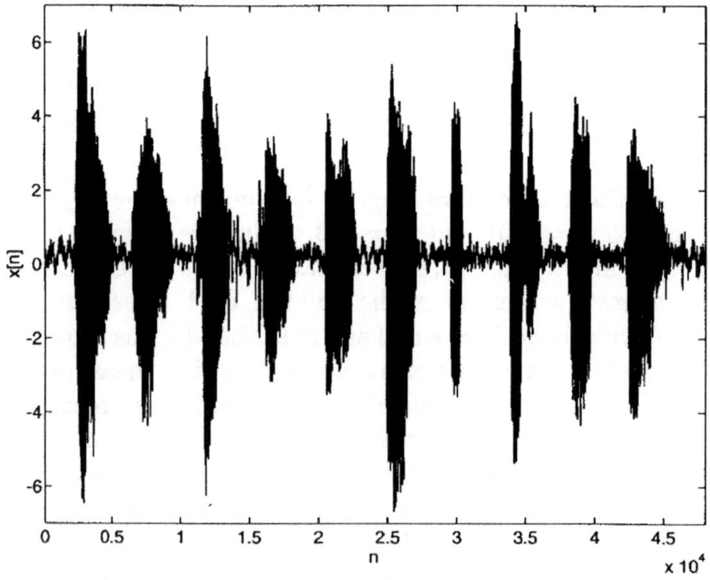

Figure 8 Input speech signal.

6 CRITICALLY SAMPLED SYSTEMS

The first experiments in subband adaptive filtering used critically sampled subbands for maximum efficiency. The primary result of these experiments is the gain in computational efficiency. However, as can be seen in Figure 9, the subband MSEs quickly level off due to the inability to properly model the aliasing, which is present in the critically sampled subbands. The point at which convergence levels off is usually referred to as the asymptotic level.

The high asymptotic levels in the subband MSEs, result in a high asymptotic level for the reconstructed full-band MSE, as shown in Figure 10. Included on this plot for comparison is the MSE for the equivalent full-band system. During the initial convergence the subband system performs equally well as compared to the full-band system. However, as mentioned earlier, the effects of aliasing in the subbands hinders the ability to properly model and the performance tapers off.

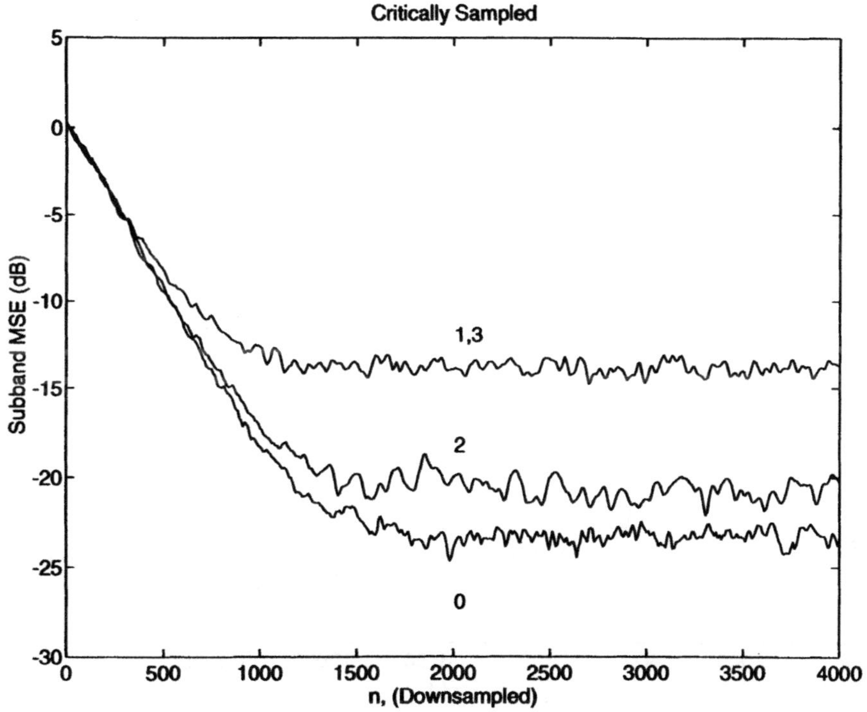

Figure 9 Subband MSEs for the critically sampled subband adaptive filtering system.

Spectral analysis of the reconstructed full-band error signal (Figure 11) shows marked spectral peaks at the crossover frequencies (corresponding exactly to the aliased parts of the spectrum) in the filter bank indicating the inability of the system to properly model in these frequency regions.

We next compare the performance of the critically sampled subband system with the full-band system using speech as the input. As mentioned earlier, the subband system performs a spectral decomposition of the input and for speech, this should be beneficial. Figure 12 illustrates the MSE of both systems and the results show very comparable performance between the two systems.

In an attempt to retain the critical sampling rate and have more accurate modeling (smaller output error), a slightly different subband system was proposed and is illustrated in Figure 13 [165].

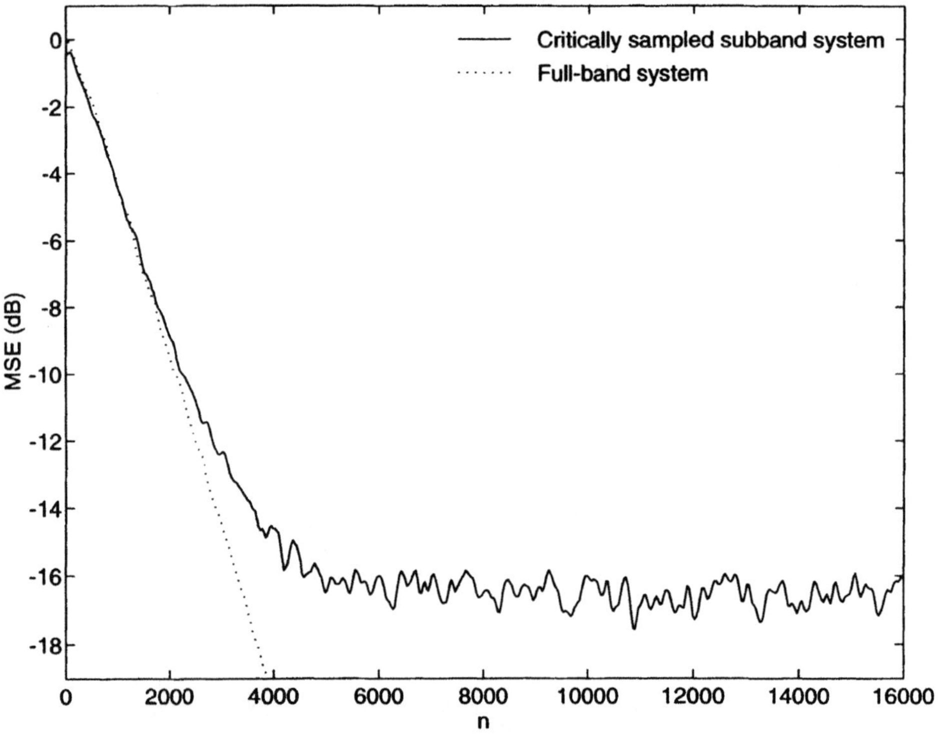

Figure 10 MSE for full-band and critically sampled subband adaptive filtering systems.

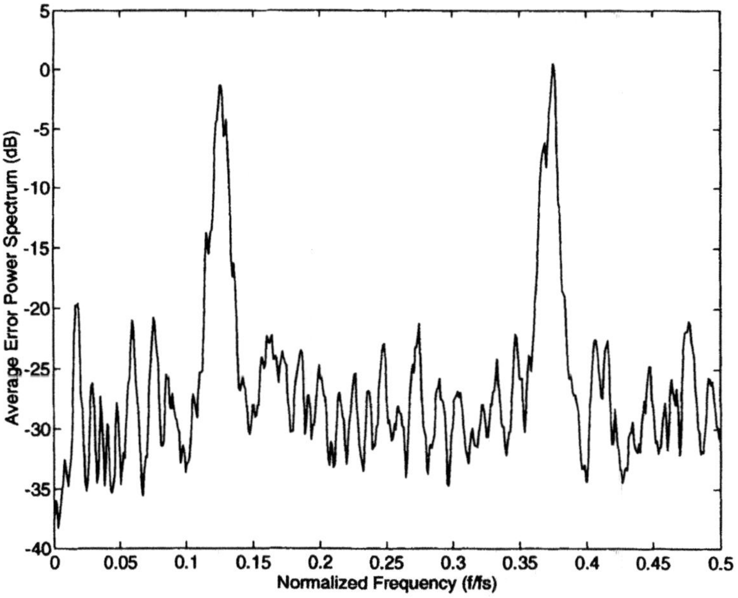

Figure 11 Average power spectrum of the error signal for the critically sampled subband adaptive filtering system.

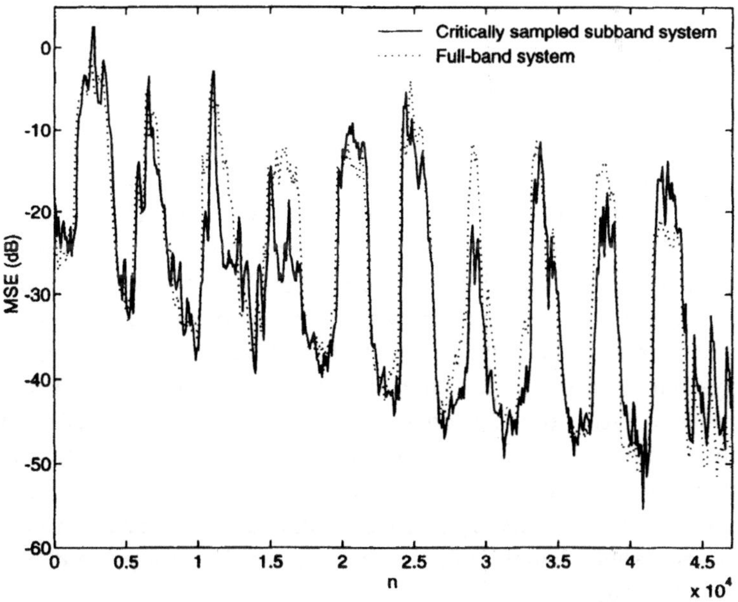

Figure 12 MSE for full-band and critically sampled subband adaptive filtering systems under speech input.

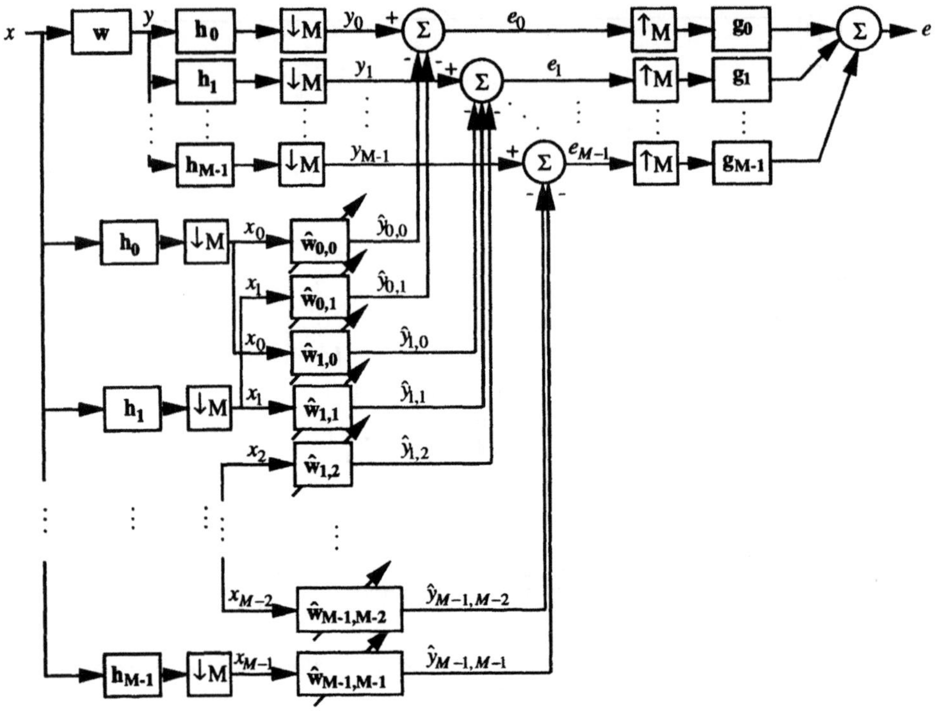

Figure 13 Critically sampled subband adaptive filtering system with adaptive cross filters.

In this system, adaptive cross filters are used in which the input is from one subband and the error signal used for update is from another subband. In experiments with input noise with a speech-like spectrum (USASI noise), it was demonstrated by Gilloire and Vetterli [165] that the critically sampled, eight-band subband system with adaptive cross filters had a lower asymptotic level than the two-band subband system without cross filters. However, there was no significant gain in the convergence rate for either system as compared to the full-band system. In a separate experiment with white noise input, a two-band subband system with adaptive cross filters and perfect reconstruction filter banks achieved perfect modeling, although the convergence rate was slower. It was ultimately concluded, however, that oversampled systems provided the best method of adaptive filtering in subbands, both in terms of convergence and computational efficiency [165]. In the next section, we present results from oversampled subband systems.

7 OVERSAMPLED SYSTEMS

Due to the inability of critically sampled subband systems without cross filters to adequately model in the presence of aliasing, we consider oversampling the subband signals as a means of eliminating the effects of aliasing. This of course, reduces the computational efficiency as compared to the critically sampled system. In these experiments, the subbands are 2× oversampled and from a practical standpoint, this is more oversampling than actually needs to be done. As can be seen in Figure 14, the subband MSEs do not have the characteristic high asymptote that the critically sampled subband MSEs have, but nonetheless exhibit an asymptotic convergence after a fast initial convergence [279].

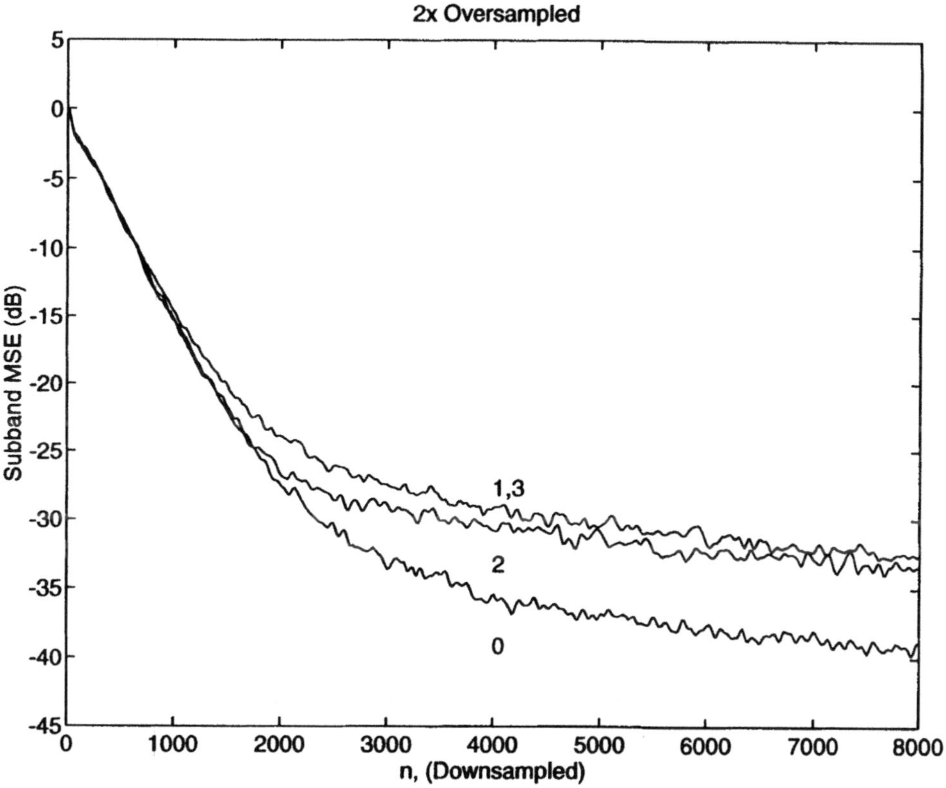

Figure 14 Subband MSEs for 2× oversampled subband adaptive filtering system.

Consequently, the reconstructed full-band MSE for the oversampled subband system (Figure 15) exhibits this slow asymptotic convergence as well. Included on this plot for comparison is the MSE for the full-band system. During the initial convergence the subband system performs better than the full-band system but is less effective afterward. The oversampled subband system does, however, have better convergence as compared to the critically sampled system. In an environment where **w** is changing over a period of time shorter than the initial convergence period (as in the case of acoustic echo cancellation), initial convergence will most affect cancellation quality.

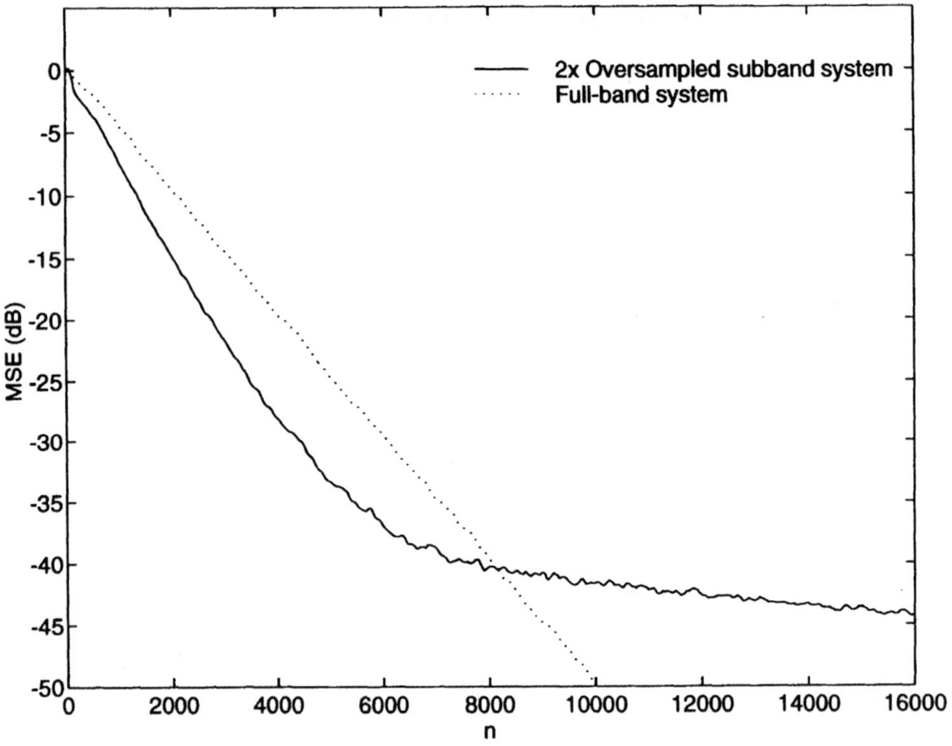

Figure 15 MSE for full-band and 2× oversampled subband adaptive filtering systems.

Spectral analysis of the error signal (Figure 16) also shows marked peaks located in the vicinity of the crossover frequencies. These peaks are not due to aliasing but rather due to the small eigenvalues associated with the analysis filter roll-off in the oversampled case. This characteristic is the basis of the theory of slow asymptotic convergence of LMS acoustic echo cancelers [279, 119].

Figure 17 contains the MSE plot for the full-band and subband systems with speech input. In this plot, the oversampled subband system achieves 0-20 dB increase in convergence over the full-band system as well as the critically sampled subband system but asymptotic convergence is still present.

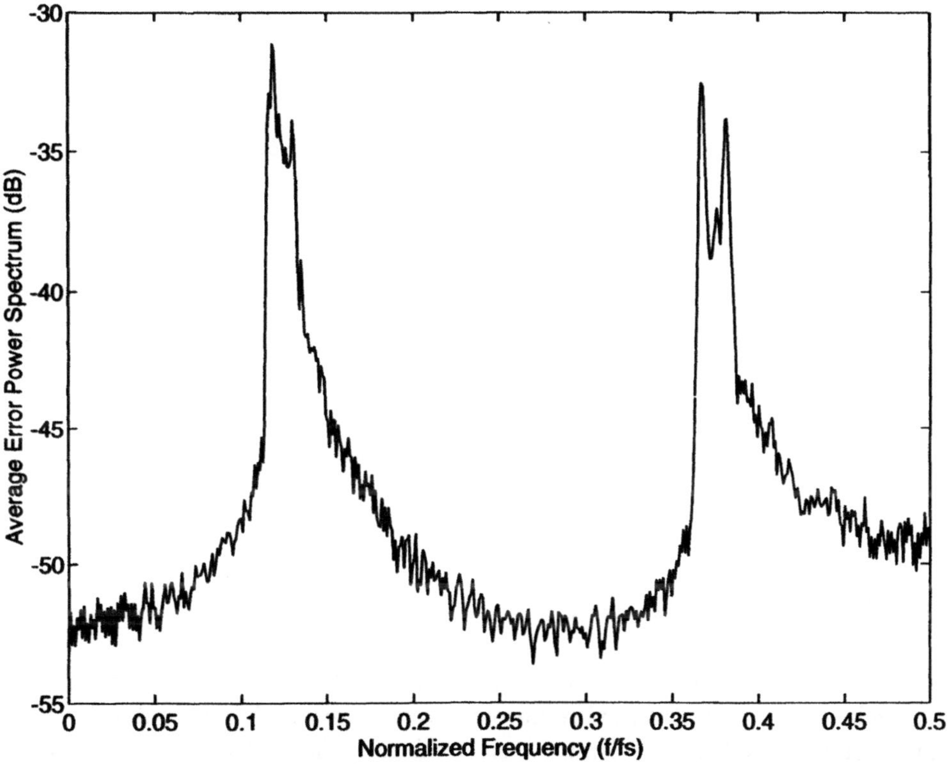

Figure 16 Average power spectrum of the error signal for the 2× oversampled subband adaptive filtering system.

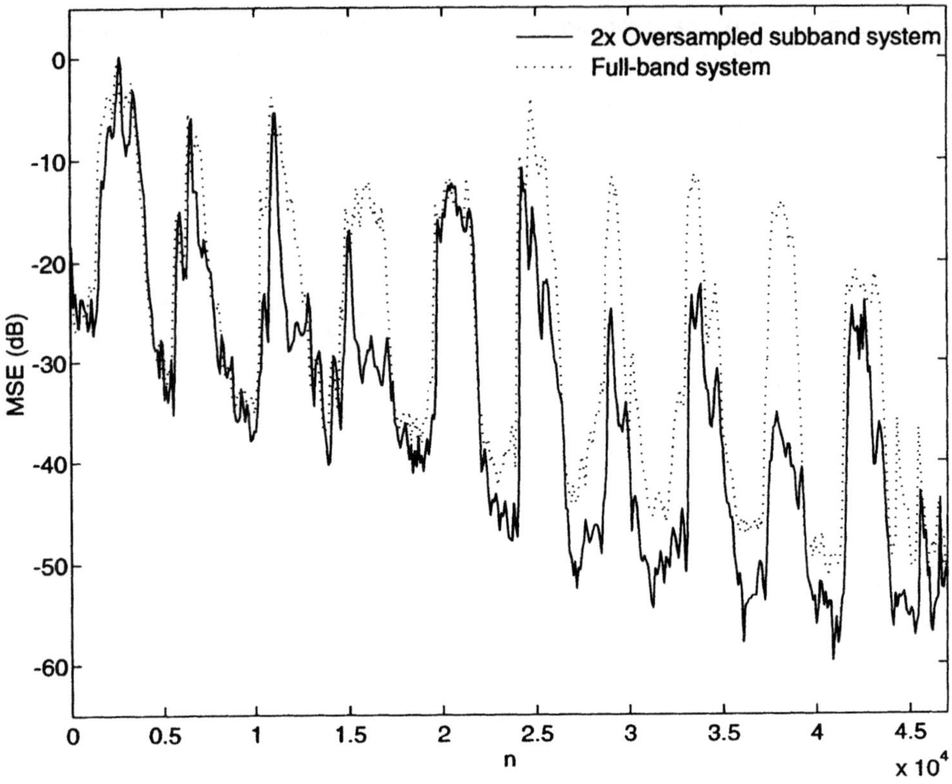

Figure 17 MSE for full-band and 2× oversampled subband adaptive filtering systems under speech input.

8 CONCLUSION

In this chapter we have seen how using multiple adaptive filters in a filter bank can reduce the computational requirements of a very high-order adaptive filter and reduce the effects of disparate eigenvalues. The performance in terms of MSE under white noise input is comparable to the full-band system during the initial convergence but degrades due to aliasing effects in the critically sampled system and slow asymptotic convergence effects in the oversampled system. Under speech input, the critically sampled subband system performs comparably to a full-band system. The 2× oversampled subband system with speech input, however, offers as much as a 20 dB reduction in MSE (in our experiments) when compared to the full-band system or critically sampled subband system.

FURTHER READING

J. Chen, H. Bes, J. Vandewalle, and P. Janssens, "A New Structure for Sub-band Acoustic Echo Canceler," in *Proc. IEEE ICASSP '88* (New York, NY) pp. 2574-2577.

A. Gilloire and M. Vetterli, "Adaptive Filtering in Sub-bands," in *Proc. IEEE ICASSP '88* (New York, NY) pp. 1572-1575.

W. Kellermann, "Analysis and Design of Multirate Systems for Cancellation of Acoustical Echoes," in *Proc. IEEE ICASSP '88* (New York, NY) pp. 2570-2573.

S. Mitra, M. Petraglia, and A. Mahalanobis, "Structural Subband Implementation of Adaptive Filters," in *Proc. 24th Asilomar Conference* (Pacific Grove, CA), Nov. 1990, pp. 232-236.

J. Shynk, "Frequency-domain and Multirate Adaptive Filtering," IEEE ASSP Magazine, vol. 9, no. 1, pp. 14-37, Jan. 1992.

H. Yasukawa, S. Shimada, and I. Furukawa, "Acoustic Echo Canceler with High Speech Quality," in *Proc. IEEE ICASSP '87* (Dallas, TX) pp. 2125-2128.

12

ANALOG SCRAMBLERS

Richard V. Cox

Speech Coding Research Department,
AT&T Bell Laboratories, Murray Hill, NJ 07974

1 INTRODUCTION

The need for secure communications has existed for many centuries. From the earliest times people have entrusted messages with messengers. Even today a head of state will send a personal message to another head of state by sending their Foreign Secretary or Ambassador. In this method of communication the security is entirely dependent on the security of the messenger. If the messenger is not trustworthy or is intercepted, then the message is compromised.

As a further example, recall that in the middle ages the nobility used wax seals for their letters or other important documents. Each such person had their own seal for sealing documents. If the document were opened en route the seal could not be restored. The recipient would know that the message had been intercepted. Of course, the security of this method was really more dependent on the messengers and the routes they took than the seal. Later, invisible inks were invented. The paper was heated or treated in order to make the message legible. Today, governments routinely send written messages via a diplomatic pouch. For highly important messages, the message is encrypted as well. Clearly the security of the communication is still dependent on the security of the pouch.

The above examples all involved securing forms of written communication. The best example of secure communication involving speech was the use of other languages. A monarch could speak to his own advisers in a language different than that of anyone else in the room. This technique has even survived until fairly recent times. During World War II the US military used Navajo Indians to relay messages in the Pacific because they knew that the Japanese military did not have anyone who could speak Navajo to translate the messages. Another technique is to use codewords in place of certain important words.

In the above examples, only the wealthy and powerful had access to secure communications. With the advent of the telephone came the need for devising means to secure speech carried by wire. Later, there was a similar need for radio as well. Both of these media are used by common people. For example, there have been numerous examples of people whose conversations have been overheard while talking over cellular or cordless telephones. Some of these occurrences have caused embarrassment while others had significant legal ramifications. Clearly the need for secure communications extends to the common man now.

Speech and our ability to recognize it are remarkably resilient. If we reduce the bandwidth, the intelligibility loss is often less than the percentage of bandwidth removed [146]. If we add noise to mask the speech signal, fairly large amounts must be added before the intelligibility is substantially reduced. Even adding noise with the same power as the speech, making the SNR of the transmitted signal 0 dB, does not reduce the speech intelligibility sufficiently. This is as compared with digital communications where the loss of a single bit could render an encrypted message indecipherable.

In the above paragraph we have mentioned two methods of communications. Before we go further, we should define some important terms to be used throughout this chapter. Communications can be conducted by either *analog* or *digital* means. For example, if we wish to transmit speech, we might use a digital speech coder to represent the speech waveform by a sequence of bits, known as a bit stream and then transmit them via a digital channel. Or we might wish to transmit them as a continuous waveform having a range of continuous values on a channel which permits this, an analog channel. This is not necessarily the original waveform of the message. For example, we can transmit audio signals via Frequency Modulation (FM). If we intercepted the transmission and reduced the signal to baseband, we still could not understand it because the signal has been transformed to FM. It is not encrypted, though. FM receivers are quite common and someone else with an FM radio can receive the same message. FM is considered an analog means of transmission, as is Amplitude Modulation.

By contrast, in digital communications the message is first transformed into a series of bits which are then transmitted. When the message is in this form, the bit stream can be encrypted using digital encryption techniques. Typically, the encryption technique depends upon a sequence of bits referred to as a *key.* The key determines which one of a large number of possible alternative transformations is made to the bit stream. The security of the message will depend upon the ***cryptanalytical strength*** of these techniques (usually just how many possible keys there are) and the level of security used in distributing the keys which will unlock the encrypted transmission.

Analog scrambling seeks to protect the transmission of the message if it is intercepted by making a transformation of the signal. Like FM it is a reversible transformation from the normal signal domain to another domain in which the signal is rendered unintelligible. Unlike FM, there should be a large number of possible transformations, so that person who intercepts it (the ***adversary***) will not be able to try enough of them to unscramble the signal. If we number the possible alternatives, we can call the index assigned to each alternative the key. For the analog scrambler to have a good cryptanalytical strength, there must be a very large number of keys, as is also the case for digital encryption.

It is possible to rate various scrambling techniques on the basis of four classes of attributes: security, quality, complexity and delay. By ***security*** we mean the cryptanalytical strength of the technique. How easy is it to defeat? As we will see, some techniques leave enough residual intelligibility in the scrambled speech that an adversary can understand the speech without a decoder. Other techniques have too few keys and are easily defeated by an adversary who can try the different possible decoders.

By ***quality*** we mean the quality of the unscrambled speech. Many of the techniques are not completely reversible. As a result, the unscrambled speech has one or more forms of distortion. Some typical forms of distortion are limited bandwidth and additive noise. Is the user willing to tolerate the amount of added distortion for the increased security that the scrambler provides? Only in the marketplace is this question answered.

Complexity directly translates into the amount of physical hardware needed to implement a given technique. Although we are interested in analog scrambling, we will be using digital signal processing techniques to implement the scrambler. The complexity can be measured as the amount of instructions per second which must be executed and the amount of memory (both RAM and ROM) required to implement the complete scrambler. In turn these quantities translate into two other important quantities, power consumption and cost. In many applications power consumption is a real consideration, especially if the scrambler is to be small and portable. It is difficult to imagine an application where cost is not an important consideration.

All of the techniques to be considered here introduce *delay* into the communications path. Large amounts of delay can cause real difficulties in holding a conversation. A general rule of thumb is that more than 250 ms delay make simultaneous two-way conversation difficult. In addition, large delays can lengthen echoes, making echo cancellation more difficult. The attribute of delay is often confused with quality. Certainly a large, onerous delay will not give the user the impression of a high quality system.

A basic question that should be answered at this point is why one should use analog scrambling rather than the combination of a digital speech coder, digital encryption and digital modulation. There are several good reasons. First, the nature of the message is usually such that its value has a relatively short lifetime. The niche for analog scrambling is to protect messages from casual eavesdropping and to provide protection from more sophisticated adversaries for a limited amount of time. It is usually assumed that digital encryption is more secure than analog scrambling and will withstand an adversary's attempts to break the system for a longer period of time. (We generally assume that adversaries know the method of encryption or scrambling and only lack the key used for the message under attack.) Second, analog scrambling should be less expensive than an equivalent digital system. (Otherwise the customer would be paying more to get less security.) Third, analog scrambling may provide a more robust system than the equivalent digital system. For example, HF radio systems used in the military are often used on very hostile channels which experience fading and low signal to noise ratios. Under such conditions, only a low bit rate digital signal can be supported with the usual modem technology or very expensive modems may be necessary. Analog scrambling techniques can give more robust performance on these channels, as well as being less expensive. Finally, we note that although our primary emphasis is on scrambling speech, the techniques used are not really speech specific and could be used to scramble any audio signal. By contrast, most low bit rate speech coders are speech specific and often have difficulty with encoding non-speech signals, such as music or environmental noises.

In the following section we give some examples of analog scrambling systems, starting with the simplest. In each case, we comment on how the scrambler measures up in each of the four attributes of security, complexity, quality and delay. Our purpose is to lead up to the two principal examples of analog scrambling using filter banks. For a more complete treatment of analog scrambling techniques, the reader is referred to [42].

In the third section we review the basic digital filter bank analysis used for frequency domain analog scrambling, including a derivation of how the channel phase equalizer works. Section 4 presents a brief summary.

2 FOUR TECHNIQUES FOR ANALOG SCRAMBLING

2.1 Simple Frequency Inversion

Figure 1 is a block diagram for a simple frequency inverter. Frequency inversion was first accomplished by analog means as shown in Figure 1a. The speech is first passed through a bandpass filter and has an absolute bandwidth of W. A template for the frequency response of this bandpass filter is shown in Figure 1b. Note that the speech has no energy content at dc. It is then modulated by frequency F, then demodulated by frequency $F - W$. The first modulation puts images of the speech at F and $-F$. The second modulation puts images at $-2F + W, -W, +W$, and $2F - W$. If we bandpass filter the result, the output signal has the inverse frequency spectrum of the original speech. Figure 1c shows how we would implement frequency inversion with a digital circuit. After bandpass filtering the analog signal, we sample it with a sampling frequency of $2W$ and multiply every other digital sample by -1. We then convert back to analog and bandpass filter once more.

The problems with this simple frequency inversion are two-fold. First, some humans can be trained to recognize inverted speech. In that case, your adversary doesn't even need a key at all. Second, there is only a single key. Your adversaries need only to recognize that the speech is inverted. They can re-invert the speech to recover the original speech. Clearly this technique fails on security. It is also clear that it is very low in complexity and has low delay. Its fourth attribute, quality, depends on the center frequency of the inverter. Normal telephone channels have a minimum bandwidth of 300 to 3200 Hz. Many have larger bandwidths, perhaps 200 to 3600 Hz. Consider what happens if the center frequency is 2000 Hz. The low frequency speech is translated to the high end. A 300 Hz tone becomes a 3700 Hz tone. If the bandwidth of the channel is 300 to 3200 Hz, the tone will be substantially attenuated in the channel. The same thing would happen to the low end of the speech spectrum. A listener would probably describe the resulting speech quality as weak, because speech tends to have more energy in the low frequencies, especially for vowel sounds.

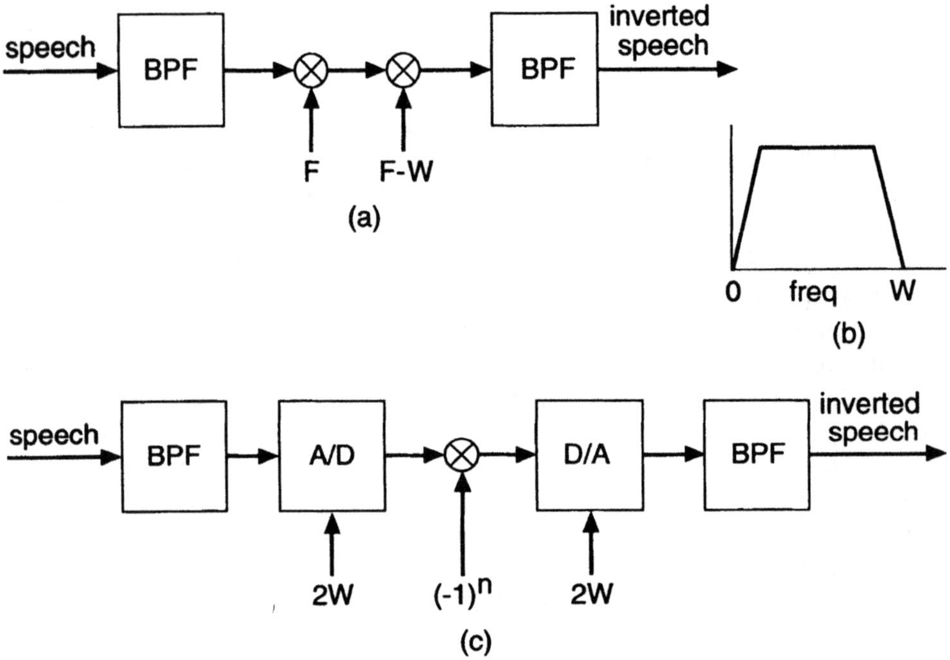

Figure 1 (a) Block diagram for an analog frequency inverter; (b) Frequency response of the bandpass filter in the analog frequency inverter; (c) Block diagram for a digital frequency inverter.

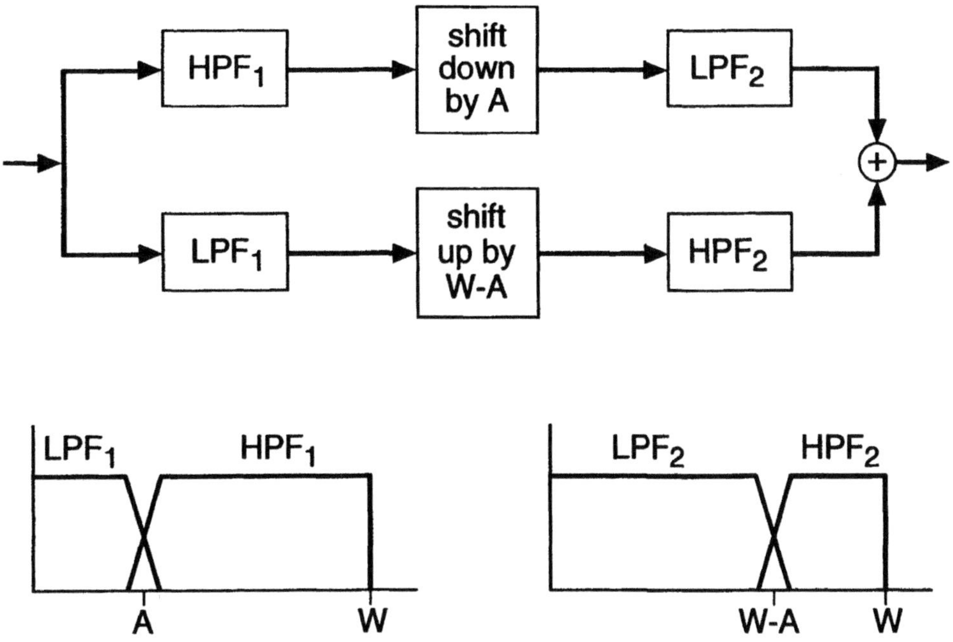

Figure 2 Block diagram for a two-band bandsplitter including frequency responses for the filters.

2.2 Band Splitting

Figure 2 gives an example of a simple bandsplitter. Once again the speech has a bandwidth of W. Two complementary filters are used to divide the speech into two bands. The lowpass filter can be of any bandwidth, A. The upper speech band then has bandwidth $W - A$. The two bands are obtained and then they are permuted. To recover the original speech, just use the complementary frequency, $W - A$, as defined by the key.

The above system is relatively easy to break because it has a very simple key. Your adversary doesn't know A, but can find it by looking at a histogram plot of the long-term spectrum. There will be a notch at A in the scrambled speech. Having found A, the means to unscramble the speech are assumed to be available. Even if there were no notch filter to give away A, it's still pretty easy for the adversary. Because speech can be frequency shifted and still retain intelligibility, the adversary can just make a tunable bandpass frequency filter. Listening to the output as the frequency is slowly changed, the adversary tunes in the unscrambled speech in much the same way as tuning for the carrier frequency of an AM radio station, stopping when the output speech quality is the highest.

Once again, the problem is that there are not enough keys and this scrambler receives low marks on security. Having an approximate key is sufficient for your adversary. Inserting a notch between the two halves of the spectrum means that the unscrambled speech will have less bandwidth than the channel. This scrambler does have relatively low complexity and delay.

Figure 3 gives an example of a multiple bandsplitter with 3 bands. In principle, the 2 band splitter can be extended to many bands. This increases the number of keys. However, this is still an easy scrambler to defeat. Your adversary needs only an equal number of tunable bandpass filters. He finds the lowest band first, which will give the speech intelligibility. Then the next lowest band, etc. can be found. So, the problem is that even though there are more dimensions to the effort, there are still not enough keys. To combat such an attack digital filter banks with more than 100 different bands can be used. The number of keys is greatly increased and the intelligibility of an individual band is greatly decreased. This makes the coder much more difficult for an adversary to attack, but relative to digital encryption techniques, this is still a relatively easy code to decipher.

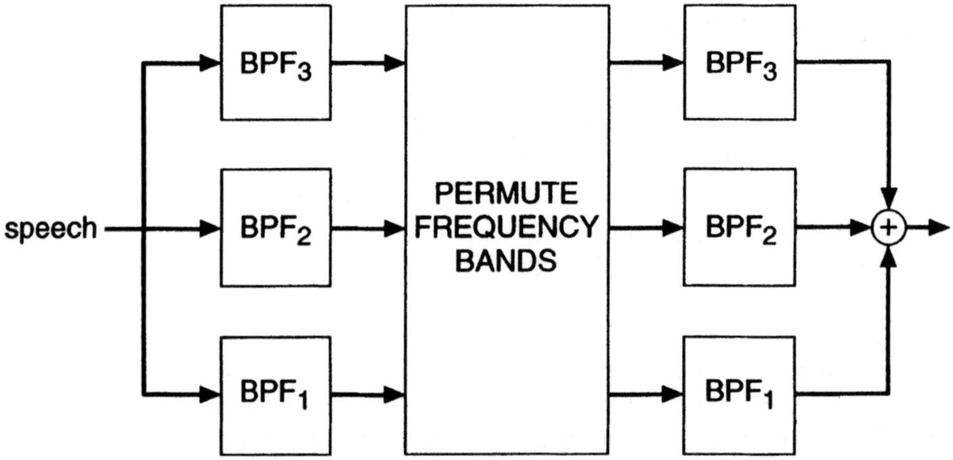

Figure 3 Block diagram for a multiband scrambler with 3 bands.

Another way to extend bandsplitters is to introduce the concept of synchronization. The splitting frequencies could be changed periodically. In the case of multi-bandsplitters, the permutation of the bands could be changed. These are called rolling code scramblers. They introduce the concept of synchronization between the encoder and the decoder.

In evaluating this scrambler we see a range of possible levels of security. Each of the proposed extensions to increase security increases delay and complexity. If the bands are kept separated in the channel to prevent aliasing, then more bands means less bandwidth for the unscrambled speech. This becomes a severe limitation if a system with 100 or more bands is to be designed. Digital filter banks with overlapping bands can be designed such that in the synthesis filter bank, the aliasing terms are cancelled. In Section 2 we discuss how to adapt these techniques for analog channels. For now we will continue the discussion of large multiband scramblers and will assume that such filter banks are available.

Figure 4 is a block diagram for a multiband scrambler. It is a generalization of Figure 3. At the transmitting or encryption end, the digitized speech is first filtered by a multiband filter bank. In order to have many keys, a high resolution filter bank is needed. This could be accomplished in several ways. The filter bank could be composed of smaller bandsplitting filters, such as

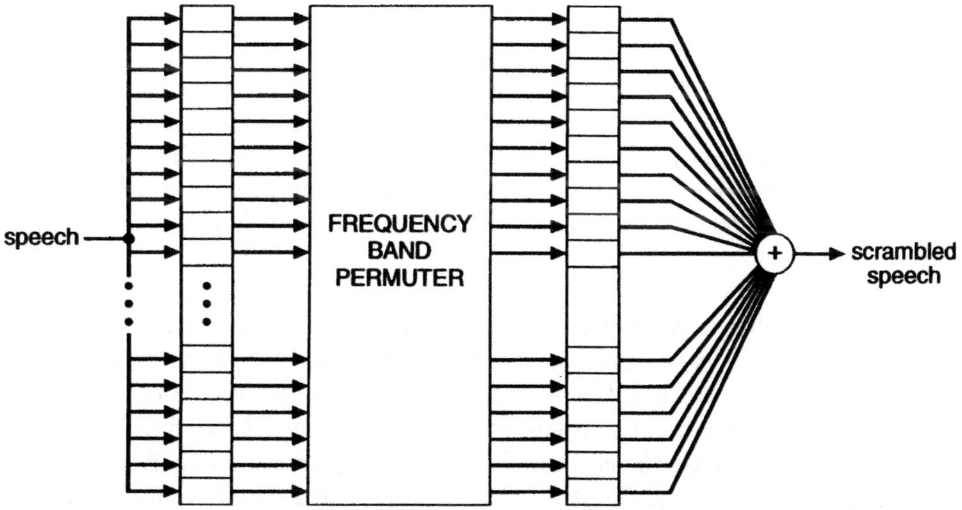

Figure 4 Block diagram for a multiband scrambler with many bands.

wavelets or quadrature mirror filter banks. A tree structure of these filter banks would result in the fine grained frequency resolution which is needed to have a large number of keys as in [211]. Or the filter bank could be transform based, such as the polyphase QMFs of Rothweiler [345] or the Time Domain Alias Cancellation filters of Princen and Bradley [324]. (Truncated ideal filters were used for an FFT-based scrambler by Lee, Chou and Chang [241] which did not require synchronization or phase equalization.)

It is important to note that these frequency bands are critically sampled. This means that for each N samples that are input, there will be N frequency domain samples which result. Next, the frequency band samples are permuted. Because of the assumption of critical sampling, the N samples can each be treated independently. Therefore a total of $N!$ permutations are possible.

In practice, the minimum telephone bandwidth is 300–3200 Hz. It is often the case that the amount is larger than this minimum, but we can't count on that. Consequently, all of the information must be transmitted over this bandwidth. It is a good precaution to filter the data to this bandwidth before beginning the scrambling process so that we have the same amount of information to scramble that we will be able to transmit. The permutation should be limited to those bands which contain information. This is another reason to bandpass filter the speech signal before performing the analysis filter bank.

For example, suppose that we wish to transmit information in the band 250–3250 Hz and we are using a filter bank with a resolution of 50 Hz. If we filter the speech with a 300–3200 Hz bandpass filter, there may still be some information in the 250–300 Hz and the 3200–3250 Hz bands. However, we should be able to ignore any information below 250 Hz or above 3250 Hz. In this example, N is equal to $4000/50 = 80$, but we are restricted to transmit the information in just 60 bands. Thus, the actual number of possible permutations is 60!, not 80!.

There is a further way to increase the number of permutations. When the analysis filter bank is critically sampled, half the bands are frequency inverted at baseband. For example, in a simple QMF filter, when brought down to baseband, the upper band samples are frequency inverted. This property can be exploited to increase the cryptanalytical strength by randomly frequency inverting channels. Ideally, about half the channels should be frequency inverted to give the maximum key strength. All possibilities, including no inversions, should be possible. As discussed earlier, the actual number of permutations which are truly different enough is orders of magnitude smaller. Once the permutation is completed, we can then synthesize the channel signal using a synthesis filter bank. At the receiver, the analysis filter bank must be the inverse of the synthesis filter of the transmitter. For the case of an ideal channel with synchronous sampling, exactly the same signal values will be obtained at the receiver. The inverse permutation is performed and then the speech can be resynthesized using the synthesis filter that is the inverse of the analysis filter at the transmitter.

In order to have a large number of possible keys, a large number of bands are necessary. A typical prototype filter might have 6 to 10 times as many taps as the number of bands. If the desired number of bands to permute is about 100, then the number of total bands would be around 128 and the total delay due to the filter banks would be between 6*4*(128/2) and 10*4*(128/2) samples, or 192 to 320 ms.

To summarize the most salient points from this section, the multiband scrambler is based on two sets of filter banks, both of which should have either perfect or near-perfect reconstruction. The analysis filter bank of the transmitter and the synthesis filter bank of the receiver are one pair. The second pair is in reverse order and consists of the synthesis filter bank of the transmitter and the analysis filter bank of the receiver. These digital filters provide great numbers of possible scrambler designs. This flexibility comes with a high price, however. The filters only work if the channel adds little or no distortion. While the filters can overlap, we need to use some channel bandwidth for sampling clock recovery and still more to achieve synchronization and sample-to-sample integrity. The most important channel problems (other than noise) are synchronization of the sampling clocks of the transmitter and receiver and compensating for nonlinear phase in the channel.

2.3 Time Segment Permutation

Figure 5 is a block diagram illustrating how time segment permutation (TSP) can be performed. TSP involves buffering in a large frame of speech, dividing the frame into segments and then permuting the segments. If there are N segments, then there are $N!$ possible permutations. However, in studies of this technique, it was found that the residual intelligibility still remained high when the task was identifying the spoken digits 0–9 [212]. Frame sizes of more than 256 msec were necessary to achieve low intelligibility, making a communication delay of twice this amount.

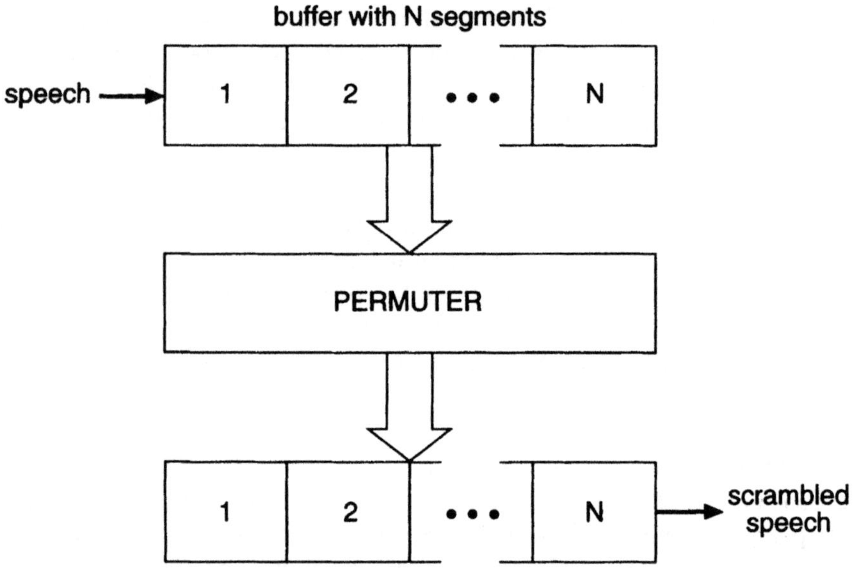

Figure 5 Block diagram for time segment permutation.

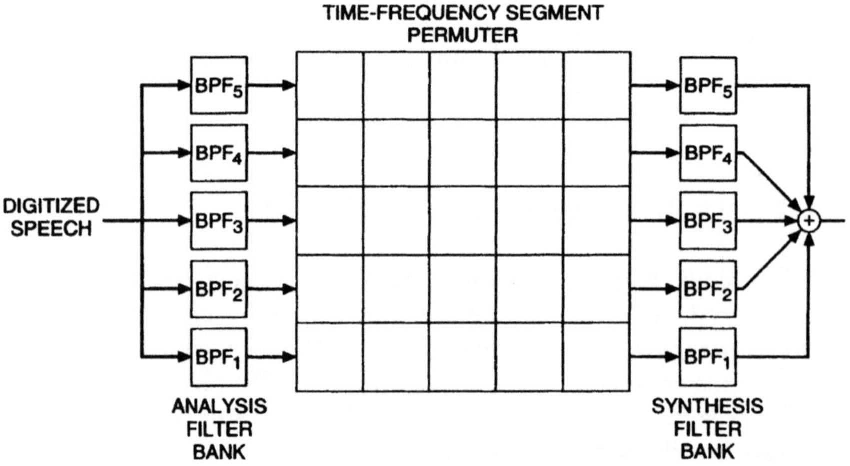

Figure 6 Block diagram for time-frequency segment permutation.

In practice, this technique adds noise to the decrypted speech. This is because the communication channel usually has both non-linear phase and bandpass characteristics. Thus, the digital samples at the decoder are not the same as those at the encoder. At the point where two permuted segments join, the bandwidth of the scrambled signal is greatest, typically equal to the Nyquist frequency (half the sampling rate) which is higher than the bandwidth of the channel. These samples will be bandpass filtered by the channel. At the decoder when the decoded speech is reassembled, the perceptible errors will be concentrated at the ends of each time segment. This will introduce a harmonically related noise structure which extends throughout the entire speech spectrum.

This technique introduces the concept that even with the proper key, we may not be able to recover the quality of the original speech. The nature of the technique is such that information is lost, resulting in lower quality output speech. The shortest time segment which could be permuted is a single sample. However, if the channel is bandlimited and has non-linear phase, then these characteristics, which introduce the noise, will result in extremely noisy, and possibly unintelligible, output speech. However, scrambling individual samples also results in the greatest amount of encryption strength.

In evaluating this coder, we see that the complexity is low in terms of the number of instructions to be executed, but high in memory (RAM). The output quality has a noise associated with it. In addition, there is substantial delay. The security of this scrambler is still relatively low unless there is a very large buffer with many segments.

2.4 Time-Frequency Segment Permutation

Time-frequency segment permutation (TFSP) was first introduced by [211]. We begin with the notion of a two-dimensional permutation. Suppose that the speech signal has been broken into both time and frequency segments, as shown in Figure 6. We will call these segments blocks, rather than segments to indicate that they are limited in both time and frequency. We can choose any permutation of the blocks. For example, suppose that we had 5 frequency bands and 5 time segments. Then there are 25! permutations.

In general, this two-dimensional permutation has been found to be more difficult to understand for casual eavesdropping. It is not clear that it is of any greater cryptanalytical strength than a multiband scrambler having the same number of permutations. In the case of the multiband scrambler the cryptanalytical strength was dependent on the number of frequency bands. For TFSP the cryptanalytical strength depends on the product of the number of frequency bands and the number of time segments. In [106] using the synchronization methods described above it was possible to unscramble permutations based on individual samples. This not only greatly increased the number of possible keys, but also made the problem more difficult for an adversary to attack because of the shortness of the segments.

Since the security is dependent upon the product of the number of bands and time segments, the number of frequency bands can be reduced compared to a multiband scrambler having the same security. Generally, this means the complexity of the arithmetic in the filter bank can be reduced. For example, the system described in [106] had 125! possible keys. For a multiband scrambler to have this many keys, there would need to be a total of 160 to 170 bands in the full bandwidth filter bank. Some 35 to 45 would not be used because they covered the portion of the spectrum that is normally bandpass filtered by the telephone system. If the ratio of the number of taps to the number of bands is in the typical range of 6 to 10, then the filters would have from 1000 to 1700 taps. By comparison, the filter size in [106] was only 48. While the actual reduction in computation is not as great as these two ratios might suggest—the TFSP scrambler must produce 25 samples in each band—there is still a considerable savings in memory storage for the filter taps and in the delay of the filter bank. Smaller filter sizes also help make the implementation details easier.

However, if there are $N!$ possible permutations, the number of significantly different keys is far fewer [210]. A periodic sequence of different permutations could be used for each segment, but this may not result in greater encryption strength. Each decrypted permutation will increase the intelligibility of the signal for your adversary.

As in the case of TSP, TFSP has the problem of non-linear phase in the channel. However, digital signal processing techniques provide a superior answer to this problem. The use of digital filter banks provides us with a means to utilize more of the channel bandwidth which in turn allows us to obtain better performance because the unscrambled speech will have more bandwidth.

Note that TFSP requires two types of synchronization as compared to the large multiband scramblers. Multiband scramblers with critical sampling require sample synchronization and phase equalization in order to make their multiband filter banks perform. TFSP requires this type of synchronization and also requires synchronization for the time segment permutation. In practice this can be handled by using in-band signaling at the end of a permutation frame. A match filter looks for the synchronization symbol which denotes the beginning of a new frame.

For the reasons given above TFSP has some computational and delay advantages when compared with large size multiband scramblers having similar security. There are two main differences between the large size multiband scramblers and TFSP which affect the speech quality. With narrower frequency bands, the large multiband scrambler can occupy a slightly greater percentage of the channel bandwidth, leading to wider bandwidth unscrambled speech. If the large size multiband scrambler does not use a rolling code, then it will have no time segments. This should reduce the noise in the unscrambled speech slightly when compared to TFSP, depending on the effectiveness of the phase equalizer for a particular channel.

TFSP and the large multiband scrambler both have relatively high delay and complexity in order to achieve high degrees of security and speech quality. The complexity is inherent in the critically sampled filter bank structure that is needed to conserve bandwidth, which is related to speech quality. This complexity is still modest by today's standards of signal processing capability. For example, the TFSP system described in [106] required less than 10 fixed point MIPs for encoder or decoder using the DSP chips of a decade ago. Total RAM usage was less than 1K words. This combination is less than the amount used for most speech coders or most modems today. The delay is a byproduct of both the filter banks and the buffer sizes required to give a high degree of security. In [106] the factors contributing to the delay were filter banks (16 ms), buffers (25 ms), the phase equalizer (8 ms) and a voice activity detector (10 ms). By contrast, a speech coder with 20 ms frame size would have a minimum delay of at least 60 ms in the same type of application. The two delays are really quite equivalent. Since the devices must typically operate in half-duplex mode anyway, this amount of delay is not a serious difficulty.

3 FILTER BANK ANALYSIS FOR ANALOG SCRAMBLERS

In the remainder of this chapter, the filtering requirements for some of these scrambling techniques will be discussed. Of the scrambling techniques described above, this discussion will be limited to those which have a meaningfully large number of keys and involve filter bank analysis. This limits the discussion to two basic types of scramblers: multiband scramblers and time-frequency segment permuters.

An important aspect of filter bank analysis is the assumption that reconstruction is either perfect or near-perfect in the absence of noise or other distortion. As alluded to above, telephone channels introduce both noise and distortion. Consequently, it is relevant to discuss the problems and potential solutions that are required to make such real-world systems work.

In both of the scramblers to be discussed in detail, we use two pairs of filter banks. The two pairs of filter banks are the analysis filter bank of the transmitter with the synthesis filter bank in the receiver and the synthesis filter bank in the transmitter with the analysis filter bank in the receiver. The first pair of filter banks is in the usual order, as in Figure 7a. The second pair of filter banks is in the reverse of the usual order, as in the example of Figure 7b. In principle,

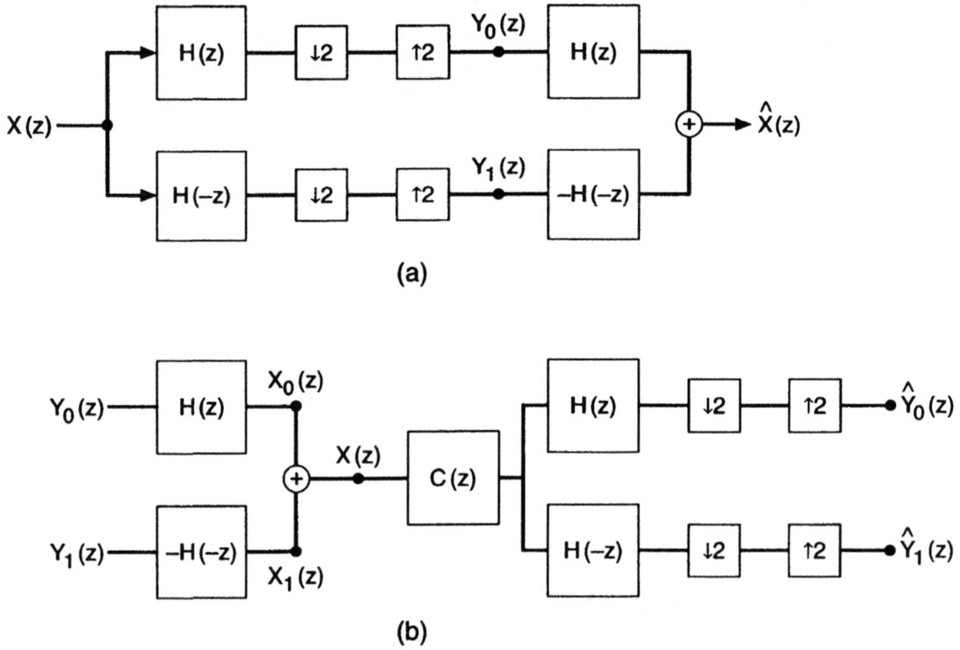

Figure 7 (a) Block diagram for analysis/synthesis filter bank pair used in time-frequency segment permutation; (b) Block diagram for synthesis/analysis filter bank pair used in time-frequency segment permutation.

this means that any set of samples we input to the synthesis filter bank in the transmitter should be recovered at the analysis filter bank in the receiver. This is the reason that critical band sampling was desired in these filter banks. It allows us the freedom of having completely independent samples.

These two filter bank pairs need not have any particular relation other than that the analysis filter bank outputs the same number of samples which the synthesis filter bank requires as input. A wavelet (i.e., octave band) filter bank could be used for the analysis in the receiver and a polyphase QMF could be used for the synthesis of the receiver. The wavelet filter bank would need to be used for the synthesis in the receiver and the polyphase QMF would need to be used for the analysis. However, we are unaware of any system which has actually mixed two different types of filter banks in this way.

Let us start with simple QMF filter bank analysis to demonstrate both the strength and weakness of using critically sampled digital filter banks for analog channels. First, lets review the conventional QMF analysis as first derived in [140]. Consider Figure 7a. The input signal is given by $x(n)$ with z-transform $X(z)$. The QMF filter bank H has the following properties:

$$h(n) = h(N-1-n) \text{ for } n = 0, 1, ..., N-1, \tag{12.1}$$

and $h(n) = 0$ for all other values of n and N is even. The transfer function of the lowpass filter is $H(z)$, while the highpass filter transfer function is given by $H(-z)$.

Therefore, the z-transform of the lower band signal is given by $H(z)X(z)$ and the upper band signal is given by $H(-z)X(z)$. After decimation and interpolation by zeros, any signal $G(z)$ is given by $G(z) + G(-z)$ where the second term is due to aliasing. In this instance we will define the upper band signal as $W_u(z) = H(-z)X(z) + H(z)X(-z)$ and the lower band signal as $W_l(z) = H(z)X(z) + H(-z)X(-z)$.

The synthesis filter bank is composed of the same lowpass filter as before, $H(z)$ and the highpass filter is now given by $-H(-z)$. Filtering $W_l(z)$ and $W_u(z)$ with these filters yields

$$Y_l(z) = H^2(z)X(z) + H(z)H(-z)X(-z) \tag{12.2}$$

and

$$Y_u(z) = -H^2(-z)X(z) - H(z)H(-z)X(-z)\,. \tag{12.3}$$

Finally, if we sum these to get $Y(z)$, we find that the aliasing terms have cancelled out and we are left with

$$Y(z) = [H^2(z) - H^2(-z)]X(z) \; . \tag{12.4}$$

This leads to the familiar constraint of QMF filters that

$$|H^2(z) - H^2(-z)| = 1. \tag{12.5}$$

We further note that the delay of $h(n)$ is $(N-1)/2$, so that the delay of the complete analysis/synthesis filter bank system is $N-1$. We are about to analyze the inner filter bank system used for scrambling where the analysis filter bank follows the synthesis filter bank, as shown in Figure 7b. In order to recover the same decimated samples that are input to the synthesis filter bank, we will insert a one sample delay in the channel. In this way the decimation phase comes out correctly.

Let $W_l(z)$ and $W_u(z)$ be the transforms of the lower and upper band input signals. Note that these signals have already been decimated and interpolated with zeros. Since all odd samples are 0, $W(z) = W(-z)$ for both the lower and upper bands. This fact will be used later.

After filtering and summation we get

$$T(z) = H(z)W_l(z) - H(-z)W_u(z). \tag{12.6}$$

$T(z)$ is first passed through the channel, $C(z)$, and then filtered, critically sampled and interpolated in each band. This yields $V_l(z)$ and $V_u(z)$. Our analysis for $V_l(z)$ yields the following results:

$$V_l(z) = H(z)C(z)T(z) - H(-z)C(-z)T(-z) \; . \tag{12.7}$$

If $C(z)$ is ideal, meaning $C(z) = z^{-1}$, then we get

$$V_l(z) = z^{-1}[H^2(z)W_l(z) - H^2(-z)W_l(-z) - H(z)H(-z)[W_u(z) - W_u(-z)]] \; . \tag{12.8}$$

Since $W_u(z) = W_u(-z)$, the $H(z)H(-z)$ terms cancel. Also, $W_l(z) = W_l(-z)$, which reduces Equation (12.8) to

$$V_l(z) = z^{-1}[H^2(z) - H^2(-z)]W_l(z) \; . \tag{12.9}$$

A similar derivation yields

$$V_u(z) = z^{-1}[H^2(z) - H^2(-z)]W_u(z) \; . \tag{12.10}$$

Next lets examine what can happen with a real channel, as opposed to an ideal one. First, the channel will not have linear phase. The channel is presumed to be analog, at least in some portions. Nonlinear phase will be contributed by the anti-aliasing filters associated with the D/A and A/D converters at the transmitter and receiver, respectively. This means that the original samples will not be recovered in any of the bands.

If the channel, $C(z)$, is not ideal, but is an odd function of z, i.e.,

$$C(z) = -C(-z) \tag{12.11}$$

then the aliasing terms involving $W_l(-z)$ and $W_u(-z)$ will cancel. This property is equivalent to

$$C(z) = z^{-k}E(z) \tag{12.12}$$

where k is an odd integer and $E(z)$ is an even function of z. This combination is just a description of a linear phase channel with a delay that is an odd integer. After cancelling out the the aliasing terms we are left with

$$V_l(z) = z^{-k}E(z)[H^2(z) - H^2(-z)]W_l(z)\ . \tag{12.13}$$

Apart from the delay term, we are also left with the effects of $E(z)$. For a good channel, this will mainly consist of ripple.

In practice, one way to make the channel linear phase was first proposed in [106]. First measure the impulse response of the channel. This can be done in a number of ways during the setup time, during blanked out intervals or during otherwise silent intervals. Once the impulse response is obtained, use the time reversed version as a pre-filter to the analysis filter in the decoder. This will have the effect of linearizing the phase of the channel. This will also have the effect of doubling the ripple in the channel, an undesirable effect. However, it does eliminate the "aliasing" terms in the two sub-channels.

It is worth noting here that this technique might fail if we were transmitting something other than frequency bands. For example, if the input to the permuter in the scrambler were the time samples of a narrow band version of the speech itself, the changes produced by $E(z)$ would result in changing the values of the samples in such a way that in the reassembled speech it would be perceived as noise. The point is that we cannot use any arbitrarily derived set of samples which are related to the original speech waveform.

The second point about a real channel is that it will have some noise. The source of this noise could come from many places. In particular, it might be quantization noise from PCM quantization of the analog signal. This noise will be added to the scrambled signal. At the receiver, the noise will be input to the analysis filter bank and its narrowband components will then be permuted using the inverse permutation of the scrambler. In the unscrambled signal, this noise will appear in all of the bands. If the noise of the channel is not too great when the speech is not scrambled, then it will not have a large effect on the quality of the unscrambled speech either. The unscrambling process will tend to produce white noise, even if the noise was colored when it was in the channel.

Third, we may have slightly different sampling frequencies at the transmitter and receiver. This is a serious problem. If left unattended its consequence is to severely damage the quality of the speech. What would happen in this instance is that the bands of one filter bank would be slightly wider than those of the other. Suppose that the receiver's bands were wider. The first band in the receiver filter bank would actually include a little of the information from the second band. The second band would have some from the third. As we go up in frequency, each successive band will have lost a larger portion of its own information to the next lower band and will have a proportionally larger amount from the next higher band. When the inverse permutation is applied, the information which is in the wrong bands will be permuted into the wrong synthesis band. This will cause a great amount of aliasing noise throughout the spectrum.

This problem was attacked by [106] by transmitting two side tones, one on either side of the scrambled speech bandwidth. Both frequencies were derived from the sampling clock of the encoder. When modulated, their difference-frequency could be used to control the sampling clock for the decoder. In this way, the sampling clock of the decoder could be linked to that of the encoder. The disadvantage of this method is that an analog guard band is required around both of these tones, thereby reducing the speech bandwidth. Another way to deal with this problem is to allow smaller tolerances on the sampling clocks. If the relative error between the two sampling clocks is 0.1%, this means that at 3 kHz, there will be a shift of 3 Hz. This will probably result in audible aliasing effects, but they will be small. Combined with periodic synchronization, there should not be a significant problem.

This section has attempted to show in detail both the strength and the weakness of using critically sampled digital filter banks for analog scrambling. The strength is that they can be used to avoid increasing the bandwidth of the signal. For an ideal channel they can provide great versatility to the designer. Their weakness is that all of this versatility is dependent on having an ideal channel. For real (non-ideal) channels, measures must be taken to correct for this as much as possible. The solution of making the channel linear phase helps cancel the aliasing terms at the expense of increasing distortion due to ripple in the channel—the ripple is doubled. Fortunately, many common telephone channels are fairly flat and the ripple is not a serious problem.

Note that a multiband system with synchronous sampling, linear phase and a flat gain response in the passband is, in effect, a truly equalized channel. By comparison echo cancelers and modem equalizers work with a model of the channel. Rather than trying to invert the effects of the channel, they attempt to compensate for them by filtering a signal with them and then comparing it to the channel output. For TFSP the channel output must be modified to match the output of the synthesis filter at the transmitter. In other words, for all the elegant mathematics of the filter banks to work we need an inverse filter for the channel response.

4 CONCLUSIONS

In this chapter we have reviewed several analog scrambling methods and studied how digital filter banks could be applied to improve their performance. Using critically sampled multiband filter banks, it is possible to produce scrambling systems that have a much larger number of possible keys, thus giving them significant cryptanalytical strength and retaining good speech quality. To achieve this level of performance, the user must tolerate a significant amount of delay and a higher amount of complexity. The two most promising methods are multiband scramblers with a large number of frequency bands and time-frequency segment permutation. Of these, TFSP requires a greater degree of synchronization, but otherwise has slightly less complexity and delay while large size multiband scramblers may have slightly better speech quality. These systems can presently be implemented using commercially available fixed point DSP chips and are generally less complex than equivalent digital encryption systems comprised of a speech coder, a digital encryption algorithm and a modem.

13

MULTITONE MODULATION

Irving Kalet

RAKEFET Consulting Co., Haifa, Israel

1 INTRODUCTION

In this chapter we discuss the application of DFT basis representation—a special case of DFT filter banks—to the problem of multitone modulation for data transmission. The concept of multitone modulation (MTM) has been known since the middle 1960's [77, 349], but has attracted much interest since the late 1980's [189, 219, 54, 144, 346, 83]. MTM is also referred to as Orthogonal Frequency Division Multiplexing (OFDM), Multi-Carrier Modulation (MCM), and Discrete Multitone (DMT). It has been proposed as a standard for the High-Speed Digital Subscriber Line (HDSL)[80] and Asymmetric Digital Subscriber Line (ADSL)[81] transmission over twisted-pair cable at data rates of 1.544 Mbps and above (as much as six or seven megabits/sec.). It has also been used as a modulation technique for high-speed data transmission [398] over the telephone switched network. In recent years MTM has also attracted much interest in wireless transmission for Direct Audio Broadcasting (DAB) [451], and for Personal Communication Services (PCS) [82]. We will concentrate on the use of MTM for data transmission over physical wire lines, e.g., the twisted-pair cables, which introduce amplitude and phase distortion in the signal. HDSL and ADSL are transmitted over these lines and, at the time of this writing, appear to be the major utilization of MTM. The main purpose of this chapter is to show that uncoded multitone transmission allows us to approach the Shannon capacity [365] of any linear time-invariant channel to within 8 to 9 dB, whether the channel be an ideal perfectly bandlimited channel or not. The fact that this is possible for any channel with general frequency transfer function, $H(f)$, is what makes MTM so important. Here we will use the cyclic frequency variable f (in Hertz) instead of the radian frequency variable Ω. As far as the author knows only the Decision Feedback Equalization

(DFE) [323, 350] technique achieves a similar result. In fact, the competition between these techniques for use in HDSL and ADSL has increased the interest in MTM [454]. After using MTM or DFE as the basic communications technique, capacity is more closely approached through the use of coding.

In order to understand the reasoning behind the use of multitone transmission one must have some background in information theory. We will now present a short summary of some of the important results of classical Shannon Information Theory [365]. This will then be followed by a discussion of the quadrature-amplitude modulation (QAM) method, which is a basic building block of MTM. The description of multitone modulation follows with some discussion of one implementation for use in HDSL and ADSL type systems. A performance comparison of MTM and DFE is also made, showing the similarity of the results of both concepts.

Most of the DMT applications in the literature have utilized the Discrete Fourier Transform (DFT) basis functions as orthogonal carriers. A DMT transceiver implementation based on the DFT basis is presented in Section 6. More recently however, subband transforms with apparently better frequency localized basis functions than the DFT basis, have been suggested as an orthogonal set for the DMT transceivers [408, 50].

2 INFORMATION THEORY

Information theory is a field based on Claude Shannon's classical work originally published in 1948 [365, 154]. In this work Shannon showed that given a transmission source with power, P, and any linear time-invariant channel with transfer function, $H(f)$, with additive Gaussian noise of power spectral density, $P_n(f)$, (see Figure 1) it is possible to send a finite number of bits of information per second, R_b bps, over this channel with a probability of making an error at the receiver approaching zero, or as close to zero as desired, as long as R_b is less than a certain value called the capacity, C, of the channel, i.e., as long as

$$C > R_b. \tag{13.1}$$

Furthermore if the rate R_b is greater than C, the probability of error becomes very large.

The above is a basic result in communications. It is a phenomenal result because it means that even with a very weak power source, it is still possible

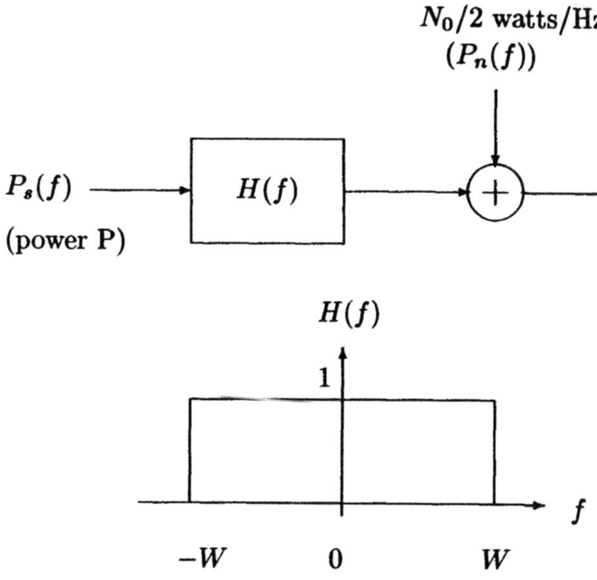

Figure 1 Channel model.

to transmit information at a reasonable bit rate with as low a probability of error as desired, as long as the bit rate R_b is less than C.

The most famous analytic result of [365] is for a perfectly bandlimited channel with bandwidth W Hz (see Figure 1) in which P/N_0W represents the signal-to-noise ratio in the given bandwidth, W Hz. In this case the capacity C is given by

$$C = W \log_2 \left[1 + \frac{P}{N_0 W}\right] \text{ bps.} \tag{13.2}$$

It can very easily be shown that C increases with increased bandwidth and that in the limit as W approaches infinity, C approaches $(1/\ln 2) \cdot (P/N_0)$ bps.

For example, for an approximate model of the telephone channel as it existed at the time of the publication of Shannon's original work, the received signal-to-noise ratio was required to be approximately 30 dB, and the bandwidth was equal to about 3 kHz. Using Equation (13.2) this gives a capacity of about 30 kbps. At that time the transmitted bit rate over this channel was on the order of hundreds of bits per second, way below the capacity. In the last forty-five years the bit rate has risen to values of 19.2 kbps and above with the use of advanced coding and modulation techniques [149, 150], developed over this period.

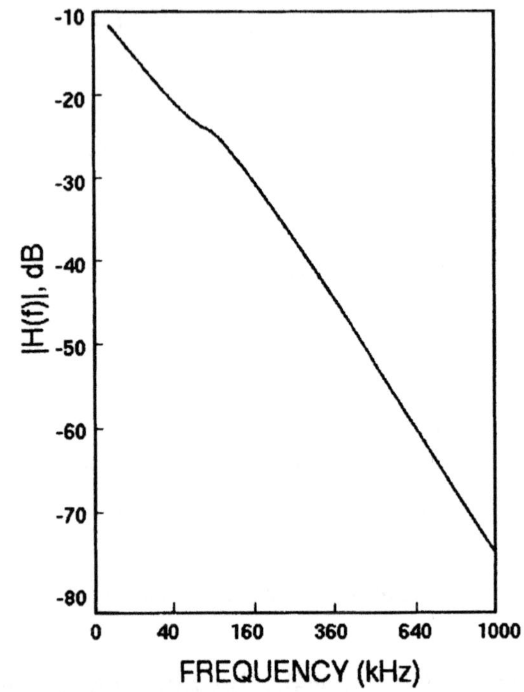

Figure 2 Attenuation characteristic $|H(f)|$ in dB for a typical twisted-pair (24 gauge, 12 Kft)—no bridge taps.

The Shannon result, which is of particular importance in the concept of MTM, is related to a general linear time-invariant channel with any frequency response, $H(f)$. For example, the $|H(f)|$ of a twisted-pair channel used for HDSL and ADSL is shown in Figure 2. In this case Shannon showed that the capacity is given by

$$C = \int_0^\infty df \log_2 \left[1 + \frac{|H(f)|^2 P_s(f)}{N_0/2}\right] \text{bps}, \tag{13.3}$$

where $P_s(f)$ is the power spectral density of the transmitted signal, $s(t)$, and the total power, P, is given by

$$P = 2\int_0^\infty P_s(f)df. \tag{13.4}$$

The capacity for a non-white Gaussian noise signal with general noise spectral density, $P_n(f)$ watts/Hz is given by the equation above with $N_0/2$ replaced by $P_n(f)$. However, it is possible to maximize the capacity, C, for a given value of P, by optimizing $P_s(f)$. The solution is the well-known "water-pouring" solution, [365, 154] shown below. The optimum $P_s(f)_{opt}$ is given by

$$P_s(f)_{opt} = \begin{cases} \lambda - \frac{N_0/2}{|H(f)|^2} & \text{for f} \in \text{A} \\ 0 & \text{otherwise,} \end{cases} \tag{13.5}$$

where λ is a Lagrangian multiplier. (For a general $P_n(f)$, replace $N_0/2$ by $P_n(f)$ in the equation above.) The spectral region, A, contains all frequencies for which $P_s(f)_{opt} \geq 0$.

The maximum capacity, C_{max}, is found by replacing Equation (13.5) in Equation (13.2) and is given by

$$C_{max} = \int_{f \in A} df \log_2 \frac{\lambda |H(f)|^2}{N_0/2} \text{bps.} \tag{13.6}$$

The reason for the name "water-pouring" or "water-filling" as the optimum solution may be seen in Figure 3. The "cup" formed by $\frac{N_0}{|H(f)|^2}$ is "filled-up" until the level λ, i.e., until the area under the λ level equals, P, the total transmitted power. The essence of the "water-pouring" solution is to utilize as much bandwidth as possible, as in an ideal bandlimited channel, but to shape the power spectral density such that $P_s(f)$ is strong where $H(f)$ is not attenuated and weak where $H(f)$ is attenuated.

The result above has been known since Shannon's work was originally published. Price, in his classic work [323] on Decision Feedback Equalization (DFE), showed that a system using DFE could approach to within 8 to 9 dB (for an error probability of 10^{-5}) of the capacity for any white Gaussian noise channel with transfer function $H(f)$. In other words the DFE technique requires 8 to 9 dB more power than that predicted by Shannon to achieve a bit rate R_b equal to the value of C (calculated with the original power, P). This result is of much interest since it means that with proper coding (to improve performance by a few dB) it is possible to approach channel capacity for any channel characteristic, $H(f)$.

In the last few years it has also been shown, that multitone modulation essentially achieves the same result as the DFE [219, 454]. Therefore, theoretically

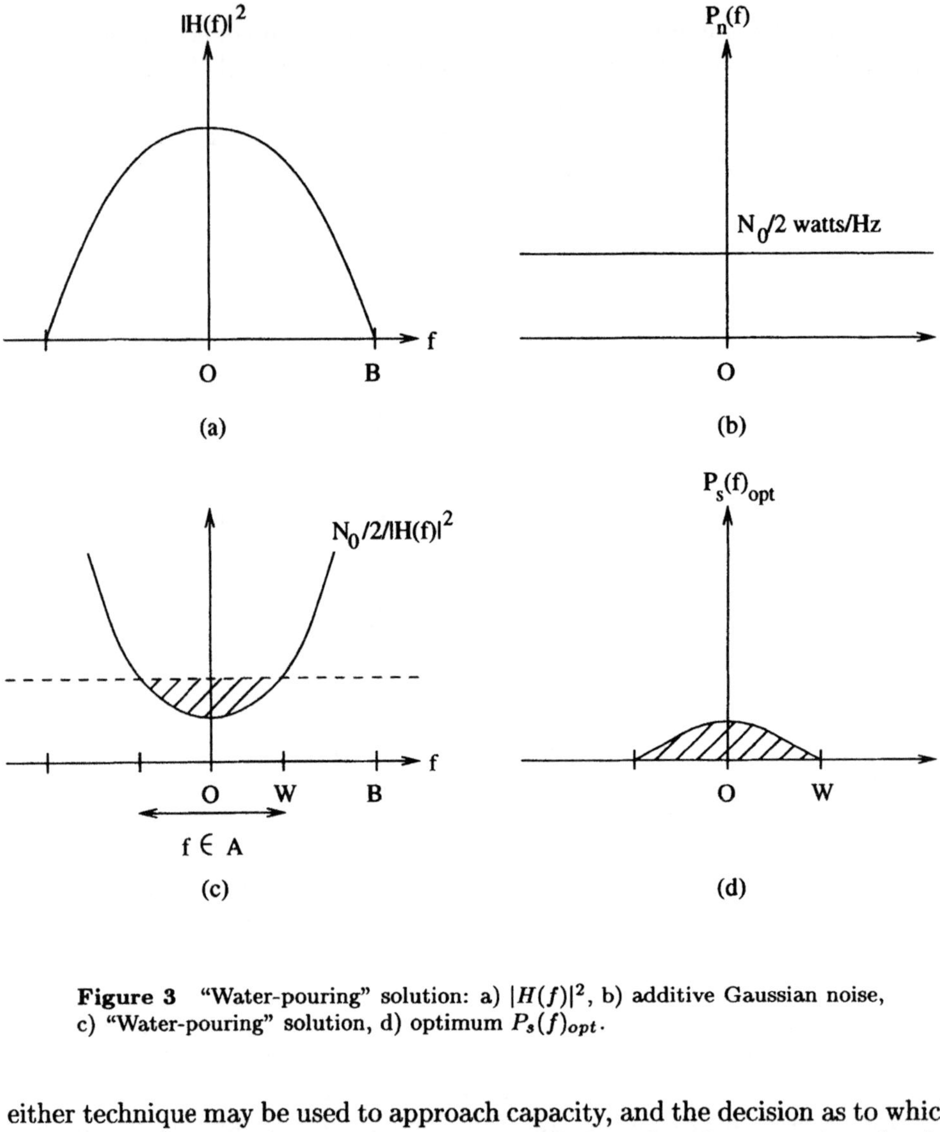

Figure 3 "Water-pouring" solution: a) $|H(f)|^2$, b) additive Gaussian noise, c) "Water-pouring" solution, d) optimum $P_s(f)_{opt}$.

either technique may be used to approach capacity, and the decision as to which to use in the real-world environment is actually based on issues such as implementation and performance in the face of impulse noise and other non-Gaussian interferences.

We will now discuss multitone modulation, including a description of quadrature amplitude modulation (QAM), which is a basic building block of the multitone concept.

The basic philosophy behind MTM is to "imitate" Shannon's idea of water-pouring, i.e., put a lot of power where the channel attenuation is small, but still use as much bandwidth as possible. This is accomplished by spreading out the transmitted power over a wide bandwidth but with decreased strength where the attenuation is strong. Many QAM signals of different powers and bits per symbol are used in order to implement this concept.

3 THE QAM MULTITONE MODULATION

The multitone QAM system under consideration is shown in Figure 4. $H(f)$ represents a general channel transfer function, which will be approximated by a staircase frequency transfer characteristic, $H_i(f)$, as shown in Figure 5. The total transmitted power, P, is divided up into N QAM signals of different powers and bits per symbol, each fitting into one frequency band, W_i. Although each QAM signal may in general be of any bandwidth W_i Hz, we will assume, without any loss of generality, that they are of equal bandwidth, Δf Hz. (In the limit as $W_i \to 0$, $H_i(f)$ approaches the continuous function $H(f)$.) The transmitted signal consists of N QAM signal tones, each with a rectangular Nyquist spectrum ([167], Chap. 4) of bandwidth equal to W_i Hz. As $W_i \to 0$, N becomes large and it becomes an "infinite" multitone signal.

Each transmitted signal tone, $x_i(t)$, is given by

$$\begin{aligned} x_i(t) &= \sum_{n=-\infty}^{\infty} a_n s(t - nT_i)\cos(2\pi f_i t) \\ &+ \sum_{n=-\infty}^{\infty} b_n s(t - nT_i)\sin(2\pi f_i t), \end{aligned} \tag{13.7}$$

where the symbol rate (and ideal Nyquist bandwidth) is W_i Hz, $T_i = 1/W_i$, and f_i is the carrier frequency of the ith signal. The baseband signal $s(t)$ is a minimum bandwidth Nyquist signal, free of intersymbol interference (ISI) and is written as

$$s(t) = 2W_i \frac{\sin 2\pi W_i t}{2\pi W_i t}\,; \quad T_i = 1/2W_i \text{ sec.} \tag{13.8}$$

The symbols a_n and b_n take on values of $\pm1, \pm3, \ldots$ depending on the number of signal points in the symmetric QAM signal (e.g., for 16 QAM, a_n, and b_n equal $\pm1, \pm3$). The signal, $s(t)$, is normalized so that the average energy per symbol equals E. The individual QAM signals have rectangular (or cross)

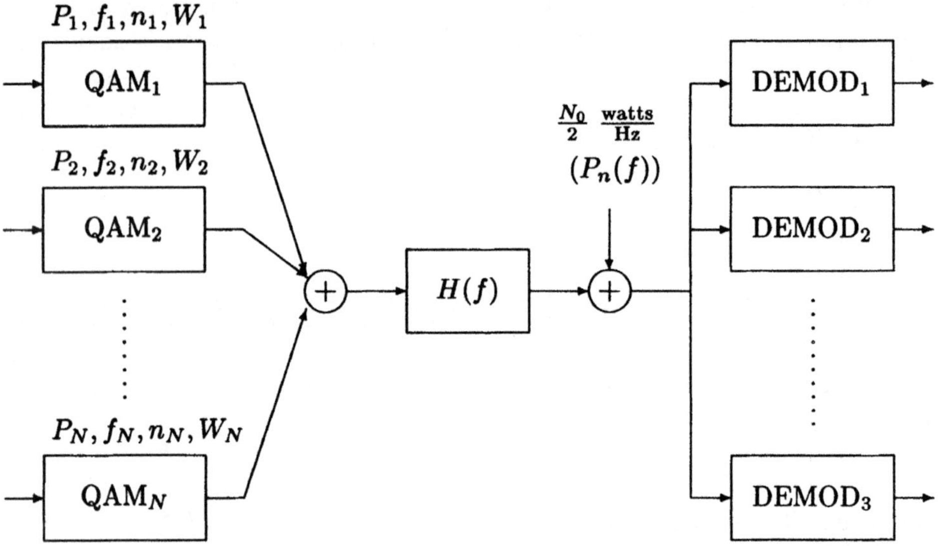

Figure 4 The multitone QAM system. Here P_i is the power transmitted in the ith tone, f_i is the center frequency of the ith tone, n_i is the number of bits per symbol for the ith tone, W_i is the bandwidth of the ith tone, and $R_i = n_i \cdot W_i$ bps is the bit rate of the ith tone.

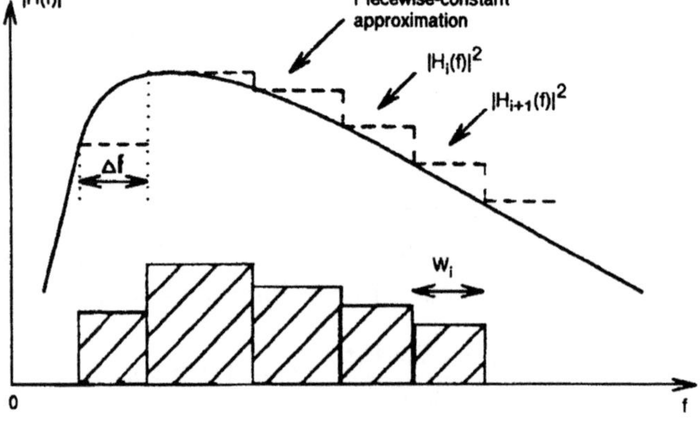

Figure 5 General channel transfer function, $|H(f)|^2$ (with piecewise-constant approximation), and MTM signal (cross-hatched).

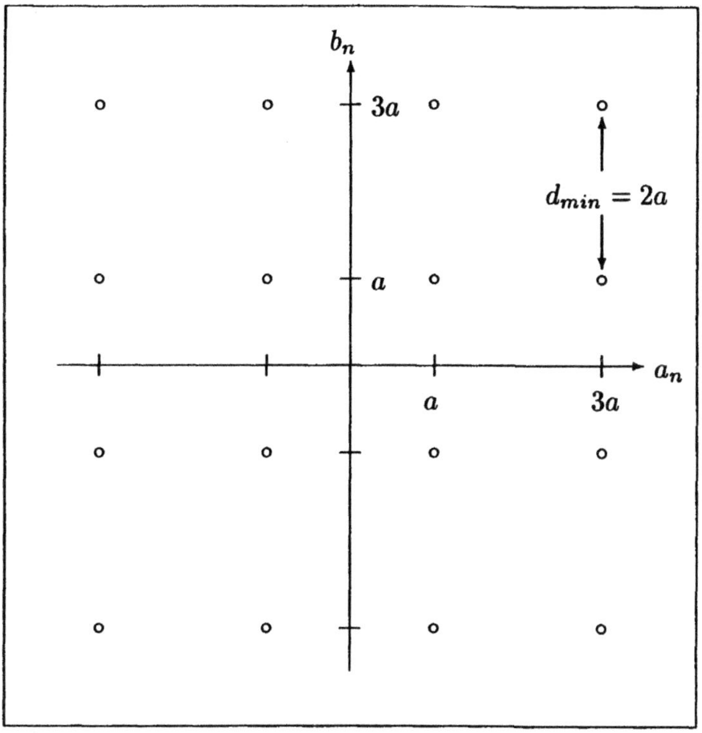

Figure 6 16-QAM constellation.

signal constellations of the type shown in Figure 6. The minimum distance between points is equal to $2a$.

The transmitted power of each QAM signal tone is equal to P_i and the total transmitted power is equal to P watts, i.e.,

$$P = \sum_{i=1}^{N} P_i. \tag{13.9}$$

Each M-ary QAM tone may have a different number of bits per symbol. M is the number of points in the QAM constellation, where $M = 2^{n_i}$, n_i being the number of bits per symbol for the ith signal, e.g., for 16-QAM, $n_i = 4$. The overall bit rate R_b, is given by

$$R_b = \sum_{i=1}^{N} n_i W_i \text{ bps.} \tag{13.10}$$

The receiver consists of bandpass filters of width W_i followed by coherent demodulators (with perfect phase recovery) and samplers at the instant of maximum signal voltage.

For even values of n, the probability of symbol error, $Pr_s\{\epsilon\}$ for a QAM signal for the case of high signal-to-noise (see [219]) for a constellation of size $M = 2^n$ signals, is given by

$$\begin{aligned} Pr_s\{\epsilon\} = & \; 4\left(1 - \tfrac{1}{\sqrt{2^n}}\right) Q\left[a\sqrt{\tfrac{2}{N_0}}\right] \\ \simeq & \; 4Q\left[a\sqrt{\tfrac{2}{N_0}}\right], \end{aligned} \tag{13.11}$$

where the Q function is defined by

$$Q(x) \triangleq \frac{1}{\sqrt{2\pi}} \int_x^{\infty} e^{-y^2/2}\, dy. \tag{13.12}$$

The value a is related to the average symbol energy E, for n even, by

$$a^2 = \frac{3E}{2}/(2^n - 1). \tag{13.13}$$

For odd values of n, the probability of error and the average energy per symbol can be closely approximated by Equation (13.11) and Equation (13.13). These expressions are also used for noninteger values of n. Although a noninteger value of n is usually impractical, such values will be considered in our systems for analytic convenience during the optimization. In real implementations in which n is an integer, the loss in optimal bit rate is very small [84]. The optimal multitone system maximizes R_b (through optimal power division among the tones and the optimal choice of bits per symbol for each tone) under the restriction that the total transmitted power is limited to P, and that the symbol error probabilities for all tones are equal.

We now turn our attention to maximizing R_b for a general linear channel with channel characteristic $|H(f)|^2$. First, we will find R_b for a general distribution of power among the different tones and then we will optimize this distribution. The magnitude square function $|H(f)|^2$ of the channel is first approximated, as mentioned earlier, by a staircase function, as shown in Figure 5. The transmitted power P, is divided into N tones, each tone with power $k_i P$ $(0 \leq k_i \leq 1)$ watts and carrying n_i bits per symbol, where

$$\sum_{i=0}^{N-1} k_i = 1. \tag{13.14}$$

Transmission is allowed only in subchannels for which $n_i \geq 2$. This requirement assumes that QAM requires at least one bit in each dimension. Later on we will eliminate this requirement. Taking the summation over all subchannels for which $n_i \geq 2$, we find the total bit rate R_b for a general distribution of k_i's. We will now assume that all channels have the same bandwidth, i.e., $W_i = \Delta f$ Hz. Therefore,

$$R_b = \sum_{i=0}^{N-1} n_i \cdot \Delta f. \tag{13.15}$$

Using Equations (13.11) and (13.13) to find n_i, we have

$$R_b = \sum_{k=0}^{N-1} \Delta f \cdot \log_2 \left[1 + \frac{(3k_i P/N_0 \Delta f)|H_i(f)|^2}{\left[Q^{-1}\left\{\frac{Pr\{\epsilon\}}{4}\right\}\right]^2} \right]. \tag{13.16}$$

We use a factor of 4 in the inverse Q function, $Q^{-1}\{\cdot\}$, as in Equation (13.11), so that our results are actually a lower bound on R_b. However, for large signal-to-noise ratios and large QAM constellations ($n >> 2$), the lower bound will be very tight.

As Δf becomes smaller the staircase approximation of the channel transfer function approaches the continuous $|H(f)|^2$. In this case, we have an "infinite" multitone system and the bit rate, R_b, is given by

$$R_b = \int_{f \in F_A} df \log_2 \left[1 + \frac{3}{[Q^{-1}\{\cdot\}]^2} \frac{P}{N_0} k(f) |H(f)|^2 \right], \tag{13.17}$$

where $k(f)$ is the continuous limit of k_i and

$$\int_{f \in F_A} k(f)\, df = 1,\ k(f) \geq 0. \tag{13.18}$$

F_A is the frequency range ($f \geq 0$) in which the integrand is greater than two (the QAM definition criterion).

In order to maximize R_b, the normalized power spectral distribution $k(f)$ is optimized. This problem is very similar to the information theory problem of maximization of the capacity of a linear channel [365]. The solution has the same form as the "water-pouring" solution. The optimum normalized power distribution, $k_o(f)$, is found [219] to be

$$k_o(f) = \begin{cases} \lambda_M - \frac{1}{K_M|H(f)|^2} & \text{for f} \in F_M \\ 0 & \text{otherwise,} \end{cases} \tag{13.19}$$

where

$$\int_{f \in F_M} k_o(f)\, df = 1, \tag{13.20}$$

$$K_M = \frac{P}{N_0} \frac{3}{[Q^{-1}\{\cdot\}]^2}, \tag{13.21}$$

and λ_M is a Lagrange multiplier.

The maximum bit rate, $R_{b_{max}}$, is given by

$$R_{b_{max}} = \int_{f \in F_A} df \ \log_2\{\lambda_M K_M |H(f)|^2\}. \tag{13.22}$$

From the condition that the integrand must be greater than two, we have that $f \in F_A$ if

$$|H(f)|^2 > \frac{4}{K_M \lambda}. \tag{13.23}$$

An interactive process is required, to jointly solve Equations (13.22) and (13.23), to find $k_0(f)$ and λ_M.

The water-pouring solution for optimizing channel capacity is a very satisfying result because it agrees with the philosophical argument of Shannon for achieving capacity. In fact, if we relax the restriction of Equation (13.23) to log $\{\cdot\} > 0$, we find that for $Pr_s\{\epsilon\} = 10^{-5}$, there is about an 8.4 dB degradation in performance when comparing an "infinite" multitone modulation to the capacity for any $|H(f)|^2$, independent of its shape.

Interestingly enough, for large signal-to-noise ratios it can easily be shown [219, 454] that the optimal $k_o(f)$ may be approximated by a uniform distribution. This approximation simplifies the physical implementation of the multitone modulation.

The optimum MTM technique is in essence a direct result of Shannon's ideas on distribution of power in the frequency domain. The difference of 8 to 9 dB between capacity and MTM can be reduced, as mentioned earlier, by proper coding. The usual coding methods are based on Ungerboeck trellis codes [411,

412, 53]. It is theoretically possible to gain as much as 6 dB with these codes, and approach to within two or three dB of capacity. In present systems the actual gain is usually between 3.5 to 4.5 dB.

The results for maximum bit rate, $R_{b_{max}}$, of Equation (13.22) are based on an "almost" infinite number of tones of "almost" zero bandwidth. In reality the number of tones must be finite, but numbers as high as 256 and 512 tones [83, 80] have been implemented. Furthermore, the number of bits per symbol, n_i, which is used in the analytic derivation of Equation (13.22), is not an integer variable, but rather a continuous variable. In most real implementations only integer values are used for n_i. However, this will not affect $R_{b_{max}}$ in more than a minor way. The reason being that if a particular n_i is supposed to equal 2.32 bits per symbol, this will in reality be reduced to 2 bits per symbol. However, some of its power will also be taken away (because n has been reduced) and given to another tone for which n_i should be 3.89 bits per symbol, thereby increasing its power so that four bits per symbol may be used. In this manner, $R_{b_{max}}$ will not be reduced by more than a few percent. There are optimal methods for redistributing the powers, and these are described in [84] and [196].

One of the most important uses for MTM is on the twisted-pair cable for HDSL and ADSL transmission of data at rates above 1 Mbps. The twisted-pair cable will be briefly discussed in the next section, including results for $R_{b_{max}}$ and C. This will be followed by a description of the DMT implementation of a multitone system, including one quantitative example.

4 THE TWISTED-PAIR CHANNEL

The twisted-pair cable is a very important element of telephony systems, local area networks (LAN), and integrated services data networks (ISDN). The cable consists of many twisted-pair wires in very close proximity to each other. Transmission may be in either direction for a particular pair of wires. In HDSL, the case we will consider, the dominating influence on the receiver detection performance for a given pair is the near-end crosstalk (NEXT) [437, 243, 107, 101] generated by transmitters located at the same end of the cable as the receiver. This is the transmission mode used in the HDSL transmission. ADSL is essentially high-speed transmission in only one direction. The transmitted signals from other twisted-pairs feed into a receiver located at the same end as the transmitters, and the NEXT interfering signal is the dominant noise factor for

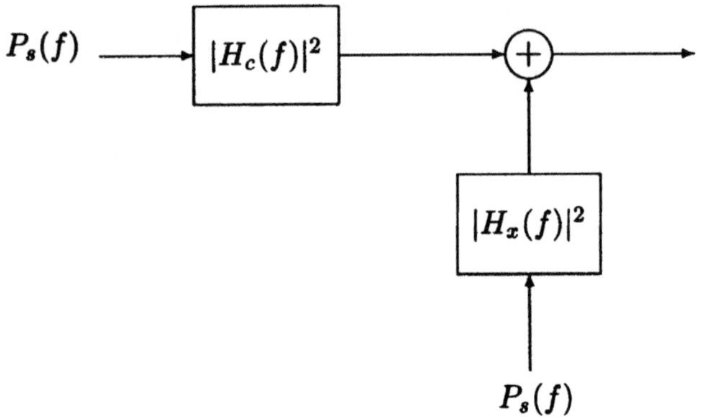

Figure 7 Twisted-pair channel model.

determining detectability performance. It is usually assumed that the nearest five to seven twisted-pairs cause most of the crosstalk and that all the users are using the same type of modems and therefore have the same power spectral density, $P_s(f)$.

A model for the twisted-pair channel including the NEXT effect has been proposed [243, 107, 101] and is shown in Figure 7. The model ignores the inherent white Gaussian noise (WGN) of the channel because the NEXT dominates the performance, but the WGN can easily be taken into account.

The model includes the twisted-pair wire attenuation function $|H_c(f)|^2$. A typical $|H_c(f)|^2$ is shown in Figure 2 [454].

The NEXT effect is modeled [437, 243, 107, 101] as a signal having the same spectral density, $P_s(f)$, as the transmitted signal (since all pairs are assumed to be transmitting similar signals), passing through a crosstalk transfer function, $|H_x(f)|^2$, (given below), which increases with increasing frequency,

$$|H_x(f)|^2 = \beta f^{3/2}, \tag{13.24}$$

where β is a constant of the cable. The interesting fact about this channel is that both the transmitted signal and the interfering NEXT signal are generated from the same spectral density, $P_s(f)$. Both $|H_x(f)|^2$ and $|H_c(f)|^2$ tend to stop transmission at higher frequencies.

The NEXT-dominated channel model used in the analysis of twisted-pair cables for HDSL has a capacity [220] and maximum multitone bit rate [219, 54] that may be found using the results of the last section. The capacity, C_{max}, is given by [141]

$$C_{max} = \int_0^{\infty} df \log_2 \left[\frac{|H_c(f)|^2}{|H_x(f)|^2}\right] \text{ bps.} \tag{13.25}$$

$R_{b_{max}}$ is given by [219]

$$R_{b_{max}} = \int_{f \in F_{NEXT}} df \log_2 \left[1 + M_n \frac{|H_c(f)|^2}{|H_x(f)|^2}\right], \tag{13.26}$$

where F_{NEXT} is the frequency range in which

$$|H_c(f)|^2 > \frac{3}{M_n}|H_x(f)|^2, \tag{13.27}$$

similar to the "QAM" condition of Equation (13.23) and

$$M_n = 3/\left(\left[Q^{-1}\left\{\frac{Pr\{\epsilon\}}{4}\right\}\right]^2\right). \tag{13.28}$$

The channel capacity C_{max}, and the maximum bit rate, $R_{b_{max}}$, almost have the same form with the exception of the constants inside the arguments of the log functions, and the bounds of integration. For most of the frequency range in Equation (13.26) the M_n term is much greater than the factor one. The difference is the constant $3/[Q^{-1}\{\frac{Pr\{\epsilon\}}{4}\}]^2$ which equals about -8.4 dB for an error probability equal to 10^{-5}. This is the same value found earlier for the AWGN channel.

Again this means that even in the NEXT case no matter what the channel characteristic, the multitone modulation allows us to be within 8-9 dB of capacity (for $Pr\{\epsilon\} = 10^{-5}$) before coding. Ungerboeck trellis coding is also applied to these channels to improve detectability performance in an effort to approach Shannon's information theory results.

Results for capacity and multitone are shown in Figure 8, for an HDSL transmission over a 26-gauge untapped twisted-pair. The curves are plots of bit rates as a function of the length of twisted-pair lines. Most HDSL users are within 12,000 feet of a central office, therefore transmission at rates approaching 3 Mbps should be possible for these users.

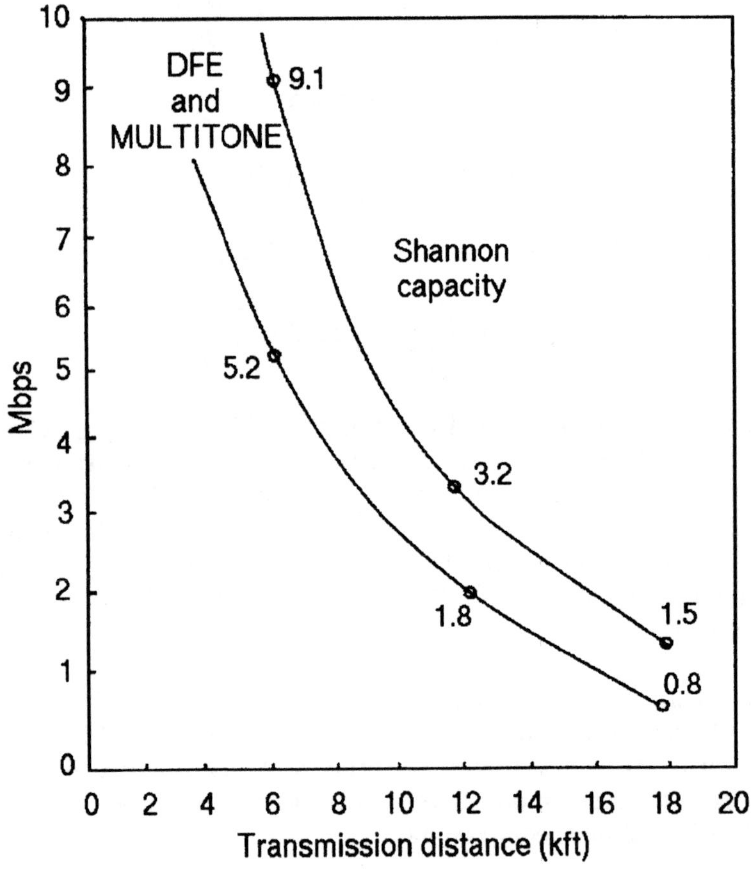

Figure 8 Comparison of capacity, multitone and DFE optimum bit rates for a 24-gauge twisted-pair.

5 DECISION FEEDBACK EQUALIZATION COMPARISON

There is a communications technique called Decision Feedback Equalization (DFE) [323, 350] which utilizes one wideband QAM signal, and a DFE equalizer at the receiver, which eliminates part (but not all) of the intersymbol interference created by the general linear channel transfer function, $H(f)$. This technique, when optimized, results in a maximum bit rate, $R_{b_{max,DFE}}$, which is similar in form to the $R_{b_{max}}$ of multitone transmission and again is about 8 to 9 dB away from the maximum capacity, C_{max}.

The equation for $R_{b_{max,DFE}}$, assuming the ideal case (for DFE) with a uniform power spectrum over the frequency range F_{DFE}, can be expressed as

$$R_{b_{max,DFE-ZF}} = \int_{f \in F_{DFE}} \log_2 \frac{1}{K_{DFE}} \frac{P}{N_0 W} |H(f)|^2 \, df, \tag{13.29}$$

$$K_{DFE} = \frac{\left[Q^{-1}\left\{\frac{Pr\{\epsilon\}}{4}\right\}\right]^2}{3}, \tag{13.30}$$

where ZF stands for "zero-forcing" as a special type of DFE.

Again this is similar to Equation (13.21) and Equation (13.22). In fact using a probability of error of 10^{-5} results in a difference of about 8 dB between capacity and $R_{b_{max,DFE-ZF}}$, a result similar to that of multitone transmission.

Equation (13.29) is similar and almost equal to Equation (13.22) for MTM and, with the exception of the constant in the argument of the log function, is also similar to Equation (13.6) for the capacity of a channel. The difference is due to the constant being about 8 to 9 dB, as it is for multitone modulation. It is this similarity in performance for both the MTM and DFE techniques that has led industry to consider both, especially for HDSL and ADSL technologies. In fact, the DFE and multitone curves appear as one curve in Figure 8.

At this point it should be mentioned that the choice between MTM and DFE is based on a number of factors besides performance over a white Gaussian noise channel. A number of these factors will now be described.

1. Implementation: The complexity of implementing either of the two techniques is a major element in the choice between the two. There has been much heated debate on this subject, with no clear choice between the two.

2. Peak-to-Average Ratio: The MTM technique is in essence the summation of a large number of signals, and its statistics tend to become Gaussian. This in turn means a high peak-to-average ratio which requires amplifiers and analog-to-digital converters, with large dynamic ranges. The DFE technique makes use of a QAM-like signal, called a carrierless AM-PM (CAP) signal [141, 388] which has a smaller peak-to-average ratio (increasing with decreasing Nyquist roll-off factor). However, it is thought that MTM may be amplitude limited (reducing the peak-to-average ratio) with minimal degradation of spectrum and detectability performance.
3. Impulsive Noise: Impulses due to switching transients and other sources are an important source of performance degradation in HDSL and ADSL systems. Each tone of a multitone modulation, by the very nature of MTM, only sees a smart part of the total wideband impulse energy. This tends to make each MTM tone much less prone to degradation due to impulse noise than a signal in a DFE-type system. However with Reed-Solomon [335] burst-error correcting block codes, the difference in performance may not be as great.

There are other issues such as delay and echo cancellation that must also be addressed. There is a very nice summary of all of these issues in [54].

6 DISCRETE MULTITONE (DMT)—AN IMPLEMENTATION

The theoretical work described in the previous sections justifies the interest in MTM. However it is still important to find a reasonable technique for implementing the MTM concept. The generation of many QAM signals at the transmitter, and then the need for many QAM demodulators at the receiver is not very appealing to system designers. However efficient DFT filter banks provide an attractive solution. An approach of this type based on the DFT has been described and implemented [346, 84]. In this design the individual QAM signals are generated as points in the DFT frequency domain in the following manner. If a 16-QAM signal is chosen by the optimization process to be transmitted at frequency f_1, it will appear as a complex number at this frequency point in the DFT frequency domain (see Figure 9). For example if the outermost corner point in the first quadrant of Figure 6 were to be transmitted, it would be written as the complex number, $3a + j3a$, at this DFT frequency. In

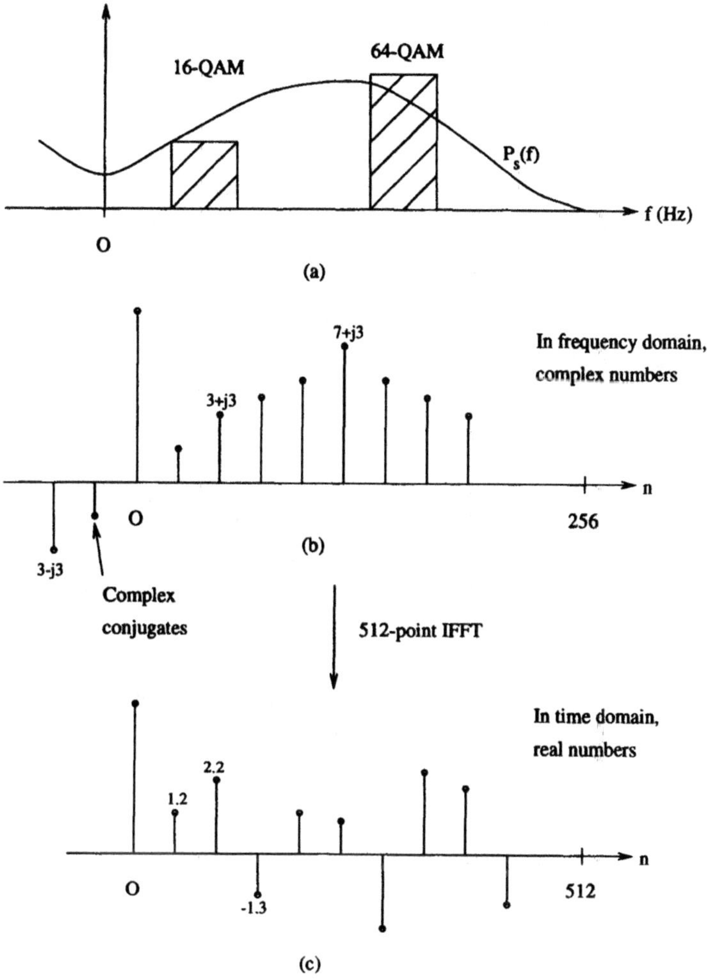

Figure 9 DMT Implementation, a) optimal $P_s(f)$, and finite multitone approximation, b) DFT (FFT) implementation in frequency-domain (512 points), c) IDFT (IFFT) in time-domain.

this manner, the different QAM signals are converted into complex points in the frequency domain. Since it is not possible to generate an infinite multitone signal, a finite number of tones, e.g., 256, may be used to approximate the optimal continuous power spectral density.

In order to transmit the signal it is necessary to transform it into a purely real sequence by exploiting Hermitian symmetry.

To guarantee this complex-to-real transformation, it is necessary to complement the original 256 complex numbers with their complex conjugates as shown in Figure 9, to generate a 512 point complex-to-real inverse FFT (IFFT). The IFFT will generate real values in the time domain. Ideally these real sample values are passed through an ideal lowpass Nyquist filter ([167], Chap. 4) with sampling period, T_s, equal to the sampling period of the DFT. This would require an ideal Nyquist bandwidth of $1/2T$ Hz. In a real implementation this signal would have a somewhat larger bandwidth (using a Nyquist raised-cosine spectrum). The above procedure guarantees an output signal whose sample values are equal to the DFT sample values. If this signal passes through an ideal distortionless channel, it can then be sampled at the receiver to recover the IDFT samples. Then a DFT is performed, to recover the original complex samples, and the entire multitone signal.

If the channel is not distortionless it is relatively easy to build a series of simple one-tap equalizers in the frequency domain to adjust for amplitude and phase distortion in every frequency bin. The technique used above has been implemented and been proposed as a standard for HDSL and ADSL transmission in the United States.

As a quantitative example of one use of this DMT concept we present the table below based on an HDSL design, which appeared in [80], with a target bit rate of 1.6 Mbps.

Table 1: HDSL—Example

- $R_{bit} = 1.6$ Mbps ($Pr_b\{\epsilon\} = 10^{-7}$)
- N = 256 independent QAM signal channels (512 - total)
- $f_{sampling} = 640$k samples/sec
- Number of FFT Blocks/sec = $\frac{640k}{512} = 1250^*$
  (Actually 640k/520 = 1230.7 DFT Blocks/sec)
- One Block (Actual = 812.5 sec)
- Number of bits/block = 1300
- Number of bits/sec = $\frac{1300 bits}{812.5 \mu s} = 1.6$ Mbps

* Eight samples are added to 512 to reduce the interblock interference effect.

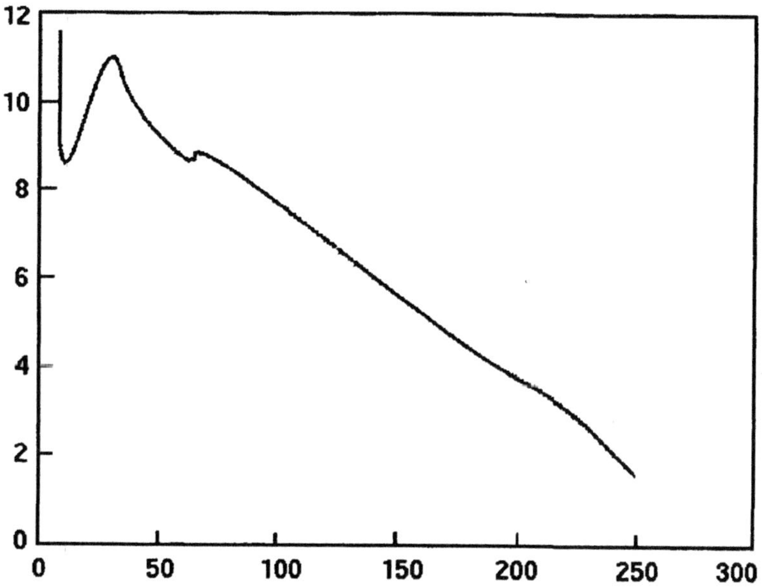

Figure 10 DMT bit allocation. The number of bits assigned per frequency bin is shown on the ordinate. Subchannel number is shown on the abscissa [80].

Because the sampling rate is 640k samples/sec, the ideal Nyquist bandwidth is about 320 kHz. This means that the multitone signal is being transmitted at frequencies at which the channel attenuation is down by as much as tens of dBs. Again, this fits the Shannon concept of using as much bandwidth as possible, but with the water pouring concept of power distribution. In Figure 10, (taken from [80]) the actual bit distribution per tone is shown. Notice the decreasing constellation size as the channel attenuation increases.

The DFT basis functions used in the work just described are naturally limited by the block size. These limitations may be overcome by using filter banks with better frequency localized bases [48]. More studies on subband transform based implementations of DMT transceivers are expected in the future.

7 SUMMARY AND THE FUTURE

In this chapter we have attempted to describe Multitone Modulation (MTM). This technique, which uses many narrowband QAM signals instead of one wide-band signal, attempts to follow Shannon's water-pouring ideas for optimizing bit transmission rates over general time-invariant channels. The concept of water-pouring, essentially means using as much bandwidth as possible, but also weighing the transmitted power spectral density, so that narrowband QAM signals with lots of power are sent where the channel shows little attenuation, and signals with less power are transmitted where the channel attenuates very strongly. At the same time the number of bits per symbol (or constellation size) is adjusted to guarantee a common error probability for all the signals. The multitone technique performs to within 8 to 9 dB (for an error probability of 10^{-5}) of the channel capacity, independent of the channel characteristic, $H(f)$. It is this property of MTM that gives it such importance, and has made it a candidate for certain kinds of data transmission over physical wire connections.

Multitone modulation is now being considered for wireless communication as well. Direct Audio Broadcasting (DAB) in Europe uses the idea of multitone, because of its slow block transmission rate so that simultaneous broadcasting can be used from many antennas with a minimum amount of multipath or delay spread interference.

MTM is also being considered for wireless Personal Communication Systems (PCS) as well. In this case it may be possible, especially for mobile users like pedestrians walking in a mall or shopping center, to learn the characteristics of this dynamic, time-varying channel, and transmit this information back to the talker (on a reverse channel) so that the transmitting device can readjust its power spectral density, and bit allocation according to the water-pouring concept, or according to the uniform power spectral density approximation. As long as the channel is quasi-stationary, (or slowly varying) information can be transmitted at maximum bit rates.

BIBLIOGRAPHY

[1] E. H. Adelson and E. P. Simoncelli. Subband image coding with three-tap pyramids. In *Proceedings Picture Coding Symposium*, 1990.

[2] N. Ahmed, T. Natarajan, and K. Rao. Discrete cosine transform. *IEEE Transactions Comput.*, C-23:90–93, January 1974.

[3] A. N. Akansu. Multiplierless suboptimal PR-QMF design. In *Proc. SPIE Visual Communication and Image Processing*, pages 723–734, November 1992.

[4] A. N. Akansu. Wavelets and filter banks: A signal processing perspective. *IEEE Circuits and Devices Magazine*, pages 14–18, November 1994.

[5] A. N. Akansu and H. Caglar. A measure of aliasing energy in multiresolution signal decomposition. In *Proceedings of the International Conference on Acoustics, Speech, and Signal Processing*, pages IV 621–624, March 1992.

[6] A. N. Akansu, H. Caglar, and Y. Liu. An objective performance measure in multiresolution signal decomposition. *IEEE Trans. Acoust. Speech Signal Process.*, ASSP-41:3153–3157, November 1993.

[7] A. N. Akansu and R. A. Haddad. *Multiresolution Signal Decomposition.* Academic Press, 1992.

[8] A. N. Akansu, R. A. Haddad, and H. Caglar. Perfect reconstruction binomial QMF-wavelet transform. In *Proc. SPIE Visual Communication and Image Processing*, volume 1360, pages 609–618, October 1990.

[9] A. N. Akansu, R. A. Haddad, and H. Caglar. Binomial QMF-wavelet transform for multiresolution signal decomposition. *IEEE Trans. on Signal Processing*, 41(1):13–19, January 1993.

[10] A. N. Akansu and Y. Liu. On signal decomposition techniques. *Optical Engineering*, pages 912–920, July 1991.

[11] A. N. Akansu and Y. Liu. Tree structures for time-frequency signal analysis. In *Proceedings of the International Conference on Acoustics, Speech, and Signal Processing*, pages 2037–2040, 1991.

[12] J. B. Allen. Personal communication.

[13] J. P. Andrew. *Subband Analysis Structures and Filters for Still Image Compression.* PhD thesis, University of Wollongong, Australia, 1994.

[14] R. Ansari. Elliptic filter design for a class of generalized halfband filters. *IEEE Trans. on Signal Processing*, ASSP-33(5):1146–1150, October 1985.

[15] R. Ansari. Two-dimensional nonrectangular interpolation, decimation, and filter banks. In *Proceedings of the International Conference on Acoustics, Speech, and Signal Processing*, 1988.

[16] R. Ansari, A. E. Cetin, and S. H. Lee. Subband coding of images using nonrectangular filter banks. In *Proc. 32nd Annual International Symposium on Optical and Optoelectronic Applied Science Engineering*, August 1988.

[17] R. Ansari and C. Guillemot. Hierarchical scheme for coding video at 4-10Mbps. In J. Vandewalle et al, editor, *Signal Processing VI: Theories and Applications, EUSIPCO '92*, pages 227–230, August 1992.

[18] R. Ansari, C. Guillemot, and J. F. Kaiser. Wavelet construction using Lagrange halfband filters. *IEEE Trans. Circuits and Systems*, CAS-38:1116–1118, September 1991.

[19] R. Ansari and C.-L. Lau. Two-dimensional IIR filters for exact reconstruction in tree-structured subband decomposition. *Electronics Letters*, 23(12):633–634, March 1987.

[20] R. Ansari, S. Lee, and L. Wu. Subband image coding using IIR filters. In *Proc. Conf. on Information Sciences and Systems*, pages 16–21, March 1988.

[21] R. Ansari and D. LeGall. Advanced television coding using exact reconstruction filter banks. In J. W. Woods, editor, *Subband Image Coding*, pages 273–318. Kluwer Academic Publishers, 1990.

[22] R. Ansari and B. Liu. A class of low-noise computationally efficient recursive digital filters. In *Proceedings of the International Conference on Acoustics, Speech, and Signal Processing*, pages 550–553, April 1981.

[23] R. Ansari and B. Liu. Interpolators and decimators as periodically time-varying filters. In *Proc. IEEE Int. Sym. Circuits and Systems*, pages 447–450, April 1981.

[24] R. Ansari and B. Liu. Transmultiplexer design using allpass filters. *IEEE Trans. Comm.*, 30(7):1569–1574, July 1982.

[25] R. Ansari and B. Liu. Efficient sampling rate alteration using recursive (IIR) digital filters. *IEEE Trans. on Signal Processing*, ASSP-31(6):1366–1373, December 1983.

[26] R. Ansari and B. Liu. Multirate signal processing and its applications. In S. K. Mitra and J. F. Kaiser, editors, *Handbook on Digital Signal Processing*. John Wiley, 1993.

[27] J-P. Antoine, P. Carrette, and R. Murenzi. The scale-angle representation in image analysis with 2D wavelet transform. In *Signal Processing VI: Theory and Applications (Proc. EUSIPCO 92, Brussels)*, pages 701–704, 1992.

[28] J-P. Antoine, P. Carrette, R. Murenzi, and B. Piette. Image analysis with 2D continuous wavelet transform. *Signal Proc.*, 31:241–272, 1993.

[29] M. Antonini, M. Barlaud, P. Mathieu, and I. Daubechies. Image coding using wavelet transform. *IEEE Trans. on Image Processing*, 1:205–220, April 1992.

[30] J. G. Apostolopoulos and J. S. Lim. Video compression for digital advanced television systems. In M. I. Sezan and R. L. Lagendijk, editors, *Motion Analysis and Image Sequence Processing*. Kluwer Academic publishers, Boston, 1992.

[31] F. Argoul, A. Arnéodo, J. Elezgaray, G. Grasseau, and R. Murenzi. Wavelet analysis of the self-similarity of diffusion-limited aggregates and electrodeposition clusters. *Phys. Rev. A*, 41:5537–5560, 1990.

[32] A. Arnéodo, F. Argoul, E. Bacry, J. Elezgaray, E. Freysz, G. Grasseau, J. F. Muzy, and B. Pouligny. Wavelet transform of fractals. In Y. Meyer, editor, *Wavelets and Applications (Proc. Marseille 1989)[264]*, pages 286–352. Springer-Verlag, Berlin, and Masson, Paris, 1991.

[33] J. L. Arrowood and M. J. T. Smith. Gibbs phenomenon suppression using FIR time-varying filter banks. In *IEEE DSP Workshop*, pages 2.1.1–2, Chicago, Illinois, September 1992.

[34] J. L. Arrowood and M. J. T. Smith. Exact reconstruction analysis/synthesis filter banks with time-varying filters. In *Proceedings of the International Conference on Acoustics, Speech, and Signal Processing*, pages III 233–236, April 1993.

[35] D. Astruc, L. Plantié, R. Murenzi, Y. Lebret, and D. Vandromme. On the use of the 3D wavelet transform for the analysis of computational fluid dynamics results. In Y. Meyer and S. Roques, editors, *Progress in Wavelet Analysis and Applications (Proc. Toulouse 1992)[266]*, pages 463–470. Frontières, 1993.

[36] H. Babic, S. K. Mitra, C. Creusere, and A. Das. Perfect reconstruction recursive QMF banks for image subband coding. In *Proc. 25th Annual Asilomar Conference on Signals, Systems, and Computers*, 1991.

[37] R. Bamberger and M. J. T. Smith. A filter bank for the directional decomposition of images: Theory and design. *IEEE Trans. on Signal Processing*, SP-40, April 1992.

[38] R. H. Bamberger and M. J. T. Smith. Efficient 2-D analysis/synthesis filter banks for directional image component representation. In *Proc. IEEE Int. Sym. Circuits and Systems*, pages 2009–2012, April 1990.

[39] M. Barlaud, P. Solé, M. Antonini, and P. Mathieu. A pyramidal scheme for lattice vector quantization of wavelet coefficients applied to image coding. In *Proceedings of the International Conference on Acoustics, Speech, and Signal Processing*, March 1992.

[40] M. Barlaud, P. Solé, T. Gaidon, M. Antonini, and P. Mathieu. Pyramidal lattice vector quantization for multiscale image coding. *IEEE Transactions on Image Processing*, 3(4):367–381, July 1994.

[41] T. Barnwell. Subband coding design incorporating recursive QMFs and optimal ADPCM coders. *IEEE Trans. Acoust. Speech Signal Process.*, ASSP-30(5):751–765, October 1982.

[42] H. J. Beker and F. C. Piper. *Secure Speech Communications*. Academic Press, London, 1985.

[43] M. Bellanger, G. Bonnerot, and M. Coudreuse. Digital filtering by polyphase network: Application to sample rate alteration and filter banks. *IEEE Trans. Acoust. Speech Signal Process.*, ASSP-24:109–114, April 1976.

[44] M. G. Bellanger. Computation rate and storage estimation in multirate digital filtering with half-band filters. *IEEE Trans. Acoust. Speech Signal Process.*, ASSP-25(4):344–346, August 1977.

[45] M. G. Bellanger, J. L. Daguet, and G. P. Lepagnol. Interpolation, extrapolation and reduction of computation speed in digital filters. *IEEE Trans. on Signal Processing*, ASSP-22(4):231–235, August 1974.

[46] F. Bellifemine, C. Cafforio, A. Chimienti, and R. Picco. Combining DCT and subband coding into an intraframe coder. *Signal Processing: Image Communication*, 5(3):235–248, May 1993.

[47] B. Belzer, J. Liao, and J. D. Villasenor. Adaptive video coding for mobile wireless networks. In *International Conference on Image Processing*, pages 972–976, November 1994.

[48] A. Benyassine. *Theory, Design and Applications of Linear Transforms for Information Transmission.* PhD thesis, New Jersey Institute of Technology, 1995.

[49] A. Benyassine and A. N. Akansu. Optimal subchannel structuring and basis selection for discrete multicarrier modulation. In *Proc. IEEE Globecom*, 1995.

[50] A. Benyassine and A. N. Akansu. Performance analysis and optimal structuring of subchannels for discrete multitone transceivers. In *Proceedings of the International Conference on Acoustics, Speech, and Signal Processing*, volume II, pages 1456–1459, April 1995.

[51] T. Berger. *Rate Distortion Theory.* Prentice-Hall Inc., Englewood Cliffs,New Jersey, 1971.

[52] M. Bierling. Displacement estimation by hierarchical blockmatching. In *SPIE Proc. Visual Communications and Image Processing*, pages 942–951, 1988.

[53] Biglieri, Divsalar, McLane, and Simon. *Introduction to trellis-coded modulation with applications.* MacMillan, 1991.

[54] J. A. C. Bingham. Multicarrier modulation for data transmission: an idea whose time has come. *IEEE Communications Magazine*, 28(5):5–14, May 1990.

[55] R. E. Blahut. Computation of channel capacity and rate-distortion functions. *IEEE Transactions on Information Theory*, IT-18:460–473, July 1972.

[56] M. E. Blain and T. R. Fischer. A comparison of vector quantization techniques in transform and subband coding of imagery. *Signal Processing: Image Communication*, 3(1):91–105, February 1991.

[57] F. Bosveld, R. L. Lagendijk, and J. Biemond. Compatible spatio-temporal subband encoding of HDTV. *Signal Processing*, 28:271–289, September 1992.

[58] F. Bosveld, R. L. Lagendijk, and J. Biemond. Hierarchical coding of HDTV. *Signal Processing: Image Communication*, 4:195–225, July 1992.

[59] F. Bosveld, R. L. Lagendijk, and J. Biemond. Compatible video compression using subband and motion compensation techniques. In *Proc. of the Int. Workshop on HDTV*, October 1993.

[60] F. Bosveld, R. L. Lagendijk, and J. Biemond. Hierarchical coding. In M. Hang and J. W. Woods, editors, *Handbook of Visual Communications*. Kluwer Academic Publishers, 1995. to appear.

[61] F. Boucherok and J. F. Vial. Compatible multiresolution coding scheme. In *Int. workshop on HDTV*, November 1992.

[62] M. Bove and A. B. Lippman. Scalable open architecture for TV. *SMPTE Journal*, 101:2–5, January 1992.

[63] K. Brandenburg and J. D. Johnston. Second generation perceptual audio coding: The hybrid coder. In *88th AES Convention*, 1990. preprint 2937(H3).

[64] M. Breeuwer and P. H. N. de With. Source coding of HDTV with compatibility to TV. In *SPIE Proc. Visual Communications and Image Processing*, pages 765–776, 1990.

[65] T. A. Brown, G. J. Saulnier, and P. Das. Direct-sequence spread spectrum acquisition using transform domain processing. In *IEEE MILCOM*, pages 1018–1022, October 1993.

[66] H. Brusewitz. *Quantization and Entropy Coding*. PhD thesis, The Royal Institute of Technology, Stockholm, Sweden, March 1987.

[67] P. J. Burt and E. H. Adelson. The Laplacian pyramid as a compact image code. *IEEE Trans. Comm.*, COM-31:552–540, April 1983.

[68] H. Caglar. *A Generalized, Parametric PR-QMF/Wavelet Transform Design Approach for Multiresolution Signal Decomposition*. PhD thesis, New Jersey Institute of Technology, 1991.

[69] H. Caglar and A. N. Akansu. A generalized, parametric PR-QMF design technique based on Bernstein polynomial approximation. *IEEE Trans. on Signal Processing*, 41(7):2314–2321, July 1993.

[70] H. Caglar, Y. Liu, and A. N. Akansu. Statistically optimized PR-QMF design. In *SPIE Proc. Visual Communications and Image Processing*, pages 86–94, November 1991.

[71] L. C. Calvez and P. Vilbe. On the uncertainty principle in discrete signals. *IEEE Trans. Circuits and Systems*, 36:394–395, June 1992.

[72] E. Carterette and M. Friedman. *Handbook of Perception*, volume IV. Academic Press, 1978.

[73] P. Cassereau. A new class of orthogonal transforms for image processing. Master's thesis, MIT, Cambridge, MA, May 1985.

[74] P. M. Cassereau, D. H. Staelin, and G. de Jager. Encoding of images based on a lapped orthogonal transform. *IEEE Trans. Comm.*, 37(2):189–193, February 1989.

[75] CCIR. Encoding parameters of digital television for studios, CCIR Recommendation 601, 1982.

[76] CCITT. Video codec for audiovisual services at p x 64 kbit/s, CCITT recommendation H.261, 1990.

[77] R. W. Chang. High-speed multichannel data transmission with bandlimited orthogonal signals. *Bell Syst. Tech. J.*, 45:1775–1796, 1966.

[78] B. Chitprasert and K. R. Rao. Discrete cosine transform filtering. In *Proceedings of the International Conference on Acoustics, Speech, and Signal Processing*, pages 1281–1284, 1990.

[79] P. A. Chou, T. Lookabaugh, and R. M. Gray. Entropy-constrained vector quantization. *IEEE Transactions on Acoustics, Speech and Signal Processing*, ASSP-37(1):31–42, January 1989.

[80] J. S. Chow, J. C. Tu, and J. M. Cioffi. A discrete multitone transceiver for HDSL applications. *IEEE J. on Sel. Areas in Commun.*, 9(6):895–908, August 1991.

[81] J. S. Chow, J. C. Tu, and J. M. Cioffi. Performance evaluation of a multichannel transceiver system for ADSL and VHDSL services. *IEEE J. on Sel. Areas in Commun.*, 9(6):909–919, August 1991.

[82] J. S. Chow, J. C. Tu, and J. M. Cioffi. Multi-channel modulation as a technique for transmission in radio channel. In *IEEE Veh. Tech. Conf. Rec.*, pages 29–33, May 1993.

[83] P. S. Chow, N. Al-Dhahir, J. M. Cioffi, and J. A. C. Bingham. A multi-carrier E1-HDSL transceiver system with coded modulation. *Eur. Trans. on Telecommun.*, 4(3):23–32, May 1993.

[84] P. S. Chow, J. M. Cioffi, and J. A. C. Bingham. A practical discrete multitone transceiver loading algorithm for data transmission over spectrally shaped channels. *IEEE Trans. Commun.*, April 1994. submitted for publication.

[85] P. Chu. Quadrature mirror filter design for an arbitrary number of equal bandwidth channels. *IEEE Trans. Acoust. Speech Signal Process.*, ASSP-33:203–218, February 1985.

[86] C. K. Chui. *An Introduction to Wavelets.* Academic Press, 1992.

[87] C. K. Chui, editor. *Wavelets: A Tutorial in Theory and Applications.* Academic Press, 1992.

[88] W. Chung and M. J. T. Smith. Spatially-varying IIR filter banks for image coding. In *Proceedings of the International Conference on Acoustics, Speech, and Signal Processing*, pages V570–573, April 1993.

[89] D. Cochran and C. Wei. Bandlimited orthogonal wavelet symbols. In *The Conference Record of the 27th Asilomar Conference on Signals, Systems and Computers*, pages 528–532, November 1993.

[90] D. Cochran and C. Wei. Wavelet-based spread spectrum communications. In *The Proceedings of the Dual-Use Technologies and Applications Conference*, pages 392–401, May 1994.

[91] A. Cohen. *Wavelets and Digital Signal Processing.* Chapman & Hall, 1994.

[92] A. Cohen and I. Daubechies. On the instability of arbitrary biorthogonal wavelet packets. *SIAM J. Math. Anal.*, 24(5):1340–1354, 1993.

[93] A. Cohen, I. Daubechies, and P. Vial. Wavelets and fast wavelet transforms on an interval. *Applied and Computational Harmonic Analysis*, 1993. to appear.

[94] A. Cohen and N. Dyn. Nonstationary subdivision schemes and multiresolution analysis. Technical report, CEREMADE, Université Paris-IX Dauphine, 1994. preprint.

[95] R. Coifman and Y. Meyer. Orthonormal wave packet bases. Dept. of Math., Yale Univ., preprint.

[96] R. Coifman, Y. Meyer, and V. Wickerhauser. Size properties of wavelet packets. In *Wavelets and Their Applications*, pages 453–470. Jones and Bartlett, 1991.

[97] R. R. Coifman and M. V. Wickerhauser. Entropy-based algorithms for best basis selection. *IEEE Trans. Inform. Theory*, March 1992.

[98] J. M. Combes, A. Grossmann, and Ph. Tchamitchian, editors. *Wavelets, Time-Frequency Methods and Phase Space*. Springer-Verlag, Berlin, second edition, 1990.

[99] A. G. Constantinides and R. A. Valenzuela. A class of efficient interpolators and decimators with application in transmultiplexers. In *Proc. IEEE Int. Sym. Circuits and Systems*, May 1982.

[100] A. G. Constantinides and R. A. Valenzuela. An efficient and modular transmultiplexer design. *IEEE Trans. Comm.*, COM-30(7):1629–1641, July 1982.

[101] R. A. Conte. A crosstalk model for balanced digital transmission in multipair cables. *AT&T Tech. J.*, 65(3):41–59, 1986.

[102] J. H. Conway and N. J. A. Sloane. Fast quantizing and decoding algorithms for lattice quantizers and codes. *IEEE Transactions on Information Theory*, IT-28:227–232, Mar. 1982.

[103] J. H. Conway and N. J. A. Sloane. A fast encoding method for lattice codes and quantizers. *IEEE Trans. Inform. Theory*, IT-29:820–824, 1983.

[104] C. F. N. Cowan and P. M. Grant. *Adaptive Filters*. Prentice-Hall, 1985.

[105] R. Cox. The design of uniformly and nonuniformly spaced pseudo QMFs. *IEEE Trans. Acoust. Speech Signal Process.*, ASSP-34:1090–1096, October 1986.

[106] R. V. Cox, D. E. Bock, K. B. Bauer, J. D. Johnston, and J. H. Snyder. The analog voice privacy system. *AT&T Tech. J*, 66(1):119–131, 1987.

[107] S. A. Cox and P. F. Adams. An analysis of digital transmission techniques for a local network. *Brit. Telecommun. Technol. J.*, 3(3):73–84, July 1985.

[108] R. E. Crochiere and L. R. Rabiner. *Multirate Digital Signal Processing*. Prentice-Hall, Englewood Cliffs,New Jersey, 1983.

[109] R. E. Crochiere, S. A. Weber, and F. L. Flanagan. Digital coding of speech in sub-bands. *Bell System Technical journal*, 55:1069–1085, Oct. 1976.

[110] A. Croisier, D. Esteban, and C. Galand. Perfect channel splitting by use of interpolation/decimation/tree decomposition techniques. In *Proceedings Inter. Conf. on Information Science and Systems*, Patras, Greece, 1976.

[111] E. A. daSilva and M. Ghanbari. A DCT-based aliasing cancellation method in subband coding. *IEEE Trans. on Circuits and Systems for Video Techn.*, 3(5):384–387, October 1993.

[112] I. Daubechies. private communications.

[113] I. Daubechies. Orthonormal bases of compactly supported wavelets. *Comm. in Pure and Applied Math.*, 41:909–996, 1988.

[114] I. Daubechies. The wavelet transform, time-frequency localization and signal analysis. *IEEE Trans. Inform. Theory*, 36:961–1005, September 1990.

[115] I. Daubechies. *Ten Lectures on Wavelets.* SIAM, Philadelphia, PA, 1992.

[116] I. Daubechies, A. Grossmann, and Y. Meyer. Painless nonorthogonal expansions. *J. Math. Phys.*, 27:1271–1283, 1986.

[117] I. Daubechies and J. C. Lagarias. Two-scale difference equations I. Existence and global regularity of solutions. *SIAM J. Math. Anal.*, 22-5:1388–1410, 1991.

[118] I. Daubechies and J. C. Lagarias. Two-scale difference equations II. Local regularity, infinite products of matrices and fractals. *SIAM J. Math. Anal.*, 23-4:1031–1079, 1992.

[119] P. L. De León II and D. M. Etter. Experimental results with increased bandwidth analysis filters in oversampled subband acoustic echo cancelers. *IEEE Signal Processing Letters*, 2(1):1–3, January 1995.

[120] R. de Queiroz and K. Rao. Adaptive extended lapped transforms. In *Proceedings of the International Conference on Acoustics, Speech, and Signal Processing*, pages III 217–220, 1993.

[121] R. de Quieroz. Subband processing of finite length signals without border distortions. In *Proceedings of the International Conference on Acoustics, Speech, and Signal Processing*, pages 613–616, April 1992.

[122] P. H. N de With, M. Breeuwer, and P. A. M. van Grinsven. Data compression systems for home-use digital video recording. *IEEE Journal on Selected Areas in Communications*, 10:97–121, January 1992.

[123] N. Delprat, B. Escudié, P. Guillemain, R. Kronland-Martinet, Ph. Tchamitchian, and B. Torrésani. Asymptotic wavelet and Gabor analysis: Extraction of instantaneous frequencies. *IEEE Trans. Inform. Theory*, IT-38:644–664, March 1992.

[124] P. Delsarte, B. Macq, and D. T. M. Slock. Signal-adapted multiresolution transform for image coding. *IEEE Trans. Inform. Theory*, 38:897–904, March 1992.

[125] DFDP4/+, Digital Filter Design Package 4. available through Atlanta Signal Processors, Inc.

[126] C. Diab, R. Prost, and R. Goutte. Exact image subband decomposition/reconstruction by DCT. *Signal Processing: Image Communication*, 4(6):489–496, November 1992.

[127] R. C. Dixon. *Spread Spectrum Systems*. Wiley-Interscience, 1976.

[128] I. Djokovic and P. P. Vaidyanathan. Generalized sampling theorems in multiresolution subspaces. Technical report, California Institute of Technology, August 1994.

[129] W. Drews and L. Gazsi. A new design method for polyphase filters using all-pass sections. *IEEE Trans. Circuits and Systems*, CAS-33(3):346–348, March 1986.

[130] D. E. Dudgeon and R. M. Mersereau. *Multidimensional Digital Signal Processing*. Prentice Hall, Englewood Cliffs, NJ, 1984.

[131] H. Dudley. The vocoder. *Bell Labs Record*, 17:122–126, 1939.

[132] M. Duval-Destin. *Analyse spatiale et spatio-temporelle de la stimulation visuelle à l'aide de la transformée en ondelettes*. Thèse de doctorat, Université d'Aix-Marseille II, 1991.

[133] M. Duval-Destin and R. Murenzi. Spatio-temporal wavelets: application to the analysis of moving patterns. In Y. Meyer and S. Roques, editors, *Progress in Wavelet Analysis and Applications (Proc. Toulouse 1992)[266]*, pages 399–408. Frontières, 1993.

[134] R. E. van Dyck and S. A. Rajala. Subband/VQ coding of color images with perceptually optimal bit allocation. *IEEE Trans. on Circuits and Systems for Video Techn.*, 4(1):68–82, February 1994.

[135] N. Dyn, G. Derfel, and D. Levin. Generalized functional equations and subdivision processes. preprint, Tel-Aviv University, 1992.

[136] N. Dyn and D. Levin. Stationary and non-stationary binary subdivision schemes. In T. Lyche and L. Schumaker, editors, *Mathematical methods in computer-aided geometric design*, pages 209–216. Academic Press, 1992.

[137] T. Ebrahimi. *Perceptually Derived Localized Linear Operators: Application to Image Sequence Compression.* PhD thesis, Ecole Polytechnique Federale de Lausanne, 1992.

[138] J. C. Ellershaw, M. J. Biggar, and A. W. Johnson. Interworking of video codecs with different spatial resolutions and aspect ratios. In *Picture Coding Symposium*, 1990.

[139] N. Erdol and F. Basbug. Use of the wavelet transform in optimal receiver design. *Applied Optics*, 33(23):5303–5307, 1994.

[140] D. Esteban and C. Galand. Application of quadrature mirror filters to split-band voice coding schemes. In *Proceedings of the International Conference on Acoustics, Speech, and Signal Processing*, pages 191–195, 1977.

[141] D. D. Falconer. Carrierless AM/PM. Technical report, Bell Laboratories, July 1975. internal memorandum.

[142] M. Farge. Wavelet transforms and their applications to turbulence. *Annu. Rev. Fluid Mech.*, 24:395–457, 1992.

[143] N. Farvardin and J. W. Modestino. Optimum quantizer performance for a class of non-Gaussian memoryless sources. *IEEE Transactions on Information Theory*, IT-30(3):485–497, May 1984.

[144] E. Feig. Practical aspects of DFT-based frequency division multiplying for data transmission. *IEEE Trans. Comm.*, 38(7):929–932, July 1990.

[145] A. Fettweis, J. Nossek, and K. Meerkotter. Reconstruction of signals after filtering and sampling rate reduction. *IEEE Trans. Acoust. Speech Signal Process.*, pages 893–903, August 1985.

[146] H. Fletcher. The mechanism of hearing as revealed through the experiments on the masking effect of thermal noise. In *Proc. National Academy of Science*, volume 24, pages 265–274, 1938.

[147] H. Fletcher. Auditory patterns. *Rev. Modern Phys.*, 12:47–65, January 1940.

[148] N. J. Fliege. *Multirate Digital Signal Processing*. Wiley, 1994.

[149] G. D. Forney, Jr. and M. V. Eyuboglu. Combined equalization and coding using precoding. *IEEE Commun. Mag.*, 30(12):25–34, December 1991.

[150] G. D. Forney, Jr., R. G. Gallagher, G. R. Lang, F. M. Longstaff, and S. V. Qureshi. Efficient modulation for bandlimited channels. *IEEE J. on Sel. Areas in Commun.*, SAC-2:632–647, September 1984.

[151] E. D. Frimout, J. Biemond, and R. L. Lagendijk. Trick mode solutions for MPEG tape recording. In *SPIE Proc. Visual Communications and Image Processing*, pages 311–321, September 1994.

[152] I. Furukawa, M. Nomura, N. Otha, and S. Ono. Hierarchical sub-band coding of super high definition images with adaptive block-size multi-stage VQ. *Signal Processing: Image Communication*, pages 5(5–6):527–538, 1993.

[153] D. Le Gall and A. Tabatabai. Subband coding of digital images using short kernel filters and arithmetic coding techniques. In *Proceedings of the International Conference on Acoustics, Speech, and Signal Processing*, pages 761–764, April 1988.

[154] R. G. Gallager. *Information Theory and Reliable Communication*. John Wiley & Sons, New York, 1968.

[155] P. P. Gandhi, S. S. Rao, and R. S. Pappu. Wavelets for baseband coding of waveforms. In *IEEE Globecom*, pages 363–367, November 1994.

[156] T. Gardos. *Analysis and Design of Multidimensional FIR Filter Banks*. PhD thesis, Georgia Institute of Technology, 1993.

[157] L. Gazsi. Explicit formulas for lattice wave digital filters. *IEEE Trans. Circuits and Systems*, CAS-32(1):68–88, January 1985.

[158] A. Gersho and R. M. Gray. *Vector Quantization and Signal Compression*. Kluwer Academic publishers, Boston, 1992.

[159] J. Gevargiz, M. Rosenmann, P. Das, and L. B. Milstein. A comparison of weighted and non-weighted transform domain processing systems for narrowband interference excision. In *IEEE MILCOM*, pages 32.3.1–4, October 1984.

[160] H. Gharavi. Subband coding of video signals. In J. W. Woods, editor, *Subband Image Coding*. Kluwer Academic Publishers, 1990.

[161] H. Gharavi. Multilayer subband-based video coding. *IEEE Trans. Comm.*, 39:1288–1291, 1991.

[162] H. Gharavi and A. Tabatabai. Subband coding of digital image using two-dimensional quadrature mirror filtering. In *SPIE Proc. Visual Communications and Image Processing*, pages 51–61, 1986.

[163] H. Gharavi and A. Tabatabai. Sub-band coding of monochrome and color images. *IEEE Trans. on Circuits and Systems*, 35:207–214, February 1988.

[164] A. Gilloire. Experiments with subband acoustic echo cancellers for teleconferencing. In *Proceedings of the International Conference on Acoustics, Speech, and Signal Processing*, pages 2141–2144, 1987.

[165] A. Gilloire and M. Vetterli. Adaptive filtering in subbands with critical sampling: Analysis, experiments, and application to acoustic echo cancellation. *IEEE Trans. on Signal Processing*, 40(8):1862–1875, August 1992.

[166] B. Girod. Scalable video for multimedia workstations. *Computers + Graphics*, pages 17(3):269–276, 1993.

[167] R. D. Gitlin, J. F. Hayes, and S. B. Weinstein. *Data Communications Principles.* Plenum Press, New York, 1992.

[168] R. Gold. Optimal binary sequences for spread spectrum multiplexing. *IEEE Trans. on Information Theory*, pages 619–621, October 1967.

[169] R. A. Gopinath. The wavelet transforms and time-scale analysis of signals. Master's thesis, Rice University, 1990.

[170] R. A. Gopinath. Factorization approach to time-varying filter banks and wavelets. In *Proceedings of the International Conference on Acoustics, Speech, and Signal Processing*, April 1994.

[171] R. M. Gray. On the asymptotic eigenvalue distribution of Toeplitz matrices. *IEEE Trans. Inform. Theory*, IT-18(6):725–730, November 1972.

[172] D. D. Greenwood. Critical bandwidth and frequency coordinates of the basilar membrane. *J. Acoust. Soc. Amer.*, 33:1344–1356, 1961.

[173] A. Grossmann, R. Kronland-Martinet, and J. Morlet. Reading and understanding continuous wavelet transforms. In J. M. Combes, A. Grossmann, and Ph. Tchamitchian, editors, *Wavelets, Time-Frequency Methods and Phase Space[98]*, pages 2–20. Springer Verlag, 1990.

[174] A. Grossmann, J. Morlet, and T. Paul. Transforms associated to square integrable group representations. *I, J. Math. Phys*, pages 2473–2479, 1985.

[175] M. Grundstrom and J. Nieweglowski. Simulations of a HDTV system with graceful degradations. In *Int. Workshop on HDTV*, 1993.

[176] R. A. Haddad, A. N. Akansu, and A. Benyassine. Time-frequency localization in transforms, subbands and wavelets: A critical review. *Optical Engineer*, 32(7):1411–1429, July 1993.

[177] R. A. Haddad, K. Park, and N. Uzun. Optimum (MSE) design of quantized M-band filter bank. In *Proceedings of the NJIT 94 Symposium on Applications of Subbands and Wavelets*, March 1994.

[178] R. A. Haddad and N. Uzun. Modeling, analysis, and compensation of quantization effects in M-band subband codecs. In *Proceedings of the International Conference on Acoustics, Speech, and Signal Processing*, pages 173–176, May 1993.

[179] S. Haykin. *Adaptive Filter Theory*. Prentice-Hall, Englewood Cliffs, New Jersey, 1986.

[180] R. P. Helman. Asymmetry of masking between noise and tone. *Percept. Psychophys.*, pages 241–246, 1972.

[181] C. Herley. *Wavelets and Filter Banks*. PhD thesis, Columbia University, 1993.

[182] C. Herley and M. Vetterli. Orthogonal time varying filter banks and wavelets. In *Proc. IEEE Int. Sym. Circuits and Systems*, April 1993.

[183] C. Herley and M. Vetterli. Wavelets and recursive filter banks. *IEEE Trans. Acoust. Speech Signal Process.*, ASSP-41(8):2537–2557, August 1993.

[184] O. Herrmann. On the approximation problem in nonrecursive digital filter design. *IEEE Trans. Circuit Theory*, CT-18(3):411–413, May 1971.

[185] N. Hess-Nielsen. Frequency localization of generalized wavelet packets. preprint Wash. Univ., St. Louis., 1993.

[186] K. Hetling, M. Medley, G. Saulnier, and P. Das. A PR-QMF (wavelet) coding based spread spectrum communications system. In *IEEE MILCOM*, pages 760–764, October 1994.

[187] K. Hetling, G. Saulnier, and P. Das. Optimized filter design for PR-QMF based spread spectrum communications. In *Proc. IEEE Int. Conf. on Communications*, 1995. Submitted.

[188] K. Hetling, G. Saulnier, and P. Das. Optimized PR-QMF based codes for multiuser communications. In *SPIE Proceedings - Wavelet Applications for Dual Use*, pages 248–259, April 1995.

[189] B. Hirosaki et al. Advanced groupband data model using orthogonally multiplexed QAM technique. *IEEE Trans. Comm.*, 34(6):587–592, June 1986.

[190] J. K. Holmes. *Coherent Spread Spectrum Systems.* John Wiley & Sons, Inc., 1982.

[191] M. Holschneider. On the wavelet transformation of fractal objects. *J. Stat. Phys.*, 50:963–993, 1988.

[192] M. Holschneider, R. Kronland-Martinet, J. Morlet, and Ph. Tchamitchian. A real-time algorithm for signal analysis with the help of the wavelet transform. In J. M. Combes, A. Grossmann, and Ph. Tchamitchian, editors, *Wavelets, Time-Frequency Methods and Phase Space[98]*, pages 286–297. Springer Verlag, 1990.

[193] U. Horn and B. Girod. Pyramid coding using lattice vector quantization for scalable video applications. In *Proc. PCS '94*, Sacramento, CA, September 1994.

[194] F. M. Hsu and A. A. Giordano. Digital whitening techniques for improving spread spectrum communications performance in the presence of narrowband jamming and interference. *IEEE Trans. Comm.*, COM-26:209–216, February 1978.

[195] D. A. Huffman. A method for the construction of minimum redundancy codes. *Proceedings of the IRE*, 40:1098–1101, 1952.

[196] D. Hughes-Hartogs. Ensemble modem structure for imperfect transmission media. U. S. Patent Nos. 4,679,227 (July 1987); 4,731,816 (March 1988); and 4,833,706 (May 1989).

[197] H. Husøy and S. Aase. Image subband coding with adaptive filter banks. In *SPIE Proc. Visual Communications and Image Processing*, November 1992.

[198] J. H. Husøy. Low-complexity subband coding of still images and video. *Optical Engineering*, 30(7), July 1991.

[199] J. H. Husøy. *Subband Coding of Still Images and Video.* PhD thesis, Norwegian Institute of Technology, 1991.

[200] J. H. Husøy, H. Gronning, and T. A. Ramstad. Subband coding of video employing efficient recursive filter banks and advanced motion compensation. In *Proc. SPIE Conference on Visual Comm. and Image Processing*, pages 546–557, October 1990.

[201] J. H. Husøy and T. A. Ramstad. Subband coding of images employing an efficient parallel IIR filterbank. In *Proc. SPIE Conference on Visual Comm. and Image Processing*, pages 752–763, November 1989.

[202] J. H. Husøy and T. A. Ramstad. Application of an efficient parallel IIR filter bank to image subband coding. *Signal Processing*, 20(4):279–292, August 1990.

[203] IEEE special issue on wavelet transforms and multiresolution signal analysis, IT-38, nr. 2. IEEE Transactions on Information Theory, March 1992.

[204] *IEEE Transactions on Consumer Electronics*, 35:132–248, August 1993.

[205] R. A. Iltis. A GLRT-based spread-spectrum receiver for joint channel estimation and interference suppression. *IEEE Trans. Comm.*, 37(3):277–288, March 1989.

[206] R. A. Iltis and L. B. Milstein. Performance analysis of narrow-band interference rejection techniques in DS spread-spectrum systems. *IEEE Trans. Comm.*, COM-32:1169–1177, November 1984.

[207] ISO 10918-1 JPEG draft international standard, CCITT recommendation T.81, 1992.

[208] ISO/IEC 13818-2, generic coding of moving pictures and associated audio, recommendation H.262 (MPEG-2), March 1994. Draft International Standard.

[209] Y. Itoh, Y. Maruyama, N. Furuya, and T. Araseki. An acoustic echo canceller for teleconference. In *Proceedings of the International Conference on Acoustics, Speech, and Signal Processing*, pages 1498–1501, 1985.

[210] N. S. Jayant. Effective number of keys in a voice privacy system based on permutation scrambling. *AT&T Tech. J*, 66(1):191–195, 1987.

[211] N. S. Jayant, R. V. Cox, B. J. McDermott, and A. M. Quinn. Analog scramblers for speech based on sequential permutations in time and frequency. *Bell Syst. Tech. J.*, 62(1):25–46, 1983.

[212] N. S. Jayant, B. J. McDermott, S. W. Christensen, and A. M. Quinn. A comparison of four methods for analog speech privacy. *IEEE Trans. Comm.*, 29:18–23, 1981.

[213] N. S. Jayant and P. Noll. *Digital Coding of Waveforms.* Prentice-Hall, Englewood Cliffs,New Jersey, 1984.

[214] P. Jeanrenaud and M. J. T. Smith. Recursive multirate image coding with adaptive prediction and finite state vector quantization. *Signal Processing*, 20(1):25–42, May 1990.

[215] J. H. Jeon, C. H. Ham, and J. K. Kim. On the hierarchical block motion estimation for video subband coding. In *SPIE Proc. Visual Communications and Image Processing*, volume 1605, pages 954–962, 1991.

[216] J. Johnston. A filter family designed for use in quadrature mirror filter banks. In *Proceedings of the International Conference on Acoustics, Speech, and Signal Processing*, pages 291–294, April 1980.

[217] J. D. Johnston and K. Brandenburg. Wideband coding perceptual considerations for speech and music. In S. Furui and M. M. Sondhi, editors, *Advances in Speech Signal Processing.* Marcel Dekker, NY, 1992.

[218] J. D. Johnston and A. J. Ferreira. Sum-difference stereo transform coding. In *Proceedings of the International Conference on Acoustics, Speech, and Signal Processing*, volume II, pages 569–572, 1992.

[219] I. Kalet. The multitone channel. *IEEE Trans. Comm.*, 37(2):119–124, February 1989.

[220] I. Kalet and S. Shamai (Shitz). On the capacity of a twisted-wire pair — Gaussian model. *IEEE Trans. Commun.*, 30(3):379–383, March 1990.

[221] G. Karlsson and M. Vetterli. Three-dimensional subband coding of video. In *Proceedings of the International Conference on Acoustics, Speech, and Signal Processing*, pages 1100–1103, April 1988.

[222] G. Karlsson and M. Vetterli. Extension of finite length signals for subband coding. *Signal Processing*, 17:161–168, June 1989.

[223] G. Karlsson and M. Vetterli. Theory of two-dimensional multirate filter banks. *IEEE Trans. Acoust. Speech Signal Process.*, ASSP-38:925–937, June 1990.

[224] J. Katto and Y. Yasuda. Performance evaluation of subband coding and optimization of its filter coefficients. In *SPIE Proc. Visual Communications and Image Processing*, pages 95–106, November 1991.

[225] M. S. Keshner. $1/f$ noise. *Proc. IEEE*, 70:212–218, March 1982.

[226] J. W. Ketchum and J. G. Proakis. Adaptive algorithms for estimating and suppressing narrow-band interference in PN spread-spectrum systems. *IEEE Trans. Comm.*, COM-30:913–924, May 1982.

[227] C. W. Kim. An algebraic approach to linear-phase regular wavelet construction. In *Proc. 1993 Conference Info. Sci. Syst.*, Baltimore, MD, March 1993. Johns Hopkins Univ.

[228] C. W. Kim and R. Ansari. FIR/IIR exact reconstruction filter banks with applications to subband coding of images. In *Proc. Midwest Symposium on Circuits and Systems*, pages 227–230, July 1991.

[229] C. W. Kim and R. Ansari. Subband decomposition procedure for quincunx sampling grids. In *Proc. SPIE Conference on Visual Comm. and Image Processing*, pages 112–123, November 1991.

[230] C. W. Kim, R. Ansari, and A. E. Cetin. A class of linear-phase biorthogonal wavelets. In *Proceedings of the International Conference on Acoustics, Speech, and Signal Processing*, pages 673–676, March 1992.

[231] Y. H. Kim and J. W. Modestino. Adaptive entropy coded subband coding of images. *IEEE Trans. on Image Processing*, 1(1):31–48, January 1992.

[232] J. R. Klauder and B. S. Skagerstam. Coherent States – Applications in Physics and Mathematical Physics. *World Scientific*, Singapore, 1985.

[233] R. Koilpillai and P. Vaidyanathan. New results on cosine modulated FIR filter banks satisfying perfect reconstruction. In *Proceedings of the International Conference on Acoustics, Speech, and Signal Processing*, pages 1793–1796, 1991.

[234] J. Kovacevic. *Filter Banks and Wavelets: Extensions and Applications*. PhD thesis, Columbia University, 1991.

[235] J. Kovacevic. Eliminating correlated errors in subband/wavelet coding systems with quantization. In *Asilomar Conf. on Signals, Systems and Computers*, 1993.

[236] J. Kovacevic and M. Vetterli. Perfect reconstruction filter banks with rational sampling rate changes. In *Proceedings of the International Conference on Acoustics, Speech, and Signal Processing*, pages 1785–1788, 1991.

[237] T. Kronander. A new approach to recursive mirror filters with a special application in subband coding of images. *IEEE Trans. Acoust. Speech Signal Process.*, pages 1496–1500, September 1988.

[238] M. Kuhn. Single frequency network of DBP-Telekom. In *Int. Workshop on HDTV*, 1993.

[239] R. E. Learned, H. Krim, B. Claus, A. S. Willsky, and W. C. Karl. Wavelet packet-based multiple access communication. *SPIE Proceedings - Wavelet Applications in Signal and Image Processing II*, 2303, July 1994.

[240] J. Lee and B. W. Dickinson. Subband video coding with temporally adaptive motion interpolation. In *Proceedings of the International Conference on Acoustics, Speech, and Signal Processing*, pages V 497–500, 1994.

[241] L. S. Lee, G. C. Chou, and C. S. Chang. A new frequency domain speech scrambling system which does not require frame synchronization. *IEEE Trans. Comm.*, 32:444–456, 1984.

[242] D. LeGall. MPEG: A video compression standard for multimedia applications. *Trans. of the ACM*, 34:46–58, 1991.

[243] J. W. Leichleider. High bit rate digital subscriber lines: A review in progress. *IEEE J. on Sel. Areas in Commun.*, pages 555–566, October 1991.

[244] L. M. Li and L. B. Milstein. Rejection of narrow-band interference in PN spread-spectrum systems using transversal filters. *IEEE Trans. Comm.*, COM-30:925–928, May 1982.

[245] M. Liebhold and E. M. Hoffert. Towards an open environment for digital video. *Communications of the ACM*, 34:103–112, April 1991.

[246] J. Lim. *Two Dimensional Signal and Image Processing.* Prentice Hall, Englewood Cliffs, NJ, 1990.

[247] Y. Linde, A. Buzo, and R. M. Gray. An algorithm for vector quantizer design. *IEEE Transactions on Communications*, COM-28:84–95, Jan. 1980.

[248] S. P. Lloyd. Least squares quantization in PCM. *Bell Laboratories Technical note*, 1957. Published in the March 1982 Special issue on Quantization: *IEEE Transactions on Information Theory, Part 1*, IT-28:129-137.

[249] J. Luo, C. W. Chen, K. J. Parker, and T. S. Huang. Three-dimensional subband video analysis and synthesis with adaptive clustering in high-frequency subbands. In *International Conference on Image Processing*, pages 255–259, November 1994.

[250] J. I. Makhoul and J. J. Wolf. Linear prediction and the spectral analysis of speech. Technical report, Bolt, Beranek, and Newman, Inc, 1972.

[251] S. G. Mallat. A theory for multiresolution signal decomposition: The wavelet representation. *IEEE Trans. Pattern Anal. and Mach. Int.*, 11(7):674–693, July 1989.

[252] H. S. Malvar. The LOT: A link between transform coding and multirate filter banks. In *Proc. IEEE Int. Sym. Circuits and Systems*, pages 835–838, 1988.

[253] H. S. Malvar. Modulated QMF filter banks with perfect reconstruction. *Electronics Letters*, 26:906–907, June 1990.

[254] H. S. Malvar. *Signal Processing with Lapped Transforms.* Artech House, 1991.

[255] H. S. Malvar and D. Staelin. The LOT transform coding without blocking effects. *IEEE Trans. Acoust. Speech Signal Process.*, ASSP-37:553–559, April 1989.

[256] D. Marr. *Vision.* Freeman, San Francisco, 1982.

[257] H. G. Martinez and T. W. Parks. A class of infinite-duration impulse response digital filters for sampling rate reduction. *IEEE Trans. Acoust. Speech Signal Process.*, ASSP-27(2):154–162, April 1979.

[258] J. Max. Quantizing for minimum distortion. *IRE Transactions on Information Theory*, IT-6(2):7–12, Mar. 1960.

[259] M. Medley, G. Saulnier, and P. Das. Applications of the wavelet transform in spread spectrum communications systems. In *SPIE Proceedings - Wavelet Applications*, volume 2242, pages 54–68, April 1994.

[260] M. Medley, G. Saulnier, and P. Das. Radiometric detection of direct-sequence spread spectrum signals with interference excision using the wavelet transform. In *Proceedings of Supercomm/ICC*, pages 1648–1652, May 1994.

[261] M. Medley, G. Saulnier, and P. Das. The application of wavelet-domain adaptive filtering to spread spectrum communications. In *SPIE Proceedings - Wavelet Applications for Dual-Use*, volume 2491, pages 233–247, April 1995.

[262] A. Mertins. Statistical optimization of PR-QMF filter banks and wavelets. In *Proceedings of the International Conference on Acoustics, Speech, and Signal Processing*, pages 113–116, 1994.

[263] M. Mettke, M. Medley, G. Saulnier, and P. Das. Wavelet transform excision using IIR filters in spread spectrum communication systems. In *IEEE Globecom*, pages 1627–1631, November 1994.

[264] Y. Meyer. *Wavelets and Applications (Proc. Marseille 1989)*. Springer-Verlag, Berlin, and Masson, Paris, 1991.

[265] Y. Meyer. *Wavelets: Algorithms and Applications.* SIAM, Philadelphia, PA, 1993.

[266] Y. Meyer and S. Roques (eds.). *Progress in Wavelet Analysis and Applications (Proc. Toulouse 1992)*. Frontières, Gif-sur-Yvette, 1993.

[267] C. A. Micchelli and H. Prautzsch. Uniform refinements of curves. *Linear Algebra and Its Applications*, 114-115:841–870, 1989.

[268] P. Millar. Mirror filters with minimum delay responses for use in subband coders. In *Proceedings of the International Conference on Acoustics, Speech, and Signal Processing*, page 11.5, March 1984.

[269] L. B. Milstein. Interference rejection techniques in spread spectrum communications. *Proc. IEEE*, 76(6):657–671, June 1988.

[270] L. B. Milstein and P. K. Das. An analysis of a real-time transform domain filtering digital communication system - Part I: Narrowband interference rejection. *IEEE Trans. Comm.*, 28(6):816–824, June 1980.

[271] L. B. Milstein and P. K. Das. An analysis of a real-time transform domain filtering digital communications system - Part II: Wideband interference rejection. *IEEE Trans. Comm.*, COM-31:21–27, June 1983.

[272] L. B. Milstein, P. K. Das, and J. Gevargiz. Processing gain advantage of transform filtering DS spread spectrum systems. In *IEEE MILCOM*, pages 21.2.1–4, October 1982.

[273] L. B. Milstein et al. On the feasibility of a CDMA overlay for personal communications networks. *IEEE Journal on Selected Areas in Communications*, 10:655–668, May 1992.

[274] F. Mintzer. Filters for distortion-free two-band multirate filter banks. *IEEE Trans. Acoust. Speech Signal Process.*, ASSP-33:626–630, June 1985.

[275] S. K. Mitra and K. Hirano. Digital all-pass networks. *IEEE Trans. Circuits and Systems*, CAS-21:207–214, February 1974.

[276] N. Mohsenian and N. M. Nasrabadi. Edge-based subband VQ techniques for images and video. *IEEE Trans. on Circuits and Systems for Video Techn.*, 4(1):53–67, February 1994.

[277] R. A. Monzingo and T. W. Miller. *Introduction to Adaptive Arrays*. John Wiley & Sons, Inc, 1980.

[278] B. C. J. Moore. *An introduction to the Psychology of Hearing*. Academic Press, 1989.

[279] D. R. Morgan. Slow asymptotic convergence of LMS acoustic echo cancelers. *IEEE Trans. Speech and Audio Processing*, 3(2):126–136, March 1995.

[280] A. Morgul, P. M. Grant, and C. F. N. Cowan. Wide-band hybrid analog/digital frequency domain adaptive filter. *IEEE Trans. Acoust. Speech Signal Process.*, ASSP-32(4):762–769, August 1984.

[281] R. Murenzi. *Ondelettes multidimensionnelles et applications à l'analyse d'images*. Thèse de doctorat, Univ. Cath. Louvain, Louvain-la-Neuve, 1990.

[282] R. Murenzi. Wavelet transforms associated to the n-dimensional euclidean group with dilations: signals in more than one dimension. In J. M. Combes, A. Grossmann, and Ph. Tchamitchian, editors, *Wavelets, Time-Frequency Methods and Phase Space[98]*. Springer Verlag, 1990.

[283] S. Nanda and W. Pearlman. Tree coding of image subbands. *IEEE Transactions of Image Processing*, 1(2):133–147, April 1992.

[284] M. J. Narasimha and A. M. Peterson. Design and application of uniform digital bandpass filter banks. In *Proceedings of the International Conference on Acoustics, Speech, and Signal Processing*, pages 499–503, 1978.

[285] N. Nasrabadi and R. King. Image coding using vector quantization: A review. *IEEE Trans. on Communications*, 36(8):957–971, August 1988.

[286] T. Naveen, F. Bosveld, J. W. Woods, and R. L. Lagendijk. Rate constrained multiresolution transmission of video. *IEEE Trans. on Circuits and Systems for Video Technology*, 1995. to appear.

[287] T. Naveen and J. W. Woods. Motion-compensated multiresolution transmission of high definition video. *IEEE Trans. on Circuits and Systems for Video Technology*, 4:29–41, February 1994.

[288] K. Nayebi. *A Time Domain Framework for the Analysis and Design of FIR Multirate Filter Bank Systems.* PhD thesis, Georgia Institute of Technology, Atlanta, GA, December 1990.

[289] K. Nayebi, T. Barnwell, and M. J. T. Smith. Time domain conditions for exact reconstruction in analysis/synthesis systems based on maximally decimated filter banks. In *19th Southeastern Symposium on System Theory*, pages 498–502, March 1987.

[290] K. Nayebi, T. Barnwell, and M. J. T. Smith. A general time domain analysis and design framework for exactly reconstructing FIR analysis/synthesis filter banks. In *Proc. IEEE Int. Sym. Circuits and Systems*, pages 2022–2025, May 1990.

[291] K. Nayebi, T. Barnwell, and M. J. T. Smith. The time domain analysis and design of exactly reconstructing analysis/synthesis filter banks. In *Proceedings of the International Conference on Acoustics, Speech, and Signal Processing*, pages 1735–1738, April 1990.

[292] K. Nayebi, T. Barnwell, and M. J. T. Smith. Design and implementation of computationally efficient modulated filter banks. In *Proc. IEEE Int. Sym. Circuits and Systems*, pages 650–653, June 1991.

[293] K. Nayebi, T. Barnwell, and M. J. T. Smith. Low delay coding of speech and audio using nonuniform band filter banks. In *Proceedings of the IEEE Workshop on Speech Coding for Telecommunications*, September 1991.

[294] K. Nayebi, T. Barnwell, and M. J. T. Smith. Analysis-synthesis systems based on time varying filter banks. In *Proceedings of the International Conference on Acoustics, Speech, and Signal Processing*, volume IV, pages 617–620, March 1992.

[295] K. Nayebi, T. Barnwell, and M. J. T. Smith. Block decimated analysis-synthesis filter banks. In *Proc. IEEE Int. Sym. Circuits and Systems*, pages 947–950, May 1992.

[296] K. Nayebi, T. Barnwell, and M. J. T. Smith. Time domain filter bank analysis: A new design theory. *IEEE Trans. Signal Process.*, pages 1412–1429, June 1992.

[297] K. Nayebi, T. Barnwell, and M. J. T. Smith. Low delay FIR filter banks: Design and evaluation. *IEEE Trans. on Signal Processing*, 42(1):24–32, January 1994.

[298] K. Nayebi, T. P. Barnwell, and M. J. T. Smith. Design of perfect reconstruction nonuniform band filter banks. In *Proceedings of the International Conference on Acoustics, Speech, and Signal Processing*, pages 1781–1784, May 1991.

[299] K. Nayebi, T. P. Barnwell, and M. J. T. Smith. Nonuniform filter banks : A reconstruction and design theory. *IEEE Trans. on Signal Processing*, 41(3), March 1993.

[300] T. Q. Nguyen. A quadratic-constrained least-squares approach to the design of digital filter banks. In *Proc. IEEE Int. Sym. Circuits and Systems*, pages 1344–1347, May 1992.

[301] V. Nuri. A theory for exactly reconstructing size-limited maximally decimated filter banks. Masters' thesis, Washington State University, Pullman, WA, May 1993.

[302] V. Nuri and R. H. Bamberger. A theoretical framework for the analysis and design of size-limited multirate filter banks. In *Proceedings of the IEEE-SP International Symposium on Time-Frequency and Time-Scale Analysis*, pages 311–314, Victoria, BC, Canada, October 1992.

[303] H. Nussbaumer. Pseudo QMF filter banks. *IBM Technical Disclosure Bulletin*, 24:3081–3087, November 1981.

[304] H. J. Nussbaumer. Polynomial transform implementation of digital filter banks. *IEEE Trans. Acoust. Speech Signal Process.*, ASSP-31:616–622, June 1983.

[305] J. R. Ohm. Motion-compensated 3D subband coding with multiresolution representation of motion parameters. In *International Conference on Image Processing*, pages 250–254, November 1994.

[306] J. R. Ohm. Three-dimensional subband coding with motion compensation. *IEEE Trans. on Image Processing*, 3:559–571, September 1994.

[307] A. V. Oppenheim and R. W. Schafer. *Digital Signal Processing*. Prentice Hall, Englewood Cliffs,New Jersey, 1975.

[308] R. S. Orr, C. Pike, M. Tzannes, S. Sandberg, and M. Bates. Covert communications employing wavelet technology. In *The Conference Record of the 27th Asilomar Conference on Signals, Systems and Computers*, pages 523–527, November 1993.

[309] P. F. Panter and W. Dite. Quantization in pulse-count modulation with nonuniform spacing of levels. *Proceedings of the IRE*, 39:44–48, Jan. 1951.

[310] A. Papoulis. *Signal Analysis*. McGraw-Hill, New York, 1977.

[311] A. Papoulis. *Probability, Random Variables, and Stochastic Processes*. McGraw Hill, third edition, 1991.

[312] K. Park and R. A. Haddad. Optimum subband filter bank design and compensation in presence of quantizers. In *Asilomar Conf. on Signals, Systems and Computers*, 1993.

[313] J. Pearl. On coding and filtering stationary signals by discrete Fourier transforms. *IEEE Trans. Inform. Theory*, pages 229–232, March 1973.

[314] M. Pecot, P. J. Tourtier, and Y. Thomas. Compatible coding of television images, Part 2 compatible system. *Signal Processing: Image Communications*, 2:249–268, 1990.

[315] W. B. Pennebaker and J. L. Mitchell. *JPEG Still Image Data Compression Standard*. Van Nostrand Reinhold, New York, 1993.

[316] S. M. Phoong, C. W. Kim, P. P. Vaidyanathan, and R. Ansari. A new class of two-channel biorthogonal filter banks and wavelet bases. *IEEE Trans. on Signal Processing*, pages 649–665, March 1995.

[317] S. M. Phoong and P. P. Vaidyanathan. Two-channel 1-D and 2-D biorthonormal filter banks with causal stable IIR and linear phase FIR filters. In *Proc. IEEE Int. Symp. Circuits and Systems*, May 1994.

[318] C. I. Podilchuk. Low bit rate subband video coding. In *International Conference on Image Processing*, pages 280–284, November 1994.

[319] C. I. Podilchuk and A. Jacquin. Subband video coding with a dynamic bit allocation and geometric vector quantization. In *SPIE Proc. IT&T Symp. Electr. Imaging Techn.*, pages 241–252, February 1992.

[320] M. R. Portnoff. *Time-Scale Modification of Speech Based on Short-Time Fourier Analysis*. PhD thesis, MIT, 1978.

[321] M. R. Portnoff. Time-frequency representation of digital signals and systems based on short-time Fourier analysis. *IEEE Trans. Acoust. Speech Signal Process.*, 28:55–69, February 1980.

[322] W. H. Press, S. A. Teukolsky, W. T. Vetterling, and B. P. Flannery. *Numerical Recipes in C.* Cambridge University Press, 1992.

[323] R. Price. Nonlinearly feedback-equalized PAM vs. capacity for noisy filter channels. In *International Conference on Communication*, pages 22.12–22.17, June 1972.

[324] J. P. Princen and A. B. Bradley. Analysis/synthesis filter bank design based on time domain alias cancellation. *IEEE Trans. Acoust. Speech Signal Process.*, pages 1153–1161, October 1986.

[325] H. S. Ptasinski and R. D. Fellman. Performance analysis of a fractal communication system. *SPIE Proceedings - Wavelet Applications*, 2242:78–86, April 1994.

[326] A. Puri and B. G. Haskell. Temporal resolution scalable video coding. In *International Conference on Image Processing*, pages 247–251, November 1994.

[327] L. Rabiner and R. Schafer. *Digital Processing of Speech Signals.* Prentice Hall, Englewood Cliffs, NJ., 1978.

[328] T. Ramstad and O. Foss. Subband coding design using recursive QMFs. *Signal Processing: Theories and Applications*, pages 747–752, 1980.

[329] T. Ramstad and J. Tanem. Cosine-modulated analysis synthesis filter banks with critical sampling and perfect reconstruction. In *Proceedings of the International Conference on Acoustics, Speech, and Signal Processing*, page 46.D4b.4, 1991.

[330] T. A. Ramstad. Digital two-rate IIR and hybrid IIR/FIR filters for sample rate conversion. *IEEE Trans. Comm.*, COM-30(7):1466–1476, July 1982.

[331] T. A. Ramstad. Analysis/synthesis filter banks with critical sampling. In *Proc. Int'l Conf. Digital Signal Processing*, pages 130–134, September 1984.

[332] T. A. Ramstad. IIR filter bank for subband coding of images,. In *Proc. IEEE Int. Sym. Circuits and Systems*, June 1988.

[333] K. P. Rao and P. Yip. *Discrete Cosine Transfroms:Algorithms, Advantages, Applications.* Academic Press, New York, 1990.

[334] F. A. Reed and P. L. Feintuch. A comparison of LMS adaptive cancellers implemented in the frequency domain and the time domain. *IEEE Trans. Acoust. Speech Signal Process.*, ASSP-29(3):770–775, June 1981.

[335] I. Reed and G. Solomon. Polynomial codes over certain finite fields. *SIAM J.*, 8:300–304, June 1960.

[336] P. A. Regalia, S. K. Mitra, M. Renfors P. P. Vaidyanathan, and Y. Neuvo. Tree-structured complementary filter banks using all-pass sections. *IEEE Trans. Circuits and Systems*, CAS-34(1470-1484), December 1987.

[337] M. Renfors and T. Saramaki. Recursive n-th band digital filters - part I: design and properties. *IEEE Trans. on Signal Processing*, CAS-34(1):24–39, January 1987.

[338] M. Renfors and T. Saramaki. Recursive n-th band digital filters - part II: design of multistage interpolators and decimators. *IEEE Trans. on Signal Processing*, CAS-34(1):40–51, January 1987.

[339] M. Richards. Helium speech enhancement using the short-time Fourier transform. *IEEE Trans. Acoust. Speech Signal Process.*, 30:841–853, December 1982.

[340] O. Rioul. Simple regularity criteria for subdivision schemes. *SIAM J. Math. Anal.*, 23-6:1544–1576, 1992.

[341] O. Rioul. A discrete-time multiresolution theory. *IEEE Trans. on Signal Processing*, 41(8), 1993.

[342] O. Rioul and M. Vetterli. Wavelets and signal processing. *IEEE Signal Processing Magazine*, 8(4):14–37, 1991.

[343] A. D. Rizos, J. G. Proakis, and T. Q. Nguyen. Comparison of DFT and cosine modulated filter banks in multicarrier modulation. In *Proc. IEEE Globecom*, 1994.

[344] M. Rosenmann, J. Gevargiz, P. Das, and L. B. Milstein. Probability of error measurement for an interference resistant transform domain processing receiver. In *IEEE MILCOM*, pages 638–640, October 1983.

[345] J. Rothweiler. Polyphase quadrature mirror filters – a new sub-band coding technique. In *Proceedings of the International Conference on Acoustics, Speech, and Signal Processing*, pages 285–288, 1983.

[346] A. Ruiz and J. Cioffi. A frequency domain approach to combined spectral shaping and coding. In *IEEE International Conference of Communications*, pages 49.3.1–49.3.5, June 1987.

[347] M. B. Ruskai, R. Coifman G. Beylkin, I. Daubechies, S. Mallat, Y. Meyer, and L. Raphael (eds.). *Wavelets and Their Applications.* Jones and Bartlett, Boston, 1992.

[348] R. J. Safranek and J. D. Johnston. A perceptually tuned sub-band image coder with image dependent quantization and post-quantization data compression. In *Proceedings of the International Conference on Acoustics, Speech, and Signal Processing*, pages 1945–1948, Glasgow, UK, 1989.

[349] B. R. Saltzberg. Performance of an efficient parallel data transmission system. *IEEE Trans. Comm.*, 15(6):805–811, December 1967.

[350] J. Salz. Optimum mean-square decision feedback equalization. *Bell Syst. Tech. J.*, 52(8):1341–1362, October 1973.

[351] S. Sandberg, M. Tzannes, P. Heller, R. Orr, C. Pike, and M. Bates. A family of wavelet-related sequences as a basis for an LPI/D communications system prototype. In *IEEE MILCOM*, pages 537–542, October 1993.

[352] S. D. Sandberg, S. Del Marco, K. Jagler, and M. A. Tzannes. Some alternatives in transform-domain suppression of narrow-band interference for signal detection and demodulation. Technical report, Aware Inc., 1994.

[353] S. D. Sandberg and M. A. Tzannes. Overlapped discrete multitone modulations for high speed copper wire communications. Technical report, Aware Inc., 1994.

[354] G. Saracco, A. Grossmann, and Ph. Tchamitchian. Use of wavelet transforms in the study of propagation of transient acoustic signals across a plane interface between two homogeneous media. In J. M. Combes, A. Grossmann, and Ph. Tchamitchian, editors, *Wavelets, Time-Frequency Methods and Phase Space[98].* Springer Verlag, 1990.

[355] G. J. Saulnier, P. Das, and L. B. Milstein. An adaptive digital suppression filter for direct sequence spread spectrum communications. *IEEE Journal on Selected Areas in Communications*, SAC-3(5):676–686, September 1985.

[356] R. W. Schafer and L. R. Rabiner. A digital signal processing approach to interpolation. In *Proc. IEEE*, pages 692–702, June 1973.

[357] G. Schamel. Graceful degradation and scalability in digital coding for terrestrial transmission. In *Int. Workshop on HDTV*, November 1992.

[358] B. Scharf. Critical bands. In *Foundations of Modern Auditory Theory*, pages 159–202. Academic Press, 1970.

[359] H. Scheuermann and H. Gockler. A comprehensive survey of digital transmultiplexing methods. In *Proc. IEEE*, pages 1419–1450, November 1981.

[360] G. Schuller and M. J. T. Smith. A general formulation for modulated perfect reconstruction filter banks with variable system delay. In *Proceedings of the NJIT 94 Symposium on Applications of Subbands and Wavelets*, pages 27–34, March 1994.

[361] G. Schuller and M. J. T. Smith. A new framework for modulated perfect reconstruction filter banks. *IEEE Trans. on Signal Processing*, 1996. to appear.

[362] P. Scotton and J. Menez. A low complexity video subband coder for ATM. *Signal Processing: Image Communications*, 6:421–433, 1994.

[363] T. Senoo and B. Girod. Vector quantization for entropy coding of image subbands. *IEEE Trans. on Image Processing*, 1(4):526–532, October 1992.

[364] E. Séré. Localisation fréquentielle des paquets d'ondelettes. *Revista Matemática Iberoamericana*, 1993. to appear.

[365] C. E. Shannon. A mathematical theory of communication. *Bell Syst. Tech. J.*, 27:379–423 & 623–656, July 1948.

[366] J. M. Shapiro. Application of the embedded wavelet hierarchical image coder to very low bit rate image coding. In *Proceedings of the International Conference on Acoustics, Speech, and Signal Processing*, pages 558–561, April 1993.

[367] E. Simoncelli. Orthogonal sub-band image transforms. Master's thesis, MIT, 1988.

[368] E. P. Simoncelli. Design of multi-dimensional derivative filters. In *Proc. IEEE ICIP*, pages I 790–793, 1994.

[369] E. P. Simoncelli and E. H. Adelson. Non-separable extensions of quadrature mirror filters to multiple dimensions. *Proceedings of the IEEE*, 78(4):652–664, April 1990.

[370] D. S. Sinha and A. H. Tewfik. Low bit rate transparent audio compression using adapted wavelets. *IEEE Trans. on Signal Processing*, 41(12):3643–3479, 1993.

[371] M. J. T. Smith. *Exact Reconstruction Analysis/Synthesis Systems and Their Application to Frequency Domain Coding.* PhD thesis, Georgia Institute of Technology, 1984.

[372] M. J. T. Smith. IIR analysis/synthesis systems. In J. W. Woods, editor, *Subband Image Coding.* Kluwer Academic Publishers, 1991.

[373] M. J. T. Smith and T. Barnwell. A procedure for designing exact reconstruction filter banks for tree-structured subband coders. In *Proceedings of the International Conference on Acoustics, Speech, and Signal Processing*, pages 27.1.1–27.1.4, March 1984.

[374] M. J. T. Smith and T. Barnwell. The design of digital filters for exact reconstruction in subband coding. *IEEE Trans. Acoust. Speech Signal Process.*, ASSP-34(3), June 1986.

[375] M. J. T. Smith and W. Chung. Adaptive IIR analysis-synthesis filter banks. In *IEEE DSP Workshop*, pages 2.3.1–2, September 1992.

[376] M. J. T. Smith and W. Chung. Spatially adaptive IIR analysis-synthesis filter banks for image coding. In *Asilomar Conference on Signals, Systems and Computers*, pages 1016–1020, October 1992.

[377] M. J. T. Smith and W. Chung. Recursive time-varying filter banks for subband image coding. *IEEE Trans. on Signal Processing*, July 1995.

[378] M. J. T. Smith and S. Eddins. Subband coding of images with octave band tree structures. In *Proceedings of the International Conference on Acoustics, Speech, and Signal Processing*, pages 1382–1385, April 1987.

[379] M. J. T. Smith and S. L. Eddins. Analysis/synthesis techniques for subband image coding. *IEEE Trans. on Acoustics, Speech, and Signal Processing*, 38(8):1446–1456, August 1991.

[380] M. J. T. Smith, R. M. Mersereau, and T. P. Barnwell. Exact reconstruction recursive filter banks for sub-band image coding. In *Proc of IEEE Miami Technicon 87*, pages 121–124, October 1987.

[381] I. Sodagar, K. Nayebi, and T. Barnwell. A class of time-varying wavelet transforms. In *Proceedings of the International Conference on Acoustics, Speech, and Signal Processing*, April 1993.

[382] I. Sodagar, K. Nayebi, T. Barnwell, and M. J. T. Smith. A new approach to time-varying FIR filter banks. In *Asilomar Conf. on Signals, Systems and Computers*, pages 1271–1274, November 1993.

[383] I. Sodagar, K. Nayebi, T. Barnwell, and M. J. T. Smith. A novel structure for time-varying FIR filter banks. In *Proceedings of the International Conference on Acoustics, Speech, and Signal Processing*, pages III 157–160, April 1994.

[384] I. Sodagar, K. Nayebi, and T. P. Barnwell. Time-varying filter banks and wavelets. *IEEE Trans. on Signal Processing*, pages 2983–2996, November 1994.

[385] I. Sodagar, K. Nayebi, and T. P. Barnwell. Time-varying analysis-synthesis systems based on filter banks and postfiltering. *IEEE Trans. on Signal Processing*, 1996. to appear.

[386] A. K. Soman and P. P. Vaidyanathan. Coding gain in paraunitary analysis/synthesis systems. *IEEE Trans. Acoust. Speech Signal Process.*, pages 1821–1835, May 1993.

[387] M. M. Sondhi and W. Kellermann. Adaptive echo cancellation for speech signals. In S. Furui and M. M. Sondhi, editors, *Advances in Speech Signal Processing*, pages 327–356. Marcel Dekker, Inc., 1992.

[388] M. Sorbara, J. J. Werner, and N. A. Zervos. Carrierless am/pm. Technical report, ANSI, September 1990. Contribution T1E1.4/90-154.

[389] H. Suzuki. A statistical model for urban radio propagation. *IEEE Trans. Comm.*, COM-25:673–680, July 1977.

[390] K. Swaminathan and P. P. Vaidyanathan. Theory and design of uniform DFT, parallel quadrature mirror filter banks. *IEEE Trans. Circuits and Systems*, CAS-33:1170–1191, December 1986.

[391] D. Taubman and A. Zakhor. Highly scalable, low delay video compression. In *International Conference on Image Processing*, pages 740–745, November 1994.

[392] D. Taubman and A. Zakhor. Multirate 3-D subband coding of video. *IEEE Trans. on Image Processing*, 3:572–588, September 1994.

[393] D. Taubman and A. Zakhor. Rate and resolution scalable subband coding of video. In *Proceedings of the International Conference on Acoustics, Speech, and Signal Processing*, pages V.493–496, 1994.

[394] L. Taxen. Polyphase filter banks using wave digital filters. *IEEE Trans. on Signal Processing*, ASSP-29(3):423–428, June 1981.

[395] M. V. Tazebay and A. N. Akansu. Progressive optimality in hierarchical filter banks. In *International Conference on Image Processing*, volume I, pages 825–829, November 1994.

[396] M. V. Tazebay and A. N. Akansu. Adaptive subband transforms in time-frequency excisers for DSSS communications systems. *IEEE Trans. on Signal Processing*, November 1995.

[397] M. V. Tazebay and A. N. Akansu. A smart time-frequency exciser for spread spectrum communications. In *Proceedings of the International Conference on Acoustics, Speech, and Signal Processing*, pages 1209–1212, May 1995.

[398] Telebit Corporation. Comparative performance results for asymmetrical duplex, v.32 (extended), and multicarrier modems. Technical report, CCITT, September 1989. SG XVII, contribution D56.

[399] H. Theile, G. Stoll, and M. Link. Low bit-rate coding of high quality audio signals. 82nd AES Convention, Preprint 2432.

[400] H. Theile, G. Stoll, and M. Link. Low bit-rate coding of high-quality audio signals an introduction to the mascam system, 1988. EBU Review - Technical, No. 230.

[401] J. V. Tobias. *Foundations of Modern Auditory Theory*, volume 1. Academic Press, 1970.

[402] J. V. Tobias. *Foundations of Modern Auditory Theory*, volume 2. Academic Press, 1972.

[403] T. Tuncer and T. Nguyen. General analysis of two-band QMF banks. *IEEE Trans. on Signal Processing*, 1995. to appear.

[404] T. E. Tuncer and G. V. H. Sandri. Butterworth wavelets and adaptive echo cancellation. In *Int. Symposium Time-Frequency and Time-Scale Analysis*, October 1992.

[405] G. L. Turin. Introduction to spread-spectrum antimultipath techniques and their application to urban digital radio. *Proc. IEEE*, 68:328–353, March 1980.

[406] M. A. Tzannes and M. C. Tzannes. Bit-by-bit channel coding using wavelets. In *IEEE Globecom*, December 1992.

[407] M. A. Tzannes and M. C. Tzannes. Block biorthogonal channel coding using wavelets. In *IEEE MILCOM*, October 1992.

[408] M. A. Tzannes et al. A multicarrier transceiver for ADSL using M-band wavelet transforms. Technical report, ANSI, Miami, FL, March 1993.

[409] G. Tziritas and C. Labit. *Motion Analysis for Image Sequence Coding.* Elsevier, Amsterdam, 1994.

[410] K. H. Tzou and T. C. Chen. Compatible HDTV coding for broadband ISDN. In *Proc. GLOBECOM'88*, pages 743–749, 1988.

[411] G. Ungerboeck. Channel coding with multilevel/phase signals. *IEEE Trans. Inform. Theory*, IT-28:55–67, January 1982.

[412] G. Ungerboeck. Trellis-coded modulation with redundant signal sets, Parts I and II. *IEEE Communications Magazine*, 25:5–27, February 1987.

[413] M. K. Uz, M. Vetterli, and D. J. LeGall. Interpolative multiresolution coding of advanced television with compatible subchannels. *IEEE Trans. on Circuits and Systems for Video Technology*, 1(1):86–99, March 1991.

[414] N. Uzun. *Cyclostationary Modeling, Analysis and Optimal Compensation of Quantization Effects in Subband Codecs.* PhD thesis, Polytechnic University, 1993.

[415] N. Uzun and R. A. Haddad. Modeling and analysis of quantization errors in two-channel subband filter structures. In *SPIE Proc. Visual Communications and Image Processing*, pages 1446–1457, 1992.

[416] P. P. Vaidyanathan. Quadrature mirror filter banks, M-band extensions and perfect reconstruction techniques. *IEEE Trans. Acoust. Speech Signal Process.*, pages 4–20, July 1987.

[417] P. P. Vaidyanathan. Theory and design of M-channel maximally decimated QMFs with arbitrary M, having the perfect-reconstruction property. *IEEE Trans. Acoust. Speech Signal Process.*, pages 476–492, April 1987.

[418] P. P. Vaidyanathan. *Multirate Systems and Filterbanks.* Prentice-Hall, Englewood Cliffs, New Jersey, 1993.

[419] P. P Vaidyanathan. Orthonormal and biorthonormal filter banks as convolvers, and convolutional coding gain. *IEEE Trans. on Signal Processing*, 41(6), June 1993.

[420] P. P. Vaidyanathan and T. Chen. Statistically optimal synthesis banks for subband coders. In *Asilomar Conf. on Signals, Systems and Computers*, 1994.

[421] P. P. Vaidyanathan and P. Q. Hoang. Lattice structures for optimal design and robust implementation of two-channel perfect reconstruction QMF banks. *IEEE Trans. Acoust. Speech Signal Process.*, ASSP-36:81–94, January 1988.

[422] L. Vandendorpe. Optimized quantization for image subband coding. *Signal Processing: Image Communication*, 4(1):65–80, November 1991.

[423] L. Vandendorpe. Hierarchical transform and subband coding for video signals. *Signal Processing: Image Communications*, 41:245–262, 1992.

[424] L. Vandendorpe and P. Delogne. Hierarchical encoding of HDTV by transform coefficients block splitting. In *SPIE Conf. on Image Processing Algorithms and Techniques*, pages 349–364, 1990.

[425] M. Vetterli. Multi-dimensional sub-band coding: some theory and algorithms. *Signal Processing*, 6:97–112, April 1984.

[426] M. Vetterli. *Analyse, Synthèse et Complexité de Calcul de Bancs de Filtres Numériques.* PhD thesis, Ecole Polytechnique Fédérale de Lausanne, 1985.

[427] M. Vetterli. A theory of multirate filter banks. *IEEE Trans. Acoust. Speech Signal Process.*, pages 356–372, March 1987.

[428] M. Vetterli and D. Le Gall. Perfect reconstruction FIR filter banks: Some properties and factorizations. *IEEE Trans. Acoust. Speech Signal Process.*, pages 1057–1071, July 1989.

[429] M. Vetterli and C. Herley. Wavelets and filter banks: theory and design. *IEEE Trans. on Signal Processing*, 40(9), September 1992.

[430] M. Vetterli and J. Kovacevic. *Wavelets and Subband Coding.* Prentice-Hall, 1995.

[431] M. Vetterli, J. Kovacevic, and D. J. LeGall. Perfect reconstruction filter banks for HDTV representation and coding. *Signal Processing: Image Communication*, 2(3):349–363, October 1990.

[432] M. Vetterli and K. M. Uz. Multiresolution coding techniques for digital television: A review. *Multidimensional Systems and Signal Processing*, pages 3:161–187, 1992.

[433] J. F. Vial, M. Pecot, P. J. Tourtier, and J. Thomas. In-band interpolation applied to motion-compensated subband coding. In *Int. Workshop on HDTV*, September 1991.

[434] L. Villemoes. Sobolev regularity of wavelets and stability of iterated filter banks. In Y. Meyer and S. Roques, editors, *Proc. Int. Conf. "Wavelets and Applications"*. Frontières, 1992.

[435] E. Viscito and J. Allebach. The design of equal complexity FIR perfect reconstruction filter banks incorporating symmetries. Technical report, Purdue University, 1989.

[436] E. Viscito and J. P. Allebach. The analysis and design of multidimensional FIR perfect reconstruction filter banks for arbitrary sampling lattices,. *IEEE Trans. Circuits and Systems*, CAS-38:29–41, January 1991.

[437] J. J. Werner. The HDSL environment. *IEEE J. on Sel. Areas in Commun.*, pages 333–344, October 1991.

[438] P. Westerink, J. Biemond, D. Boekee, and J. Woods. Sub-band coding of images using vector quantization. *IEEE Trans. on Communications*, 36(6):713–719, June 1988.

[439] P. H. Westerink. *Subband Coding of Images.* PhD thesis, T. U. Delft, October 1989.

[440] P. H. Westerink, J. Biemond, and D. E. Boekee. An optimal bit allocation algorithm for subband coding. In *Proceedings of the International Conference on Acoustics, Speech, and Signal Processing*, pages 757–760, April 1988.

[441] P. H. Westerink, J. Biemond, and D. E. Boekee. Scalar quantization error analysis for image subband coding using QMF's. *IEEE Trans. on Signal Processing*, pages 421–428, February 1992.

[442] P. H. Westerink, J. Biemond, and F. Muller. Subband coding of image sequences at low bit rates. *Signal Processing: Image Communications*, 2:441–448, 1990.

[443] B. Widrow and S. Stearns. *Adaptive signal processing.* Prentice Hall, Englewood Cliffs, New Jersey, 1985.

[444] W. Wisnoe. *Utilisation de la méthode de transformée en ondelettes 2D pour l'analyse de visualisation d'écoulements.* Thèse de doctorat, ENSAE, Toulouse, 1993.

[445] W. Wisnoe, P. Gajan, A. Strzelecki, C. Lempereur, and J-M. Mathé. The use of the two-dimensional wavelet transform in flow visualization processing. In Y. Meyer and S. Roques (eds.), editors, *Progress in Wavelet Analysis and Applications (Proc. Toulouse 1992)[266].* Frontières, 1993.

[446] R. C. Wood. On optimum quantization. *IEEE Transactions on Information Theory*, IT-15:248–252, Mar. 1969.

[447] J. W. Woods and S. D. O'Neil. Subband coding of images. *IEEE Trans. on Acoustics, Speech, and Signal Processing*, 34:1278–1288, October 1986.

[448] John W. Woods, editor. *Subband Image Coding*. Kluwer Academic publishers, Boston, 1991.

[449] G. W. Wornell. Communications over fractal channels. In *Proceedings of the International Conference on Acoustics, Speech, and Signal Processing*, pages 1945–1948, May 1991.

[450] G. W. Wornell and A. V. Oppenheim. Wavelet based representations for a class of self-similar signals with application to fractal modulation. *IEEE Trans. Inform. Theory*, IT-38(2):785–800, March 1992.

[451] J. M. Wozencraft and P. H. Moose. Modulation and coding for in-band DAB using multi-frequency modulation,. In *Nat. Ass. of Broadcasters 1991 Conf.*, April 1991.

[452] Y. Yamada, S. Tazaki, and R. M. Gray. Asymptotic performance of block quantizers with difference distortion measures. *IEEE Transactions on Information Theory*, IT-26(1):6–14, Jan. 1980.

[453] S. Zafar, Y. Q. Zhang, and H. Jabbari. Multiscale video representation using multiresolution motion compensation and wavelet decomposition. *IEEE Journal on Selected Areas in Communications*, 11:24–35, 1993.

[454] N. Zervos and I. Kalet. Optimized decision feedback equalization versus optimized orthogonal frequency division multiplexing for high-speed data transmission over the local cable network. In *International Conference on Communications*, pages 35.2–35.2.6, June 1989.

[455] Y. Q. Zhang and S. Zafar. Motion-compensated wavelet transform coding for color video compression. *IEEE Trans. on Circuits and Systems for Video Technology*, 2:285–296, September 1992.

[456] R. E. Ziemer and R. L. Peterson. *Digital Communications and Spread Spectrum Systems*. Macmillan Publishing Company, New York, 1985.

INDEX

GPSR Compliance

The European Union's (EU) General Product Safety Regulation (GPSR) is a set of rules that requires consumer products to be safe and our obligations to ensure this.

If you have any concerns about our products, you can contact us on ProductSafety@springernature.com

In case Publisher is established outside the EU, the EU authorized representative is:

Springer Nature Customer Service Center GmbH
Europaplatz 3
69115 Heidelberg, Germany

Batch number: 10755284

Printed by Printforce, the Netherlands